Advances in Fuzzy MCDM, Hybrid Methods, Fuzzy Number Ranking and Their Applications

Advances in Fuzzy MCDM, Hybrid Methods, Fuzzy Number Ranking and Their Applications

Guest Editors

Ta-Chung Chu
Wei-Chang Yeh

MDPI

Basel • Beijing • Wuhan • Barcelona • Belgrade • Novi Sad • Cluj • Manchester

Guest Editors
Ta-Chung Chu
Department of Industrial Management and Information
Southern Taiwan University of Science and Technology
Tainan City
Taiwan

Wei-Chang Yeh
Department of Industrial Engineering and Engineering Management
National Tsing Hua University
Hsinchu
Taiwan

Editorial Office
MDPI AG
Grosspeteranlage 5
4052 Basel, Switzerland

This is a reprint of the Special Issue, published open access by the journal *Axioms* (ISSN 2075-1680), freely accessible at: https://www.mdpi.com/journal/axioms/special_issues/7F551OEE24.

For citation purposes, cite each article independently as indicated on the article page online and as indicated below:

Lastname, A.A.; Lastname, B.B. Article Title. *Journal Name* **Year**, *Volume Number*, Page Range.

ISBN 978-3-7258-3432-7 (Hbk)
ISBN 978-3-7258-3431-0 (PDF)
https://doi.org/10.3390/books978-3-7258-3431-0

Contents

About the Editors

Ta-Chung Chu

Ta-Chung Chu received his Ph.D. degree at the Department of Industrial Engineering from the University of Texas at Arlington, USA. He served as Chairman from August 2010 to July 2016 and is currently a professor at the Department of Industrial Management and Information, Southern Taiwan University of Science and Technology, Taiwan. His research interests are in fuzzy multiple-criteria decision making, fuzzy number ranking, and their applications. Prof. Chu has published 60 journal papers and 100 conference papers. He obtained a research award from the National Science and Technology Council (NSTC), Taiwan, in the year of 2000, and obtained an NSTC Special Talent Research Award in the year of 2012. Prof. Chu was invited by the Industrial Engineering Graduate Program, Faculty of Engineering, Kasetsart University, Thailand, July 2-6, 2012, to serve as a lecturer for course 01206527. From July 3 to October 2, 2007, with the funding of the NSTC, he went to North Carolina State University in Raleigh, USA, as a visiting scholar. Prof. Chu has been invited by renowned journals as a reviewer and he has reviewed 243 journal papers. He was included in the 「Career-Long Impact List」 of the "World's Top 2% Scientists" in 2021-2024 and the 「Year Impact List」 in 2021, 2022, and 2024. Prof. Chu served as Chair for the Conference of Industrial Management and Information Application Innovation (IMIAI 2013 and 2014). He was one of the four Invited Speakers in 2024 of the 7th International Conference on Knowledge Management Systems. Currently, he serves as Academic Editor of *PLOS ONE*.

Wei-Chang Yeh

Wei-Chang Yeh is currently a chair professor of the Department of Industrial Engineering and Engineering Management at National Tsing Hua University in Taiwan. He received his M.S. and Ph.D. from the Department of Industrial Engineering at the University of Texas at Arlington. The majority of his research is focused on algorithms, including exact solution methods and soft computing. He has published more than 350 research papers in highly ranked journals and conference papers and has been awarded the Outstanding Research Award twice, the Distinguished Scholars Research Project once, and an Overseas Research Fellowship twice by the Ministry of Science and Technology (MOST) in Taiwan. He has been invited to serve as an Associate Editor of three journals, namely *IEEE Transactions on Reliability*, *IEEE Access*, and *Reliability Engineering & System Safety*. He proposed a novel soft computing algorithm called simplified swarm optimization (SSO) and a new implicit enumeration algorithm called the binary-addition-tree algorithm (BAT). Both SSO and BAT demonstrated the simplicity, effectiveness, efficiency, and flexibility of make-to-fit for solving NP-hard problems. He has been granted more than 55 patents; he is enlisted on the global list of the top 2% of scientists (2020-2023) by Stanford University; and he has been elected as International Fellow, MOST Fellow (Ministry of Science and Technology Taiwan, 2021.), and CIIE Fellow (Chinese Institute of Industrial Engineers), 2022, as well as being given the Guoguang Invention Medal and the titles of Outstanding Inventor of Taiwan and Doctor of Erudition, by the Chinese Innovation and Invention Society.

Preface

This reprint contains ten papers published in a Special Issue of the MDPI journal *Axioms*, entitled "Advances in Fuzzy MCDM, Hybrid Methods, Fuzzy Number Ranking, and Their Applications." In this volume, the authors, who are eminent mathematicians and experts in their fields, have presented their recent results on fuzzy MCDM, hybrid methods, fuzzy number ranking, and/or relevant applications, which also include open problems for further scientific research. The ten papers have been accepted after a rigorous review procedure. The objective of this Special Issue is to provide a platform for researchers to publish their recent work, delve deeper into various problems, and solve them mathematically. Finally, we would like to thank Mr. Owen Su, the Managing Editor, and his team for their attentive assistance in the review process and inviting authors.

Ta-Chung Chu and Wei-Chang Yeh
Guest Editors

axioms

MDPI

Article

A Framework for I^*-Statistical Convergence of Fuzzy Numbers

Tanushri [1], Ayaz Ahmad [1] and Ayhan Esi [2,*]

1 Department of Mathematics, National Institute of Technology Patna, Patna 800005, India; tanushri.phd22.ma@nitp.ac.in (T.); ayaz@nitp.ac.in (A.A.)
2 Department of Mathematics, Malatya Turgut Ozal University, Malatya 44040, Turkey
* Correspondence: aesi23@hotmail.com

Abstract: In this study, we investigate the concept of I^*-statistical convergence for sequences of fuzzy numbers. We establish several theorems that provide a comprehensive understanding of this notion, including the uniqueness of limits, the relationship between I^*-statistical convergence and classical convergence, and the algebraic properties of I^*-statistically convergent sequences. We also introduce the concept of I^*-statistical pre-Cauchy and I^*-statistical Cauchy sequences and explore its connection to I^*-statistical convergence. Our results show that every I^*-statistically convergent sequence is I^*-statistically pre-Cauchy, but the converse is not necessarily true. Furthermore, we provide a sufficient condition for an I^*-statistically pre-Cauchy sequence to be I^*-statistically convergent, which involves the concept of $I^* - lim\ inf$.

Keywords: I^*-statistical convergence; I^*-statistical Cauchy sequences; I^*-statistical pre-Cauchy; $I^* - lim\ inf$

MSC: 40A05; 40A35; 40D25

Citation: Tanushri; Ahmad, A.; Esi, A. A Framework for I^*-Statistical Convergence of Fuzzy Numbers. *Axioms* **2024**, *13*, 639. https://doi.org/10.3390/axioms13090639

Academic Editors: Salvatore Sessa, Ta-Chung Chu and Wei-Chang Yeh

Received: 1 August 2024
Revised: 3 September 2024
Accepted: 15 September 2024
Published: 18 September 2024

1. Introduction

There has been considerable progress in the convergence theory concerning fuzzy number sequence due to seminal works and innovative extensions that have taken place. Matloka [1] introduced the primary definition of convergence of sequences of fuzzy numbers and defined its limit and discussed its algebraic properties, while Nanda [2] studied the spaces of bounded and convergent sequences of fuzzy numbers and showed that they are complete metric spaces which furthered its theoretical background. Variations are manifested by sequences that do not converge under classical convergence conditions. Most mathematical problems involve sequences that are not convergent in the usual sense. There is now a realization of the necessity of considering more classes of sequences for determining or discussing their convergences. One of the approaches is to consider sequences that converge when we restrict our attention to large subsets of natural numbers in some meaningful sense. For example, if we define an important subset as all natural numbers apart from those with finitely many, then we get the traditional concept of convergence. On the other hand, recourse may be made to subsets having zero natural density. The natural density of a subset $\mathcal{A}$ of $\mathbb{N}$ is formally expressed as $\delta(\mathcal{A})$, and it is defined as follows:

$$\delta(\mathcal{A}) = \lim_{n\to\infty} \frac{1}{n}|\{k < n : k \in \mathcal{A}\}|,$$

which will lead us to a type of convergence namely, statistical convergence. The concept of statistical convergence for sequences of real numbers was independently introduced by Fast [3] and Schoenberg [4]. This foundational idea was later expanded by Savaş [5], who discussed alternative conditions for sequences of fuzzy numbers to be statistically Cauchy. Subsequent research further explored the nuances of this area, notably by Connor [6], who introduced the concept of statistically pre-Cauchy sequences and demonstrated that statistically convergent sequences are inherently statistically pre-Cauchy. The exploration of statistical convergence from a sequence space perspective and its connection to summability theory was advanced by researchers like Fridy [7] and Salát [8]. For a foundational understanding of statistical convergence, we recommend consulting works such as [9–13]. Some of the applications of statistical convergence can be found in [14,15]. Kostyrko et al. [16] extended the concept of statistical convergence by introducing I-convergence and I^*-convergence, which utilize ideals in metric spaces. They discussed several basic properties of these new types of convergence. For a detailed examination of I-convergence, we suggest referring to [17–21].

Kumar and Kumar [22] applied the concepts of I-convergence, I^*-convergence, and I-Cauchy sequences to sequences of fuzzy numbers, with further developments in this area discussed in works such as [23,24].

Savaş and Das [25] later extended I-convergence to I-statistical convergence, aiming to unify λ-statistical and A-statistical convergence using ideals. They introduced the notion of I-statistically pre-Cauchy sequences, which were further investigated by Debnath et al. [26]. Later on, Debnath et al. [27] discussed I-statistical convergence, introducing I-statistical limit points and cluster points, and exploring their basic properties. They extended I-statistical convergence and proved that I^*-statistical convergence implies I-statistical convergence. In recent years, various authors have studied different kinds of convergence by generalising statistical convergence via ideals in different spaces and for different types of sequences, for example, [28–30]. However, the properties and consequences of I^*-statistical convergence have not been thoroughly discussed, which motivated our current research.

This article investigates the concept of I^*-statistical convergence for sequences of fuzzy numbers in metric space. We have proved that under I^*-statistical convergence the limit of the sequence is unique. We established several theorems that comprehensively understand this notion, which include the relationship between I^*-statistical convergence and classical convergence and the algebraic properties of I^*-statistically convergent sequences. We also defined I^*-statistically pre-Cauchy sequences and I^*-statistical Cauchy sequences and explored their connection to I^*-statistical convergence. Our results show that every I^*-statistically convergent sequence is I^*-statistically pre-Cauchy, but the converse is not necessarily true. Furthermore, we provide a sufficient condition for an I^*-statistically pre-Cauchy sequence to be I^*-statistically convergent, which involves the concept of $I^*- \, lim\, inf$.

2. Preliminaries

In the theory of fuzzy numbers, we start by considering intervals denoted by $\mathcal{A}$ with endpoints $\underline{\mathcal{A}}$ and $\overline{\mathcal{A}}$. The set D comprises all closed, bounded intervals on the real line $\mathbb{R}$, represented as:

$$D = \{\mathcal{A} \subset \mathbb{R} : \mathcal{A} = [\underline{\mathcal{A}}, \overline{\mathcal{A}}]\}.$$

For any $\mathcal{A}, \mathcal{B}$ in D, we define $\mathcal{A} \leqslant \mathcal{B}$ iff $\underline{\mathcal{A}} \leqslant \underline{\mathcal{B}}$ and $\overline{\mathcal{A}} \leqslant \overline{\mathcal{B}}$, with the distance function $d(\mathcal{A}, \mathcal{B})$ being the maximum of $|\underline{\mathcal{A}} - \underline{\mathcal{B}}|$ and $|\overline{\mathcal{A}} - \overline{\mathcal{B}}|$.

The metric d establishes a Hausdorff metric on D, rendering (D, d) a complete metric space. Moreover, $\leqslant$ acts as a partial order on D.

Definition 1 ([22]). *A fuzzy number is a function X from $\mathbb{R}$ to $[0,1]$, which satisfy the following conditions:*

(i) *X is normal, i.e., there exists an $X_0 \in \mathbb{R}$ such that $X(X_0) = 1$;*
(ii) *X is fuzzy convex, i.e., for any $x, y \in \mathbb{R}$ and $\lambda \in [0,1]$, $X(\lambda x + (1-\lambda)y) \geqslant \min\{X(x), X(y)\}$;*
(iii) *X is upper semi-continuous;*
(iv) *The closure of the set $\{x \in \mathbb{R} : X(x) > 0\}$, denoted by X^0 is compact.*

The properties (i)–(iv) imply that for each $\alpha \in (0,1]$, the α-level set:

$$\mathcal{X}^{\alpha} = \{x \in \mathbb{R} : \mathcal{X}(x) \geqslant \alpha\} = \left[\underline{\mathcal{X}}^{\alpha}, \overline{\mathcal{X}}^{\alpha}\right].$$

where $\mathcal{X}^{\alpha}$ represents a non-empty, compact, and convex subset of the real numbers $\mathbb{R}$.

The set of all fuzzy numbers is denoted by $L(\mathbb{R})$, and the set comprising all sequences of fuzzy numbers is represented by $L(\$)$. We define a mapping, denoted as $\overline{d}$, which takes pairs of fuzzy numbers from $L(\mathbb{R}) \times L(\mathbb{R})$ and maps them to the real numbers $\mathbb{R}$. Formally, this mapping $\overline{d}$ can be expressed as follows:

$$\overline{d}(\mathcal{X}, \mathcal{Y}) = \sup_{\alpha \in [0,1]} d(\mathcal{X}^{\alpha}, \mathcal{Y}^{\alpha}).$$

where $\overline{d}(\mathcal{X}, \mathcal{Y})$ computes the supremum of the distance, d, between the α-level sets of fuzzy numbers $\mathcal{X}$ and $\mathcal{Y}$ across all values of α within the interval $[0, 1]$.

Puri and Ralescu [31] demonstrated that the space $(L(\mathbb{R}), \bar{d})$ constitutes a complete metric space: "We define the relation $\mathcal{X} \leqslant \mathcal{Y}$ for $\mathcal{X}, \mathcal{Y} \in L(\mathbb{R})$ if $\underline{\mathcal{X}}^{\alpha} \leqslant \underline{\mathcal{Y}}^{\alpha}$ and $\overline{\mathcal{X}}^{\alpha} \leqslant \overline{\mathcal{Y}}^{\alpha}$ for each $\alpha \in [0, 1]$. Furthermore, we define $\mathcal{X} < \mathcal{Y}$ if $\mathcal{X} \leqslant \mathcal{Y}$ and there exists some $\alpha_0 \in [0, 1]$ such that $\underline{\mathcal{X}}^{\alpha_0} < \underline{\mathcal{Y}}^{\alpha_0}$ or $\overline{\mathcal{X}}^{\alpha_0} < \overline{\mathcal{Y}}^{\alpha_0}$. If neither $\mathcal{X} \leqslant \mathcal{Y}$ nor $\mathcal{Y} \leqslant \mathcal{X}$ holds, we say that $\mathcal{X}$ and $\mathcal{Y}$ are incomparable fuzzy numbers". Moreover, they continue that in the metric space $L(\mathbb{R})$, "we can define addition $\mathcal{X} + \mathcal{Y}$ and scalar multiplication $\lambda\mathcal{X}$, where λ is a real number, in terms of α-level sets as follows:

$$[\mathcal{X} + \mathcal{Y}]^{\alpha} = [\mathcal{X}]^{\alpha} + [\mathcal{Y}]^{\alpha}$$

for each $\alpha \in [0, 1]$, and

$$[\lambda\mathcal{X}]^{\alpha} = \lambda[\mathcal{X}]^{\alpha}$$

for each $\alpha \in [0, 1]$, respectively".

Regarding fuzzy integers within a subset S of $L(\mathbb{R})$, if there exists a fuzzy integer denoted by μ such that $\mathcal{X} \leqslant \mu$ holds for every $\mathcal{X}$ in the subset S, we designate S as having an upper bound, with μ serving as the upper bound for the set. Similarly, we define the lower bound.

For each $\alpha \in [0, 1]$, if we define $\overline{\mathcal{Z}}^{\alpha} := \overline{\mathcal{X}}^{\alpha} + \overline{\mathcal{Y}}^{\alpha}$ and $\underline{\mathcal{Z}}^{\alpha} := \underline{\mathcal{X}}^{\alpha} + \underline{\mathcal{Y}}^{\alpha}$, we can express $\mathcal{Z}$ as the sum of $\mathcal{X}$ and $\mathcal{Y}$, denoted as $\mathcal{Z} = \mathcal{X} + \mathcal{Y}$. Similarly, following a comparable pattern, we represent $\mathcal{Z}$ as the difference of $\mathcal{X}$ and $\mathcal{Y}$, expressed as $\mathcal{Z} = \mathcal{X} - \mathcal{Y}$, iff $\overline{\mathcal{Z}}^{\alpha} := \overline{\mathcal{X}}^{\alpha} - \underline{\mathcal{Y}}^{\alpha}$ and $\underline{\mathcal{Z}}^{\alpha} := \underline{\mathcal{X}}^{\alpha} - \overline{\mathcal{Y}}^{\alpha}$ for each $\alpha \in [0, 1]$.

Definition 2 ([22])**.** *A sequence $\mathcal{X} = (\mathcal{X}_n)$ of fuzzy numbers are said to be convergent to a fuzzy number $\mathcal{X}_0$ if, for every $\varepsilon > 0$, there exists a positive integer m such that $\overline{d}(\mathcal{X}_n, \mathcal{X}_0) < \varepsilon$ for every $n \geqslant m$. The fuzzy number $\mathcal{X}_0$ is referred to as the ordinary limit of the sequence $(\mathcal{X}_n)$, denoted as* $\lim_{n \to \infty} \mathcal{X}_n = \mathcal{X}_0$.

Definition 3 ([22])**.** *A sequence $\mathcal{X} = (\mathcal{X}_n)$ of fuzzy numbers are regarded as a Cauchy sequence if, for every $\varepsilon > 0$, there exists a positive integer n_0 such that $\overline{d}(\mathcal{X}_n, \mathcal{X}_m) < \varepsilon$ for all $n, m \geqslant n_0$.*

Definition 4 ([22])**.** *A sequence $\mathcal{X} = (\mathcal{X}_n)$ of fuzzy numbers are categorized as a bounded sequence if the set $\{\mathcal{X}_n : n \in \mathbb{N}\}$, comprising all the fuzzy numbers in the sequence is itself a bounded set of fuzzy numbers.*

Definition 5 ([22])**.** *A sequence $\mathcal{X} = (\mathcal{X}_n)$ of fuzzy numbers are considered to be statistically convergent to a fuzzy number $\mathcal{X}_0$ if, for any $\varepsilon > 0$, the set $A(\varepsilon) = \{n \in \mathbb{N} : \overline{d}(\mathcal{X}_n, \mathcal{X}_0) \geqslant \varepsilon\}$ exhibits a natural density of zero. In this context, the natural density of a set refers to the proportion of natural numbers within the set concerning the whole set of natural numbers. The fuzzy number $\mathcal{X}_0$ is termed the statistical limit of the sequence $(\mathcal{X}_n)$, denoted as $st - \lim_{n \to \infty} \mathcal{X}_n = \mathcal{X}_0$.*

Definition 6 ([22])**.** *A sequence $\mathcal{X} = (\mathcal{X}_n)$ of fuzzy numbers are termed statistically Cauchy if, for any $\varepsilon > 0$, there exists a positive integer $m = m(\varepsilon)$ such that the set $\{n \in \mathbb{N} : \bar{d}(\mathcal{X}_n, \mathcal{X}_m) \geqslant \varepsilon\}$ has a natural density of zero. In this context, the term "natural density" pertains to the proportion of natural numbers within the set concerning the entire set of natural numbers.*

Throughout this paper, we will use $\mathbb{R}$ and $\mathbb{N}$ to represent, respectively, the set of real numbers and positive integers. We will denote the power set of any set $\mathcal{X}$ as $P(\mathcal{X})$, and the complement of the set $\mathcal{A}$ will be denoted as $\mathcal{A}^c$.

Definition 7 ([22])**.** *Let X be a non-empty set, then a collection of subsets I contained in the power set of X denoted as $P(X)$ is said to be ideal iff it satisfies the following conditions:*

(i) The empty set belongs to I, i.e., $\emptyset \in I$;
(ii) For any set $\mathcal{A}$ and $\mathcal{B}$ belonging to I, $\mathcal{A} \cup \mathcal{B}$ also belongs to I;
(iii) If $\mathcal{A} \in I$ and $\mathcal{B} \subset \mathcal{A}$ then $\mathcal{B} \in I$.

Definition 8. *Let X be a non-empty set. A non-empty family of sets F contained within the power set $P(X)$ is denoted as a filter on X iff it adheres to the following criteria:*

(i) The empty set $\emptyset$ is not an element of the filter, meaning $\emptyset \notin F$;
(ii) For any two sets $\mathcal{A}$ and $\mathcal{B}$ that belong to the filter, their intersection denoted as $\mathcal{A} \cap \mathcal{B}$ is also a part of the filter formally expressed as $\mathcal{A} \cap \mathcal{B} \in F$;
(iii) If a set $\mathcal{A}$ is a member of the filter and $\mathcal{B}$ is a super set of $\mathcal{A}$, then $\mathcal{B}$ is also an element of the filter, i.e., $\mathcal{B} \in F$.

Conditions (i), (ii), and (iii) jointly define the properties of a filter on set X.

An ideal I is termed non-trivial if it satisfies two conditions: it is not an empty set ($I \neq \emptyset$), and it does not contain the entire set X ($X \notin I$). Notably, a non-trivial ideal $I \subset P(X)$ corresponds to a filter, denoted as $F(I)$, which is formed by taking the set complement of each element of I with respect to the entire set X. The filter $F(I)$ is referred to as the filter associated with the ideal I.

An ideal I in X is considered admissible iff it includes all singleton sets, i.e.,$\{\{x\} : x \in X\}$.

Definition 9 ([22])**.** *Suppose $I \subset P(\mathbb{N})$ is a non-trivial ideal. We define a sequence $\mathcal{X} = (\mathcal{X}_n)$ of fuzzy numbers as I-convergent to a fuzzy number $\mathcal{X}_0$ if, for any ϵ, the set $A(\epsilon) = \{n \in \mathbb{N} : \bar{d}(\mathcal{X}_n, \mathcal{X}_0) \geqslant \epsilon\} \in I$.*

The fuzzy number $\mathcal{X}_0$ is then referred to as the I-limit of the sequence $(\mathcal{X}_n)$, and this is denoted as $\lim_{n\to\infty} \mathcal{X}_n = \mathcal{X}_0$.

The set of fuzzy number sequences that are both convergent and I-convergent can be denoted by ℓ_1. These sequences exhibit both conventional convergence and convergence according to the ideal I, providing a rich framework for the study of their convergence properties. Throughout the paper, we consider I as an admissible ideal.

Definition 10 ([22])**.** *A sequence $\mathcal{X} = (\mathcal{X}_n) \in L(\mathbb{S})$ of fuzzy numbers is said to be I^*-convergent to a fuzzy number $\mathcal{X}_0$ iff there exists a set $\mathcal{K} = \{m_1 < m_2 < m_3 < \cdots < m_k < \dots\} \subset \mathbb{N}$ such that $\mathcal{K} \in F(I)$ and $\bar{d}\left(\mathcal{X}_{m_k}, \mathcal{X}_0\right) \to 0$ as $n \to \infty$.*

3. I^*-Statistical Convergence of Sequence of Fuzzy Numbers

Definition 11. *A sequence $\mathcal{X} = (\mathcal{X}_n) \in L(\mathbb{S})$ is said to be I^*-statistically convergent to a fuzzy number $\mathcal{X}_0$ if and only if there exists a set $\mathcal{K} = \{m_1 < m_2 < m_3 \cdots < m_k < \dots\} \subset \mathbb{N}$ and for each $\epsilon > 0$ we have $\lim_{n\to\infty} \frac{1}{n}|\{m_k < n : \bar{d}(\mathcal{X}_{m_k}, \mathcal{X}_0) < \epsilon\} \in F(I)| = 1$. $\mathcal{X}_0$ is the I^*-statistical limit of $\mathcal{X}_n$ and is denoted by $I^* - st - lim_{n\to\infty}\mathcal{X}_n = \mathcal{X}_0$.*

Example 1. *Consider the sequence $\mathcal{X} = (\mathcal{X}_n)$, which is defined as follows:*

$$\mathcal{X} = (\mathcal{X}_n) = \begin{cases} 0 & \text{for } n = \mathcal{k}^2 \text{ where } \mathcal{k} \in \mathbb{N} \\ \frac{1}{n} & \text{otherwise} \end{cases}$$

which is I^*-statistically convergent to 0. Let $\mathcal{K} = \{m_1 < m_2 < m_3 < \cdots < m_{\mathcal{k}} < \ldots\} \subset \mathbb{N}$, where $m_1, m_2, m_3, \ldots, m_k, \ldots$ are all non-perfect square natural numbers. Then, for each $\epsilon > 0$, we have:

$$\lim_{n\to\infty} \frac{1}{n}\left|\left\{m_{\mathcal{k}} < n : \bar{d}(\mathcal{X}_{m_{\mathcal{k}}}, 0) < \epsilon\right\} \in \mathcal{K}\right| = 1.$$

It is trivial to show that I is an ideal if it is the collection of subsets of the set $X = \{n \in \mathbb{N} : n = k^2\}$. This implies that $\mathcal{K} \in F(I)$. Therefore:

$$\lim_{n\to\infty} \frac{1}{n}\left|\left\{m_{\mathcal{k}} < n : \bar{d}(\mathcal{X}_{m_{\mathcal{k}}}, 0) < \epsilon\right\} \in F(I)\right| = 1.$$

Theorem 1. *If I is an admissible ideal, then a sequence $\mathcal{X} = (\mathcal{X}_n) \in L(\$)$ that is I^*-statistically convergent will converge to a unique limit.*

Proof. Let $\mathcal{X} = (\mathcal{X}_n) \in L(\$)$ be an I^*-statistically convergent sequences to two different fuzzy numbers $\mathcal{X}_0$ and $\mathcal{Y}_0$. Without the loss of generality, suppose that $\mathcal{X}_0$ and $\mathcal{Y}_0$ are comparable fuzzy numbers. Consequently, there exists $\alpha_0 \in [0,1]$ such that:

$$\underline{\mathcal{X}}_0^{\alpha_0} < \underline{\mathcal{Y}}_0^{\alpha_0} \quad \text{and} \quad \overline{\mathcal{X}}_0^{\alpha_0} > \overline{\mathcal{Y}}_0^{\alpha_0} \tag{1}$$

or

$$\underline{\mathcal{X}_0}^{\alpha_0} > \underline{\mathcal{Y}_0}^{\alpha_0} \quad \text{and} \quad \overline{\mathcal{X}_0}^{\alpha_0} < \overline{\mathcal{Y}_0}^{\alpha_0}. \tag{2}$$

We will prove that (1) and (2) can be performed in a similar manner.

Let us assume that (1) is valid. Choose $\xi_1 = \underline{\mathcal{Y}_0}^{\alpha_0} - \underline{\mathcal{X}_0}^{\alpha_0}$ and $\xi_2 = \underline{\mathcal{X}_0}^{\alpha_0} - \underline{\mathcal{Y}_0}^{\alpha_0}$. Clearly $\xi_1 > 0$ and $\xi_2 > 0$. Let $\xi' = min\{\xi_1, \xi_2\}$. Select ϵ such that $0 < \epsilon < \xi'$. Given that $(\mathcal{X}_n)$ is I^*-statistical convergent to both $\mathcal{X}_0$ and $\mathcal{Y}_0$ therefore, we have $\mathcal{M} = \{m_1 < m_2 < m_3 < \cdots < m_{\mathcal{k}} < \ldots\} \subset \mathbb{N}$ and $\mathcal{K} = \{n_1 < n_2 < n_3 < \cdots < n_{\mathcal{k}} < \ldots\} \subset \mathbb{N}$ such that for every $\epsilon > 0$:

$$\begin{aligned} &\lim_{n\to\infty} \frac{1}{n}|\{m_{\mathcal{k}} < n : \bar{d}(\mathcal{X}_{m_{\mathcal{k}}}, \mathcal{X}_0) \leqslant \epsilon\} \in F(I)| = 1 \text{ and} \\ &\lim_{n\to\infty} \frac{1}{n}|\{n_k < n : \bar{d}(\mathcal{X}_{n_{\mathcal{k}}}, \mathcal{Y}_0) \leqslant \epsilon\} \in F(I)| = 1 \end{aligned} \tag{3}$$

since $F(I)$ is a filter on $\mathbb{N}$ therefore, by the definition of filter $\mathcal{M} \cap \mathcal{N} \neq \phi$.

Let $m \in \mathcal{M} \cap \mathcal{N}$ then by (3) there exists positive integers $\mathcal{k}_1$ and $\mathcal{k}_2$ such that:

$$\begin{aligned} &\lim_{n\to\infty} \frac{1}{n}|\{m_{\mathcal{k}} < n : \bar{d}(\mathcal{X}_{m_{\mathcal{k}}}, \mathcal{X}_0) \leqslant \epsilon\} \in F(I)| = 1 \text{ for every } m_{\mathcal{k}} \in \mathcal{M} \\ &\text{with} m_{\mathcal{k}} > \mathcal{K}_1 \text{ and} \\ &\lim_{n\to\infty} \frac{1}{n}|\{n_k < n : \bar{d}(\mathcal{X}_{n_{\mathcal{k}}}, \mathcal{Y}_0) \leqslant \epsilon\} \in F(I)| = 1 \text{ for every } n_{\mathcal{k}} \in \mathcal{N} \\ &\text{with } n_{\mathcal{k}} > \mathcal{K}_2. \end{aligned} \tag{4}$$

Let $\mathcal{k} = max\{\mathcal{k}_1, \mathcal{k}_2\}$ the (4) follows for $m \in \mathcal{M} \cap \mathcal{N}$ with $n_{\mathcal{k}}, m_{\mathcal{k}} > \mathcal{k}$. For each $\alpha \in [0,1]$ and $m = max\{m_{\mathcal{k}}, n_{\mathcal{k}}\}$ we have, $\lim_{n\to\infty} \frac{1}{n}|\{m < n : \bar{d}(\mathcal{X}_m^{\alpha_0}, \mathcal{X}_0) \leqslant \epsilon\} \in F(I)| = 1$ and $\lim_{n\to\infty} \frac{1}{n}|\{m < n : \bar{d}\ (\mathcal{X}_m^{\alpha_0}, \mathcal{Y}_0) \leqslant \epsilon\} \in F(I)| = 1$. Now the definition of d implies:

$$\begin{aligned} &\left|\underline{\mathcal{X}}_m^{\alpha_0} - \underline{\mathcal{X}}_0^{\alpha_0}\right| < \varepsilon \text{ and } \left|\underline{\mathcal{X}}_m^{\alpha_0} - \underline{\mathcal{Y}}_0^{\alpha_0}\right| < \varepsilon, \\ &\left|\overline{\mathcal{X}}_m^{\alpha_0} - \overline{\mathcal{X}}_0^{\alpha_0}\right| < \varepsilon \text{ and } \left|\overline{\mathcal{X}}_m^{\alpha_0} - \overline{\mathcal{Y}}_0^{\alpha_0}\right| < \varepsilon. \end{aligned}$$

$\underline{X}_m^{\alpha_0} \in (\underline{X}_0^{\alpha_0} - \varepsilon, \underline{X}_0^{\alpha_0} + \varepsilon) \cap (\underline{\mathcal{Y}}_0^{\alpha_0} - \varepsilon, \underline{\mathcal{Y}}_0^{\alpha_0} + \varepsilon) = \Phi$. Thus, a contradiction arises, implying the comparability of fuzzy numbers X_0 and $\mathcal{Y}_0$. Consider $X_0 \leqslant \mathcal{Y}_0$ and the neighborhoods $\mathcal{A} = \left\{n \in \mathbb{N} : \bar{d}(X_n, X_0) < \varepsilon\right\}$ and $\mathcal{B} = \left\{n \in \mathbb{N} : \bar{d}(X_n, \mathcal{Y}_0) < \varepsilon\right\}$ of X_0 and $\mathcal{Y}_0$, respectively, are disjoint for $\varepsilon = \frac{\bar{d}(X_0, \mathcal{Y}_0)}{3} > 0$. By Definition (8), both the sets $\mathcal{A}, \mathcal{B} \in F(I)$ so that $\mathcal{A} \cap \mathcal{B} \neq \Phi$. A contradiction has arrived that the neighborhoods of X_0 and $\mathcal{Y}_0$ are disjoint. Hence, X_0 is determined uniquely. □

Theorem 2. *Let $X = (X_n)$ and $\mathcal{Y} = (\mathcal{Y}_n) \in L(\$)$ then:*

(i) $lim_{n\to\infty} X_n = X_0$ implies $I^* - st - lim_{n\to\infty} X_n = X_0$;
(ii) $I^* - st - lim_{n\to\infty} X_n = X_0$ and $c \in \mathbb{R}$, then $I^* - st - lim_{n\to\infty} cX_n = cX_0$;
(iii) If $I^* - st - lim_{n\to\infty} = X_0$ and $I^* - st - lim_{n\to\infty} \mathcal{Y}_n = \mathcal{Y}_0$ then $I^* - st - lim_{n\to\infty}(X_n + \mathcal{Y}_n) = (X_0 + \mathcal{Y}_0)$.

Proof.

(i) Let $lim_{n\to\infty} X_n = X_0$, then for each $\epsilon > 0$ there exists a positive integer m(say) such that $\bar{d}(X_n, X_0) < \epsilon$ for every $n \geqslant m$. Then, for $\epsilon > 0$ let $A(\epsilon) = \{m_k : \bar{d}(X_{m_k}, X_0) < \epsilon\}$ for set $\mathcal{K} = \{m_1 < m_2 < \ldots < m_k < \ldots\} \subset \mathbb{N}$ is an infinite set then there exists a set $H = \{n_1, n_2, n_3, \ldots, n_k\}$ such that $\mathbb{N} - H = \mathcal{K}$ and H is a finite set, and therefore, $H \in I$ as I is an admissible ideal. This implies that $\mathcal{K} \in F(I)$ and $\delta(\mathcal{K}) = 1$. Thus, $\lim_{n\to\infty} \frac{1}{n}|\{m_k < n : \bar{d}(X_{m_k}, X_0) \leqslant \epsilon\} \in F(I)| = 1$. Hence, $lim_{n\to\infty} X_n = X_0$ implies $I^* - st - lim_{n\to\infty} X_n = X_0$.

(ii) Let $\alpha \in [0,1]$ and $c \in \mathbb{R}$. Let $\mathcal{K} = \{m_1 < m_2 < \ldots < m_k < \ldots\} \subset \mathbb{N}$ and $\epsilon > 0$ be given. Since $d(cX_n{}^\alpha, cX_0{}^\alpha) = |c| d(X_n{}^\alpha, X_0{}^\alpha)$. Therefore, $\bar{d}(cX_{m_k}, cX_0) = |c|\bar{d}(X_{m_k}, X_0)$. As $I^* - st - lim_{n\to\infty} X_n = X_0$. Therefore, the set $A(\epsilon) = \{m_k : \bar{d}(X_{m_k}, X_0) \leqslant \epsilon\} \in F(I)$ and $\delta(A(\epsilon)) = 1$. Let $B(\epsilon) = \{m_k : \bar{d}(cX_{m_k}, cX_0) \leqslant \epsilon\}$. We will show that $B(\epsilon)$ is contained in $A(\epsilon_1)$ for some $0 < \epsilon_1 < \epsilon$. Let $m_p \in B(\epsilon)$, then $\bar{d}(cX_{m_k}, cX_0) \leqslant \epsilon$, which implies that $|c|\bar{d}(X_{m_k}, X_0) \leqslant \epsilon$, that is, $\bar{d}(X_{m_k}, X_0) \leqslant \frac{\epsilon}{|c|} = \epsilon_1$(say). Therefore, $m \in A(\epsilon_1)$. Since X_n is I^*-statistically convergent therefore, $A(\epsilon_1) \in F(I)$ and by this $B(\epsilon) \in F(I)$. Hence, $I^* - st - lim_{n\to\infty} cX_n = cX_0$.

(iii) For $\alpha \in [0,1]$, let $X_n{}^\alpha$, $\mathcal{Y}_n{}^\alpha$, $X_0{}^\alpha$, and $\mathcal{Y}_0{}^\alpha$ be the α level sets of X_n, $\mathcal{Y}_n$, X_0, and $\mathcal{Y}_0$, respectively. Since $\bar{d}(X_n{}^\alpha + \mathcal{Y}_n{}^\alpha, X_0{}^\alpha + \mathcal{Y}_0{}^\alpha) \leqslant \bar{d}(X_n{}^\alpha, X_0{}^\alpha) + \bar{d}(\mathcal{Y}_n{}^\alpha, \mathcal{Y}_0{}^\alpha)$, therefore, $\bar{d}(X_n + \mathcal{Y}_n, X_0 + \mathcal{Y}_0) \leqslant \bar{d}(X_n, X_0) + \bar{d}(\mathcal{Y}_n, \mathcal{Y}_0)$. Let $\epsilon > 0$ be given. Since X_n and $\mathcal{Y}_n$ are I^*-statistically convergent, therefore, there exists $\mathcal{K} = \{m_1 < m_2 < \cdots < m_k < \ldots\} \subset \mathbb{N}$ such that $\lim_{n\to\infty} \frac{1}{n}|\{m_k < n : \bar{d}(X_{m_k}, X_0) \leqslant \epsilon\} \in F(I)| = 1$ and $\lim_{n\to\infty} \frac{1}{n}|\{m_k < n : \bar{d}(\mathcal{Y}_{m_k}, \mathcal{Y}_0) \leqslant \epsilon\} \in F(I)| = 1$. Take $A(\frac{\epsilon}{2}) = \{m_k : \bar{d}(X_{m_k}, X_0) < \frac{\epsilon}{2}\}$, $B(\frac{\epsilon}{2}) = \{m_k : \bar{d}(\mathcal{Y}_{m_k}, \mathcal{Y}_0) < \frac{\epsilon}{2}\}$ and $C(\epsilon) = \{m_k : \bar{d}(X_{m_k} + \mathcal{Y}_{m_k}, X_0 + \mathcal{Y}_0) < \epsilon\}$. Since, $A(\frac{\epsilon}{2}) \in F(I)$ and $B(\frac{\epsilon}{2}) \in F(I)$, therefore, $A(\frac{\epsilon}{2}) \cap B(\frac{\epsilon}{2}) \neq \phi$ and belongs to the filter; thus, we have for all $n \in A(\frac{\epsilon}{2}) \cap B(\frac{\epsilon}{2}) \subset C(\frac{\epsilon}{2}) \in F(I)$, i.e., $\lim_{n\to\infty} \frac{1}{n}|\{m_k < n : \bar{d}(X_{m_k} + \mathcal{Y}_{m_k}, X_0 + \mathcal{Y}_0) \leqslant \epsilon\} \in F(I)| = 1$. Hence, $I^* - st - lim(X_n + \mathcal{Y}_n) = (X_0 + \mathcal{Y}_0)$.

□

Theorem 3. *For any sequence $X = (X_n) \in L(\$)$ if there exists two sequences $\mathcal{Y} = (\mathcal{Y}_n)$, $Z = (Z_n) \in L(\$)$ of fuzzy numbers such that $X = \mathcal{Y} + Z$, $\bar{d}(\mathcal{Y}_n, X_0) \to 0$ as $n \to 0$ and $SuppZ = \{n \in \mathbb{N} : Z_n \neq 0\} \in I$ and $\delta(SuppZ) = 0$, then X is I^*-statistically convergent.*

Proof. Let $\mathcal{Y} = (\mathcal{Y}_n), Z = (Z_n) \in L(\$)$ such that $X = \mathcal{Y} + Z$, $\bar{d}(\mathcal{Y}_n, X_0) \to 0$ as $n \to 0$ and $\lim_{n\to\infty} \frac{1}{n}|\{n \in \mathbb{N} : Z_n \neq 0\} \in I| = 0$.

Let $\mathcal{K} = \{n \in \mathbb{N} : Z_n = 0\}$. Since $SuppZ$ belongs to I then $\mathcal{K} \in F(I)$ with $\delta(\mathcal{K}) = 1$ and also $\mathcal{K}$ is an infinite set as otherwise $\mathcal{K} \in I$. Let $\mathcal{K} = \{m_1 < m_2 < m_3 \cdots < m_k < \ldots\} \subset \mathbb{N}$ such that $\delta(k) = 1$ then $X_{m_k} = \mathcal{Y}_{m_k}$ for each $n \in \mathbb{N}$. Since $Z_n = 0$ for all $n \in K$. It is given that $\bar{d}(\mathcal{Y}_n, \mathcal{X}_0) \to 0$. Therefore, $\bar{d}(\mathcal{X}_{m_{\mathcal{K}}}, \mathcal{X}_0) \to \infty$. Thus, $\lim_{n\to\infty} \frac{1}{n}|\{m_{\mathcal{K}} < n : \bar{d}(\mathcal{X}_{m_{\mathcal{K}}}, \mathcal{X}_0) \leqslant \epsilon\} \in F(I)| = 1$. This proves that $\mathcal{X}$ is I^*-statistically convergent. □

4. I^*-Statistically Cauchy and I^*-Statistically Pre-Cauchy Sequences

Definition 12. *A sequence $\mathcal{X} = (\mathcal{X}_n)$ is said to be I^*-statistically Cauchy if there exists a set $\mathcal{K} = \{m_1 < m_2 < m_3 < \ldots < m_{\mathcal{K}} < \ldots\} \subset \mathbb{N}$ and for each $\epsilon > 0$, there exists $m_p \in \mathbb{N}(\epsilon)$ such that $\lim_{n\to\infty} \frac{1}{n}|\{m_{\mathcal{K}} < n : \bar{d}(\mathcal{X}_{m_{\mathcal{K}}}, \mathcal{X}_{m_p}) \leqslant \epsilon\} \in F(I)| = 1$. I^*_{ca} denotes the collection of all I^*-statistically Cauchy sequences.*

Definition 13. *A sequence $\mathcal{X} = (\mathcal{X}_n)$ is said to be I^*-statistically pre-Cauchy if there exists a set $\mathcal{K} = \{m_1 < m_2 < m_3 < \ldots < m_{\mathcal{K}} < \ldots\} \subset \mathbb{N}$ and for each $\epsilon > 0$ we have $\lim_{n\to\infty} \frac{1}{n^2}|\{(m_{\mathcal{K}}, m_p) : \bar{d}(\mathcal{X}_{m_{\mathcal{K}}}, \mathcal{X}_{m_p}) \leqslant \epsilon; m_{\mathcal{K}}, m_p \leqslant n\} \in F(I)| = 1$.*

Theorem 4. *Every I^*-statistically convergent sequence is I^*-statistically Cauchy.*

Proof. Let $\mathcal{X} = (\mathcal{X}_n)$ be I^*-statistically convergent to $\mathcal{X}_0$. Then, there exists a set $\mathcal{K} = \{m_1 < m_2 < m_3 \cdots < m_{\mathcal{K}} < \ldots\} \subset \mathbb{N}$ and for each $\epsilon > 0$ we have $\lim_{n\to\infty} \frac{1}{n}|\{m_{\mathcal{K}} < n : \bar{d}(\mathcal{X}_{m_{\mathcal{K}}}, \mathcal{X}_0) \leqslant \epsilon\} \in F(I)| = 1$. Let $C = \{m_{\mathcal{K}} < n : \bar{d}(\mathcal{X}_{m_{\mathcal{K}}}, \mathcal{X}_0) \leqslant \frac{\epsilon}{2}\} \in F(I)$ and $\delta(C) = 1$. Since I is an admissible ideal, therefore, we can choose $\bar{d}(\mathcal{X}_{m_{\mathcal{K}}}, \mathcal{X}_0) \leqslant \frac{\epsilon}{2}$. Define $B = \{m_{\mathcal{K}} < n : \bar{d}(\mathcal{X}_{m_{\mathcal{K}}}, \mathcal{X}_0) \leqslant \epsilon\}$. We need to show that $C \subset B$. Let $\bar{d}(\mathcal{X}_{m_{\mathcal{K}}}, \mathcal{X}_0)$ be any arbitrary element of C, then $\bar{d}(\mathcal{X}_{m_{\mathcal{K}}}, \mathcal{X}_0) < \frac{\epsilon}{2}$, $\bar{d}(\mathcal{X}_{m_{\mathcal{K}}}, \mathcal{X}_0) + \bar{d}(\mathcal{X}_{m_p}, \mathcal{X}_0) \leqslant \frac{\epsilon}{2} + \frac{\epsilon}{2}$, and $\bar{d}(\mathcal{X}_{m_{\mathcal{K}}}, \mathcal{X}_{m_p}) \leqslant \epsilon$, which shows that every element of C is as element of B. Therefore, $C \subset B$. According to the Definition (8) $B \in F(I)$ and since $\delta(C) = 1$, this implies that $\delta(B) = 1$. Hence, we have $\lim_{n\to\infty} \frac{1}{n}|\{m_{\mathcal{K}} < n : \bar{d}(\mathcal{X}_{m_{\mathcal{K}}}, \mathcal{X}_{m_p}) \leqslant \epsilon\} \in F(I)| = 1$. □

Theorem 5. *Every I^*-statistically Cauchy sequence is I^*-statistically pre-Cauchy.*

Proof. Let $\mathcal{X} = (\mathcal{X}_n)$ be any arbitrary sequence of I^*_{ca}. Then, there exists a set $\mathcal{K} = \{m_1 < m_2 < m_3 \cdots < m_{\mathcal{K}} < \ldots\} \subset \mathbb{N}$ and for each $\epsilon > 0$ we have $\lim_{n\to\infty} \frac{1}{n}|\{m_{\mathcal{K}} < n : \bar{d}(\mathcal{X}_{m_{\mathcal{K}}}, \mathcal{X}_0) \leqslant \epsilon\} \in F(I)| = 1$. Let $C = \{m_{\mathcal{K}} < n : \bar{d}(\mathcal{X}_{m_{\mathcal{K}}}, \mathcal{X}_0) \leqslant \frac{\epsilon}{2}\} \in F(I)$ and $\delta(C) = 1$. Now without any loss of generality define T such that $\mathcal{X}_{m_p}$ be any term of the sequence $\mathcal{X}_n$ and $T = \{(m_{\mathcal{K}}, m_p) : \bar{d}(\mathcal{X}_{m_{\mathcal{K}}}, \mathcal{X}_{m_p}) \leqslant \epsilon; m_{\mathcal{K}}, m_p \leqslant n\}$ and by Definition (8) $T \in F(I)$ and $\delta(K) = 1$. That is, $\lim_{n\to\infty} \frac{1}{n^2}|\{(m_{\mathcal{K}}, m_p) : \bar{d}(\mathcal{X}_{m_{\mathcal{K}}}, \mathcal{X}_{m_p}) \leqslant \epsilon; m_{\mathcal{K}}, m_p \leqslant n\} \in F(I)| = 1$, which shows that every I^*-statistically Cauchy sequence is I^*-statistically pre-Cauchy. □

Remark 1. *Every I^*-statistically pre-Cauchy sequence need not be I^*-statistically Cauchy.*

To understand this we will consider the following example.

Example 2. *Let $\mathcal{X} = (\mathcal{X}_n)$ be a sequence defined as:*

$$\mathcal{X}_n = \begin{cases} (0, 1, 2) & \text{if } n \text{ is a odd,} \\ (0, 0.5, 1) & \text{if } n \text{ is a even.} \end{cases}$$

where (a, b, c) denotes a triangular fuzzy number [32] with peak at b and support $[a, c]$. Let $\epsilon > 0$ be arbitrary. Without the loss of generality, we can choose $n_0 \in \mathbb{N}$ such that $n \geqslant n_0$, we have:

$$\frac{1}{n^2}|\{(m_{\mathcal{K}}, m_p) : \bar{d}(\mathcal{X}_{m_{\mathcal{K}}}, \mathcal{X}_{m_p}) \leqslant \epsilon; m_{\mathcal{K}}, m_p \leqslant n\}|$$

$$\geqslant \frac{1}{n^2}|\{(m_k, m_p) : \bar{d}(x_{m_k}, x_{m_p}) \leqslant \epsilon; m_k, m_p \leqslant n_0\}|$$

Let K be the collection of all odd natural numbers, $K = \{m_1 < m_2 < m_3 < \dots\} \subset \mathbb{N}$*(say). This implies* $K \in F(I)$*. Since* $m_k, m_p \leqslant n_0$ *and belongs to K implies that* m_k, m_p *are both odd, and therefore:*

$$\bar{d}(x_{m_k}, x_{m_p}) = \bar{d}((0,1,2),(0,1,2)) = 0 \leqslant \epsilon$$

Let $C = \{(m_k, m_p) : \bar{d}(x_{m_k}, x_{m_p}) \leqslant \epsilon; x_{m_k}, x_{m_p} \leqslant n_0\}$ *and* C^c *denotes the compliment of C. We will show that* $\lim_{n\to\infty} \frac{1}{n^2}|C^c| = 0$*. Since* C^c *contains all even numbers less than or equal to* n_0*. Thus, we have:*

$$\frac{1}{n^2}|C^c| \leqslant \frac{n_0/2}{n^2} \leqslant \frac{1}{2n}.$$

Since n_0 *is fixed, the right-hand side approaches 0 as* $n \to \infty$*. Therefore, we have* $\lim_{n\to\infty} \dfrac{1}{n^2}|\{(m_k, m_p) : \bar{d}(x_{m_k}, x_{m_p}) \leqslant \epsilon; m_k, m_p \leqslant n\} \in F(I)| = 1$*, which shows that x is* I^**-statistically pre-Cauchy.*

However, x is not I^**-statistically Cauchy. Suppose for the sake of contradiction that x is* I^**-statistically Cauchy. Then, there exists a set* $\mathcal{K} = \{m_1 < m_2 < m_3 < \dots\} \subset \mathbb{N}$ *and for each* $\epsilon > 0$*, there exists* $m_p \in \mathbb{N}(\epsilon)$ *such that:*

$$\lim_{n\to\infty} \frac{1}{n}|\{m_k < n : \bar{d}(x_{m_k}, x_{m_p}) \leqslant \epsilon\} \in F(I)| = 1$$

Without the loss of generality we can choose $n_0 \in \mathbb{N}$ *such that* $n \geqslant n_0$*, we have:*

$$\lim_{n\to\infty} \frac{1}{n}|\{m_k < n : \bar{d}(x_{m_k}, x_{m_p}) \leqslant \epsilon\}| \geqslant \lim_{n\to\infty} \frac{1}{n}|\{m_k < n_0 : \bar{d}(x_{m_k}, x_{m_p}) \leqslant \epsilon\}|$$

Let $D = \{m_k \leqslant n_0 : \bar{d}(x_{m_k}, x_{m_p}) \leqslant \epsilon\}$ *and* D^c *denotes the compliment of D. We will show that* $\lim_{n\to\infty} \frac{1}{n}|D^c| = 0$*. Since* D^c *contains all even numbers less than or equal to* n_0*. Thus, we have:*

$$\frac{1}{n}|D^c| \leqslant \frac{n_0/2}{n} \leqslant \frac{1}{2}$$

which is a contradiction, so x is not I^**-statistically Cauchy.*

Theorem 6. *Every* I^**- statistically convergent sequence is* I^**-statistically pre-Cauchy.*

Proof. The proof is trivial from Theorem 4 and 5. See the Appendix A. □

To illustrate the concept of a sequence that is I^*-statistically pre-Cauchy but not I^*-statistically convergent, we can consider the the following example. Understanding that any I^*-statistically convergent sequence must contain a subsequence that converges in the usual sense is crucial. Let us look at the example below.

Example 3. *Let* $x = (x_k)$ *be a sequence. Consider the sequence* $x = (x_k)$ *defined such that for* $(m_k - 1)! < k < m_k!$*, we have* $x_k = \sum_{n=1}^{m_k} \frac{1}{n}$*. This sequence* $x = (x_k)$ *does not possess any convergent subsequences, implying that x is not* I^**-statistically convergent. However, despite the lack of convergent subsequences, the sequence is* I^**-statistically pre-Cauchy. This means that while the entire sequence does not converge in the* I^**-statistical sense, it still satisfies the pre-Cauchy criterion under* I^**-statistical conditions.*

Let $\epsilon > 0$ *be given and let* $\mathcal{K} = \{m_1 < m_2 < m_3 \dots < m_k < \dots\} \subset \mathbb{N} \in F(I)$*,* $m_k \in \mathbb{N}$ *satisfy* $\frac{1}{m_k} < \epsilon$*. Now, consider the case where* $m_k! < n < (m_k + 1)!$ *and* $(m_k - 1)! < j, k < n$*, then* $d(x_{m_j}, x_{m_k}) \leqslant \frac{1}{m_k} < \epsilon$*. It follows that, for* $m_k! < n < (m_k + 1)!$*,* $\frac{1}{n^2}|\{(m_j, m_k) : d(x_{m_j}, x_{m_k}) < \epsilon, m_j, m_k \leqslant n)\}|$*, we have* $\geqslant \frac{1}{n^2}[n - (m_k - 1)!]^2$*,* $\geqslant \left[1 - \frac{(m_k-1)!}{m_k!}\right]^2$*, and* $= [1 - \frac{1}{m_k}]^2$*.*

Since $lim_{k\to\infty}(1-\frac{1}{m_k})=1$. *As a result, X is I^*-statistically pre-Cauchy.*

Before we present the next theorem, we need to introduce the definition of the $I^*-lim\ inf$. Let us first outline this concept.

Definition 14. *Let I be an admissible ideal of $\mathbb{N}$ and let $X=(X_n)\in L(\$)$. Let $\mathcal{A}_x=\{\mu\in L(\mathbb{R}):\{k:x_k<\mu\}\in F(I)\}$ then the $I^*-lim\ inf$ is given by:*

$$I^*-lim\ inf X=\begin{cases} inf\ \mathcal{A}_x & if\ \mathcal{A}_x\neq\phi \\ \infty & if\ \mathcal{A}_x=\phi \end{cases}$$

It is known that "$I^*-lim\ inf X=\eta$(finite) if and only if for arbitrary $\epsilon>0$ $\{k:x_k<\eta+\epsilon\}\in F(I)$ and $\{k:x_k<\eta-\epsilon\}\notin F(I)$".

Theorem 7. *Suppose $X=(X_k)\in L(\$)$ is I^*-statistically pre-Cauchy. If X has a subsequence (X_{p_k}) that converges to X_0 and $0<I^*-lim\ inf\frac{1}{n}|\{p_k\leqslant n:k\in\mathbb{N}\}|<\infty$ then X is I^*-statistically convergent to X_0.*

Proof. Let $\epsilon>0$ be given. Since $limX_{p_k}=X_0$ choose $r\in\mathbb{N}$ such that $\bar{d}(X_j,X_0)<\frac{\epsilon}{2}$ whenever $j>r$ and $j=p_k$ for some k. Let $\mathcal{A}=\{p_k:p_k>r,k\in\mathbb{N}\}$ and $\mathcal{A}(\epsilon)=\{k:\bar{d}(X_k,X_0)\geqslant\epsilon\}$. Now note that $\frac{1}{n^2}|\{n\in\mathbb{N}:\bar{d}(X_{m_k},X_{m_p})\leqslant\frac{\epsilon}{2},m_k,m_p<n\}|$

$\leqslant\frac{1}{n^2}\sum_{\mathcal{A}(\epsilon)\times\mathcal{A}}(m_j,m_k)$

$=\frac{1}{n}|\{p_k\leqslant n:p_k\in\mathcal{A}\}|.\frac{1}{n}|\{k\leqslant n:\bar{d}(X_k,X_0)\geqslant\epsilon\}|$.

Since X is I^*-statistically pre-Cauchy, then there exists a set $\mathcal{K}=\{m_1<m_2<\cdots<m_k<\dots\}$ $\subset\mathbb{N}$ and for each $\epsilon>0$ we have $\lim_{n\to\infty}\frac{1}{n^2}|\{(m_k,m_j):\bar{d}(X_{m_k},X_{m_j})\leqslant\epsilon;m_k,m_p<n\}\in F(I)|=1$. Let $C=\{(m_k,m_j):\bar{d}(X_{m_k},X_{m_j})\leqslant\frac{\epsilon}{2};m_k,m_p\}\in F(I)$ and $\delta(C)=1$. Again, since $I^*-lim\ inf\frac{1}{n}|\{p_k\leqslant:k\in\mathbb{N}\}|=b>0$(say). So, $\{n\in\mathbb{N}:\frac{1}{n}|\{p_k\leqslant n:k\in\mathbb{N}\}|>\frac{b}{2}\}=D$(say) $\in F(I)$. As C and D belongs to $F(I)$ so, $C\cap D\in F(I)$, i.e., consequently $\lim_{n\to\infty}\frac{1}{n}|\{m_k<n:\bar{d}(X_{m_k},X_0)\leqslant\epsilon\}\in F(I)|=1$. This shows that X is I^*-statistically convergent. □

5. Conclusions

Our study has thoroughly examined the concept of I^*-statistical convergence for sequences of fuzzy numbers within a metric space. Our investigation confirms the uniqueness of the limit under I^*-statistical convergence, establishing a firm foundation for understanding this advanced mathematical concept. Through the development of several key theorems, we have elucidated the relationship between I^*-statistical convergence and classical convergence, alongside the algebraic properties intrinsic to I^*-statistically convergent sequences.

Additionally, our work has introduced and analyzed I^*-statistically pre-Cauchy and I^*-statistically Cauchy sequences, highlighting their intricate connection to I^*-statistical convergence. Notably, we demonstrated that while every I^*-statistically convergent sequence is necessarily I^*-statistically pre-Cauchy, the reverse does not universally apply. To further enrich the theoretical framework, we provided a sufficient condition for an I^*-statistically pre-Cauchy sequence to achieve I^*-statistical convergence, utilizing the concept of I^*-lim inf. These findings contribute significantly to the broader understanding of convergence in the context of fuzzy number sequences and open avenues for future research in this area.

The future scope of this study includes examining the monotonicity and boundedness of sequences of fuzzy numbers within the framework of I^*-statistical convergence. Additionally, this concept can be extended to explore convergence in the context of double and triple sequences, broadening the applicability of I^*-statistical convergence. Further research could also investigate these convergence properties in various other mathematical spaces, potentially unveiling new theoretical insights and applications.

Author Contributions: Writing—original draft preparation, T. and A.A.; Writing—review and editing, A.E. All authors have read and agreed to the published version of the manuscript.

Funding: This research received no external funding.

Data Availability Statement: The manuscript has no associated data.

Acknowledgments: The authors would like to thank the editor and reviewers for their valuable comments and suggestions, which improved this paper significantly.

Conflicts of Interest: The authors declare no conflicts of interest.

Appendix A. Proof of Theorem 6.

Proof. From Theorem 4, we know that every I^*-statistically convergent sequence is I^*-statistically Cauchy. Additionally, Theorem 5 establishes that every I^*-statistically Cauchy sequence is I^*-statistically pre-Cauchy. Therefore, it follows that every I^*-statistically convergent sequence is also I^*-statistically pre-Cauchy. □

References

1. Matloka, M. Sequences of fuzzy numbers. *Busefal* **1986**, *28*, 28–37.
2. Nanda, S. On sequences of fuzzy numbers. *Fuzzy Sets Syst.* **1989**, *33*, 123–126. [CrossRef]
3. Fast, H. Sur la convergence statistique. *Colloq. Math.* **1951**, *2*, 241–244. [CrossRef]
4. Schoenberg, I.J. The Integrability of Certain Functions and Related Summability Methods. *Am. Math. Mon.* **1959**, *66*, 361–775. [CrossRef]
5. Savas, E. On statistically convergent sequences of fuzzy numbers. *Inf. Sci.* **2001**, *137*, 277–282. [CrossRef]
6. Connor, J.; Fridy, J.; Kline, J. Statistically pre-Cauchy sequences. *Analysis* **1994**, *14*, 311–318. [CrossRef]
7. Fridy, J.A. On statistical convergence. *Analysis* **1985**, *5*, 301–314. [CrossRef]
8. Šalát, T. On statistically convergent sequences of real numbers. *Math. Slovaca* **1980**, *30*, 139–150.
9. Mursaleen, M.A.; Serra-Capizzano, S. Statistical convergence via q-calculus and a Korovkin's type approximation theorem. *Axioms* **2022**, *11*, 70. [CrossRef]
10. Savas, E.; Mursaleen, M. Bézier type Kantorovich q-Baskakov operators via wavelets and some approximation properties. *Bull. Iran. Math. Soc.* **2023**, *49*, 68. [CrossRef]
11. Mursaleen, M.A.; Kilicman, A.; Nasiruzzaman, M. Approximation by q-Bernstein-Stancu-Kantorovich operators with shifted knots of real parameters. *arXiv* **2022**, arXiv:2201.03935. [CrossRef]
12. Nasiruzzaman, M.; Aljohani, A. Approximation by α-α-Bernstein–Schurer operators and shape preserving properties via q-analogue. *Math. Methods Appl. Sci.* **2023**, *46*, 2354–2372. [CrossRef]
13. Mursaleen, M.A.; Lamichhane, B.; Kilicman, A.; Senu, N. On q-statistical approximation of wavelets aided Kantorovich q-Baskakov operators. *Filomat* **2024**, *38*, 3261–3274.
14. Burgin, M.; Duman, O. Statistical convergence and convergence in statistics. *arXiv* **2006**, arXiv:math/0612179.
15. Cai, Q.-B. Approximation properties of Kantorovich-type q-Bernstein-Stancu-Schurer operators. *J. Comput. Anal. Appl.* **2017**, *23*, 847–859.
16. Kostyrko, P.; Wilczynski, W. I-convergence. *Real Anal. Exch.* **2000**, *26*, 669–686. [CrossRef]
17. Dems, K. On I-Cauchy sequences. *Real Anal. Exch.* **2004**, *30*, 1.
18. Kostyrko, P.; Macaj, M.; Salat, T.; Sleziak, M. I-convergence and extremal I-limit points. *Math. Slovaca* **2005**, *55*, 443–464.
19. Kumar, V. On i and i*-convergence of double sequences. *Math. Commun.* **2007**, *12*, 171–181.
20. Kumar, V.; Singh, N. I-core of double sequences. *Int. J. Contemp. Math. Sci.* **2007**, *2*, 1137–1145. [CrossRef]
21. Aytar, S.; Pehlivan, S. On i-convergent sequences of real numbers. *Ital. J. Pure Appl. Math.* **2007**, *21*, 191.
22. Kumar, V.; Kumar, K. On the ideal convergence of sequences of fuzzy numbers. *Inf. Sci.* **2008** [CrossRef]
23. Kocinac, L.D.; Rashid, M.H.M. On ideal convergence of double sequences in the topology induced by a fuzzy 2-norm. *TWMS J. Pure Appl. Math.* **2017**, *8*, 97–111.
24. Altaweel, N.H.; Rashid, M.H.M.; Albalawi, O.; Alshehri, M.G.; Eljaneid, N.H.E.; Albalawi, R. On the Ideal Convergent Sequences in Fuzzy Normed Space. *Symmetry* **2023**, *15*, 936. [CrossRef]
25. Savas, E.; Das, P. A generalized statistical convergence via ideals. *Appl. Math. Lett.* **2011**, *24*, 826–830. [CrossRef]
26. Debnath, S.; Debnath, J. On I-statistically convergent sequence spaces defined by sequences of Orlicz functions using matrix transformation. *Proyecciones (Antofagasta)* **2014**, *33*, 277–285. [CrossRef]
27. Debnath, S.; Rakshit, D. On I-statistical convergence. *Iran. J. Math. Sci. Inform.* **2017**, *13*, 101–109.

28. Rashid, M.H.M.; Al-Subh, S.A. Statistical Convergence with Rough I3-Lacunary and Wijsman Rough I3-Statistical Convergence in 2-Normed Spaces. *Int. J. Anal. Appl.* **2024**, *22*, 115. [CrossRef]
29. Choudhurya, C.; Debnatha, S.; Esib, A. Further results on I^- deferred statistical convergence. *Filomat* **2024**, *38*, 769–777. [CrossRef]
30. Kişi, Ö.; Choudhury, C. A Study on Rough I-Deferred Statistical Convergence in Gradual Normed Linear Spaces. *Proc. Natl. Acad. Sci. India Sect. A Phys. Sci.* **2024**, *94*, 113–126. [CrossRef]
31. Puri, M.L.; Ralescu, D.A. Differentials of fuzzy functions. *J. Math. Anal. Appl.* **1983**, *91*, 552–558. [CrossRef]
32. Guo, S.; Zhao, H. Fuzzy best-worst multi-criteria decision-making method and its applications. *Knowl.-Based Syst.* **2017**, *121*, 23–31. [CrossRef]

axioms

MDPI

Article

Extensions of Orders to a Power Set vs. Scores of Hesitant Fuzzy Elements: Points in Common of Two Parallel Theories

Esteban Induráin [1,*,†], Ana Munárriz [2,†] and M. Sergio Sara [3,†]

1 InaMat2 (Institute for Advanced Materials and Mathematics), Departamento de Estadística, Informática y Matemáticas, Universidad Pública de Navarra, 31006 Pamplona, Spain
2 Inarbe (Institute for Advanced Research in Business and Economics), Departamento de Estadística, Informática y Matemáticas, Universidad Pública de Navarra, 31006 Pamplona, Spain; ana.munarriz@unavarra.es
3 Departamento de Estadística, Informática y Matemáticas, Universidad Pública de Navarra, 31006 Pamplona, Spain; sara.142587@e.unavarra.es
* Correspondence: steiner@unavarra.es; Tel.: +34-948169551
† These authors contributed equally to this work.

Abstract: We deal with two apparently disparate theories. One of them studies extensions of orderings from a set to its power set. The other one defines suitable scores on hesitant fuzzy elements. We show that both theories have the same mathematical substrate. Thus, important possibility/impossibility results concerning criteria for extensions can be transferred to new results on scores. And conversely, conditions imposed a priori on scores can give rise to new extension criteria. This enhances and enriches both theories. We show examples of translations of classical results on extensions in the context of scores. Also, we state new results concerning the impossibility of finding a utility function representing some kind of extension order if some restrictions are imposed on the utility function considered as a score.

Keywords: extensions of orders from a set to its power set; criteria of extensions of orderings; possibility and impossibility results; hesitant fuzzy elements and sets; scores

MSC: 03A72; 06A06; 91B02

Citation: Induráin, E.; Munárriz, A.; Sara, M.S. Extensions of Orders to a Power Set vs. Scores of Hesitant Fuzzy Elements: Points in Common of Two Parallel Theories. *Axioms* **2024**, *13*, 549. https://doi.org/10.3390/axioms13080549

Academic Editors: Ta-Chung Chu and Wei-Chang Yeh

Received: 5 June 2024
Revised: 5 August 2024
Accepted: 8 August 2024
Published: 13 August 2024

1. Introduction

1.1. Motivation of the Manuscript

Sometimes it happens that two apparently disparate theories have a parallel development. In a sense, one could be considered a part of the other, even if they belong to totally different frameworks. When this occurs, the results of each theory can be applied to the other, thereby reinforcing both.

In the present manuscript, we compare two such parallel theories.

One theory involves extending total orders from a set to its power set, following some criteria established a priori. This had already been studied by people working in mathematical economics from 1980 on (see [1,2]) when dealing with the problem of ranking sets of objects. Depending on the chosen criteria, we may reach either possibility or impossibility results.

On the other hand, we have the theory of scores on hesitant fuzzy sets. In terms of extensions of orderings, we start with the usual order on the unit interval $[0,1]$ and, using scores (see [3]), we get a new ordering on the subsets of $[0,1]$ (also known as the hesitant fuzzy elements). Again, we seek to ensure that the scores accomplish some pre-defined criteria. Depending on these criteria, we may have a suitable score or not (possibility or impossibility, again).

At this stage, we notice that the second setting derives from the first.

In fact, some criteria used by mathematical economists may be unknown to fuzzy set theorists, and vice versa. This observation suggests the potential to broaden the scope of criteria and possibility/impossibility results.

1.2. Aim and Objectives

This work arises from our professional experiences as mathematicians working with two different collectives, namely economists and engineers, who sometimes, unfortunately, overlook each other's mathematical applications and achievements. These distinct achievements—economists have their own, and engineers have theirs—could actually be mutually beneficial and lead to future collaborations. This is especially true once we recognize and highlight that the mathematical foundation is the same, and the results obtained by economists could indeed be translated into the context used by engineers, and vice versa.

We aim to highlight the fact that problems in mathematical economics, such as extending orders from a set to its power set in the spirit of "ranking sets of objects", and problems related to defining and analyzing scores of hesitant fuzzy elements encountered by mathematicians and engineers in fuzzy set theory and artificial intelligence approaches are, in a sense, equivalent. Furthermore, technical results obtained in one approach can often be translated into new results in the other approach, enriching both theories, which often ignore each other despite sharing the same mathematical substrate.

Specifically, we want to show the analogy between the study of extensions of orderings from a set to its power set and the definition of scores of hesitant fuzzy elements.

1.3. Contents of the Manuscript

Bearing these ideas in mind, the contents of the manuscript are as follows:

After the Introduction and the section on Preliminaries, we introduce, on the one hand, classical criteria to extend orderings from a set to its power set (Section 3) and analyze their compatibility (Section 4). On the other hand, we introduce different classes of scores for hesitant fuzzy sets (Section 5) and analyze compatibility among extra required features on scores. In Section 6, we explore the analogies between both approaches, showing that some (in)compatibility results in one setting could be translated in some way to the other setting. A final section of concluding remarks and lines for future research closes the paper.

2. Preliminaries

2.1. Extension of Orderings from a Set to Its Power Set

Let X denote a nonempty set.

Definition 1. *A preorder $\precsim$ on X is a binary relation on X that is reflexive and transitive. An antisymmetric preorder is said to be an order. A complete (or total) preorder $\precsim$ on a set X is a preorder such that if $a, b \in X$, then $(a \precsim b) \vee (b \precsim a)$ holds. If $\precsim$ is a preorder on X, then as usual, we denote the associated asymmetric relation by $\prec$ and the associated equivalence relation by $\sim$; these are defined by $a \prec b \Leftrightarrow (a \precsim b) \wedge \neg(b \precsim a)$ and $a \sim b \Leftrightarrow (a \precsim b) \wedge (b \precsim a)$. If this asymmetric relation is transitive, then $\precsim$ is said to be a quasi-transitive preorder. For any preorder $\precsim$ on X, the indifference part of $\precsim$, denoted by $ind(\precsim)$ is the binary relation on X defined by $(a, b) \in ind(\precsim)$ if and only if $a \prec c \iff b \prec c$ as well as $c \prec a \iff c \prec b$, for any $c \in X$. A complete preorder $\precsim$ defined on X is usually called a preference, and it is said to be representable if there exists a real-valued function $u : X \to \mathbb{R}$ such that $a \precsim b \Leftrightarrow u(a) \leq u(b)$ holds for every $a, b \in X$. Here u is said to be a utility function representing $\precsim$.*

From now on, we will consider X endowed with a complete preorder $\precsim$. Sometimes we will work in the particular case in which $\precsim$ is a total order.

Definition 2. *An extension of the complete preorder $\precsim$ from the set X to its power set $\mathcal{P}(X)$ is another preorder $\precsim_E$, now defined in $\mathcal{P}(X)$ such that the following property holds true for any elements $a, b \in X$: $a \precsim b \Leftrightarrow \{a\} \precsim_E \{b\}$.*

Remark 1.

(i) *Among possible extensions, the literature pays special attention to the following situations: (a) Both $\precsim$ and $\precsim_E$ are total orders; (b) both $\precsim$ and $\precsim_E$ are complete preorders; (c) $\precsim$ is a total order, but $\precsim_E$ is a complete preorder.*

(ii) *Some authors do not consider $\emptyset$ in the extensions, so they define the extensions not on the whole power set $\mathcal{P}(X)$ but instead on $\pi(X) = \mathcal{P}(X) \setminus \{\emptyset\}$.*

(iii) *The most classical criteria appeared from 1950 on, mainly in several papers related to social choice and decision making (see, e.g., [1,4]). They usually consider a total order $\precsim$ defined on a finite set X, whereas the extension $\precsim_E$ can be either a total order or just a complete preorder depending on the context.*

2.2. Hesitant Fuzzy Elements, Hesitant Fuzzy Sets and Scores

Now we present some nomenclature (see, e.g., [5]) to be used henceforward.

By X, we will denote a nonempty set, also called the universe.

Definition 3 ([6])**.** *A (type-1) fuzzy subset H of X is defined as a function $\mu_H : X \to [0,1]$. Here μ_H is called the membership function of H. If μ_H takes values just in $\{0,1\}$, the corresponding subset μ_H is a classical crisp (i.e., non-fuzzy) subset of X.*

Throughout the manuscript, the following notation will be used:

- $\mathcal{P}([0,1])$ denotes the collection of all subsets of the unit interval $[0,1]$,
- $\Pi([0,1])$ denotes the collection of all nonempty subsets of $[0,1]$,
- $\mathcal{I}([0,1])$ constitutes the subset of all intervals in $[0,1]$,
- $\mathcal{I}_C([0,1])$ consists of all closed intervals in $[0,1]$,
- $\mathcal{I}^{\cup}([0,1])$ consist of all finite unions of intervals in $[0,1]$,
- $\mathcal{F}([0,1])$ denotes the collection of all nonempty finite subsets of $[0,1]$, and
- $\mathcal{F}_n([0,1])$ denotes the collection of all nonempty subsets of $[0,1]$ with at most n elements.

If $X, Y \subseteq \mathbb{R}$, we write $Y > X$ to mean $y > x$ for all $y \in Y$, $x \in X$. Notice that $Y > X$ implies $X \cap Y = \emptyset$.

The main objects to be handled in our analysis are called the hesitant fuzzy elements (HFEs for short):

Definition 4 ([7,8])**.** *A subset E of $[0,1]$ is said to be a hesitant fuzzy element (HFE). Also, a function $h \colon X \to \mathcal{P}([0,1])$ is called a hesitant fuzzy set (HFS) over X.*

Remark 2. *Classical fuzzy sets extend crisp subsets with the assistance of a first level of uncertainty: the membership function μ_H maps any element x of the universe X with its "uncertainty degree". Thus, mapping x into a number from $[0,1]$ graduates the acceptability of the claim that this element belongs to the fuzzy subset H, which therefore generalizes the idea of classical subsets. HFSs are a particular case of "type-2 fuzzy sets" [9–11].*

Basically, a type-1 fuzzy set on a universe X is a map h from X into $[0,1]$. Concerning type-2 fuzzy subsets, they are defined as functions from X to the family of type-1 fuzzy subsets of $[0,1]$; see, e.g., [12–14]. In that sense, a type-1 fuzzy set has a grade of membership that is crisp, namely a number in the unit interval. And a type-2 has grades of membership that are fuzzy subsets of $[0,1]$, that is, maps from $[0,1]$ into itself. Notice that HFSs correspond to the particular case of type-2 fuzzy subsets of X such that the grade maps $[0,1]$ into $\{0,1\}$.

In many practical applications, the HFSs considered map each element of X to a finite subset of $[0,1]$. This particular case is well known in the literature:

Definition 5 ([15])**.** *A function $h\colon X \to \mathcal{F}([0,1])$ is said to be a typical hesitant fuzzy set (THFS) of X. An element of $\mathcal{F}([0,1])$ is called a typical hesitant fuzzy element (THFE).*

Scores are now formally defined:

Definition 6. *Given a family $\mathcal{G} \subseteq \mathcal{P}([0,1])$, a score on $\mathcal{G}$ is defined as a function $s\colon \mathcal{G} \to [0,1]$ that satisfies the following properties:*

1. *The score assigned to the empty set is zero. That is, $s(\varnothing) = 0$ provided that $\varnothing \in \mathcal{G}$;*
2. *For all $E \in \mathcal{G}$, we have that the score assigned to E must lie between the infimum and the supremum of E. That is, $\inf(E) \leqslant s(E) \leqslant \sup(E)$. (In particular, $s(\{t\}) = t$ holds true for each t in the unit interval).*

A score on $\mathcal{G} = \mathcal{P}([0,1])$ (respectively, on $\mathcal{G} = \mathcal{F}([0,1])$) is called total (respectively, typical). Additionally, a score on $\mathcal{G} = \mathcal{I}([0,1])$ is said to be an interval score.

Remark 3. *One significant application of scores defined on HFEs is that they enable the reduction of uncertainty by one level. If $h\colon X \to \mathcal{P}([0,1])$ defines an HFS, and s is a score defined on $\mathcal{P}([0,1])$, then the composition $s \circ h$ directly defines a (type-1) fuzzy set over X.*

3. Classical Criteria to Extend Orderings from a Set to Its Power Set

The most classical criteria appeared from 1950 on, mainly in several papers related to social choice and decision making (see, e.g., [1,4,16]). They usually consider a total order $\precsim$ defined on a finite set X, whereas the extension $\precsim_E$ can be either a total order or just a complete preorder depending on the context.

In order to classify criteria, we may focus on several important aspects, usually related to some objective that we want to achieve from the restrictions imposed.

In the first classification of extension criteria, we may focus on the following facts:

(i) maxima and/or minima of sets; (ii) means; (iii) monotonicity.

Needless to say, there are some other possible classes, so that this is just a first approximation. In addition, these classes are not always pairwise disjoint.

In general, what we may expect a priori is that if we take criteria from different classes, these give rise to an incompatibility result. However, this is not always true.

Here are several criteria based on the maxima or minima of a finite set.

Definition 7. *Let X stand for a finite set endowed with a total order $\precsim$. An extension $\precsim_E$ satisfies the Pure Maximality Criterion [PMA] (respectively, the Pure Minimality Criterion [PMI]) if for every nonempty subsets $A, B \subseteq X$ it holds true that $\max A \prec \max B \Rightarrow A \prec_E B$ (respectively, $\min A \prec \min B \Rightarrow A \prec_E B$). Of course, here maxima and minima are taken as regards the given total order $\precsim$ on X. They exist because X is finite.*

Definition 8. *Let X stand for a finite set endowed with a total order $\precsim$. An extension $\precsim_E$ satisfies the Gärdenfors Principle [G] (see [4]) if for every nonempty subset $A \subseteq X$ and any element $x \notin A$ it holds true that $x \prec \min A \Rightarrow A \cup \{x\} \prec_E A$, and also $\max A \prec x \Rightarrow A \prec_E A \cup \{x\}$. Again, maxima and minima are taken here with respect to the given total order $\precsim$ on X.*

Definition 9. *Let X stand for a set endowed with a complete preorder $\precsim$. An extension $\precsim_E$ satisfies the Barberà-Pattanaik Property [BP] if for every $x, y \in X$ with $x \prec y$ it holds true that $\{x\} \prec_E \{x,y\} \prec_E \{y\}$.*

In addition, $\precsim_E$ satisfies the generalized Barberà-Pattanaik Property [GBP] if for every nonempty subsets $A, B \subseteq X$ such that $a \prec b$ holds for any $a \in A, b \in B$, we also have that $A \prec_E A \cup B \prec B$.

Remark 4. *Notice that, unlike the Gärdenfors principle, here we do not ask X to be finite. If X is finite, the Barberà-Pattanaik Property [BP] is much weaker than the Gärdenfors Principle [G], of which it is an immediate consequence in that case. In addition, we could also think that [BP]* *reminds us of an idea of a mean, so that if an element $y \in X$ is better than another one x, in the extension to the power set, the subset $\{x, y\}$ should lie between the subsets $\{x\}$ and $\{y\}$. This becomes much clearer when the extension is represented by a numerical function (utility) so that the value for $\{x, y\}$ represents, in a sense, a mean of the values for $\{x\}$ and $\{y\}$.*

Let us now introduce some more criteria based on the intuitive idea of a mean.

Definition 10. *Let X stand for a set endowed with a complete preorder $\precsim$. An extension $\precsim_E$ satisfies the Kelly Criterion [K] (see [17]) if given two nonempty subsets $A, B \in \mathcal{P}(X)$ with $A \neq B$ and such that $y \precsim x$ holds for every $x \in A$; $y \in B$, then $B \prec_E A$ also holds true.*

Remark 5. *Notice that this criterion [K] tells us that "the elements of A are at least as good as those in B", but there is at least one that is better since $A \neq B$. Thus, at least intuitively, an "average value" or a "mean" should be higher in A.*

Definition 11. *Let X stand for a set endowed with a complete preorder $\precsim$. An extension $\precsim_E$ satisfies the Criterion of Useful Elements [CUE]. If for every subset $A \subseteq X$ such that A has at least two elements, there is $a \in A$ such that $A \setminus \{a\} \prec_E A$. Such an element $a \in A$ is said to be a useful element for the subset A.*

Definition 12. *Let X be a set endowed with a complete preorder $\precsim$. An extension $\precsim_E$ satisfies the Criterion of Singular Elements [CSE] if for every subset $A \subseteq X$ such that A has at least two elements, there is $a \in A$ such that $A \prec_E A \setminus \{a\}$. Such an element $a \in A$ is said to be a singular element for the subset A. Also $\precsim_E$ satisfies the Robustness Criterion [R] if for every nonempty subsets $A, B, C \in \mathcal{P}(X)$ such that $y \prec x$ holds for any $x \in A \cup B$; $y \in C$, it holds that $B \prec_E A \Rightarrow B \cup C \prec_E A$.*

Let us pay attention now to criteria based on monotonicity properties.

Definition 13. *Let X stand for a set endowed with a complete preorder $\precsim$. An extension $\precsim_E$ satisfies the criterion of Monotonicity for Inferior Element [MIE] (see [1], where the terminology is different) if for every $x, y, z \in X$ such that $z \prec y \prec x$ it holds true that $\{y, z\} \prec_E \{x, z\}$. Similarly, the extension $\precsim_E$ satisfies the criterion of Monotonicity for Superior Element [MSE] (see [1]) if for every $x, y, z \in X$ such that $z \prec y \prec x$ it holds true that $\{z, x\} \prec_E \{y, x\}$.*

Definition 14. *Let X stand for a set endowed with a complete preorder $\precsim$. An extension $\precsim_E$ satisfies the criterion of Monotonicity for Sets [MS] if for every nonempty subsets $A, B, C \subseteq X$ with $C \cap (A \cup B) = \emptyset$ it holds that $A \prec_E B \Leftrightarrow A \cup C \prec_E B \cup C$. Moreover, $\precsim_E$ satisfies the criterion of Simple Monotonicity for Elements [SME] (see [1]) if for every nonempty subsets $A, B \subseteq X$ and $x \in X$ such that $x \notin (A \cup B)$ it holds true that $A \prec_E B \Rightarrow A \cup \{x\} \prec_E B \cup \{x\}$. Also, $\precsim_E$ satisfies the criterion of Strong Monotonicity for Elements [STRME] (see [1]) if for every nonempty subsets $A, B \subseteq X$ and $x \in X$ with $x \notin (A \cup B)$ it holds that $A \prec_E B \Leftrightarrow A \cup \{x\} \prec_E B \cup \{x\}$.*

Definition 15. *Let X stand for a set endowed with a complete preorder $\precsim$. An extension $\precsim_E$ satisfies the General Criterion of Monotonicity for Superior Elements [CSUP] (see [1], with a different nomenclature) if for every nonempty subsets $A, B \subseteq X$, and any element $x \in X$ such that $y \prec x$ holds true for every $y \in A \cup B$, it holds then true that $B \prec_E A \Rightarrow B \cup \{x\} \prec_E A \cup \{x\}$. Also, $\precsim_E$ satisfies the General Criterion of Monotonicity for Inferior Elements [CINF] (see [1]) if for every nonempty subsets $A, B \subseteq X$, and any element $x \in X$ such that $x \prec y$ holds true for every $y \in A \cup B$, it holds then true that $B \prec_E A \Rightarrow B \cup \{x\} \prec_E A \cup \{x\}$.*

Remark 6. *Notice that [CINF] is a consequence of [R] if we consider $C = \{x\}$ in the corresponding definition of Robustness. Therefore [R] implies [CINF].*

Definition 16. *Let X stand for a set endowed with a complete preorder* $\precsim$*. An extension* $\precsim_E$ *satisfies the Criterion of Monotonicity Relative to Disjoint Sets [MDS] (see once more [1], with a different terminology) if for every nonempty sets* $A, B \subseteq X$ *such that* $A \cap B = \emptyset$*, and any* $x \in X$ *such that* $x \notin (A \cup B)$*, it holds then true that* $B \prec_E A \Rightarrow B \cup \{x\} \prec_E A \cup \{x\}$*. Also,* $\precsim_E$ *satisfies the Criterion of Monotonicity Relative to Nested Sets [MNS] (see [1]) if for every nonempty sets* $A, B \subseteq X$ *such that* $B \subseteq A$*, and any element* $x \in X$ *such that* $x \notin A$*, it holds then true that* $A \prec_E A \cup \{x\} \Rightarrow B \prec_E B \cup \{x\}$*.*

4. Compatibility of Criteria of Extension of Orderings

4.1. Complete Preorders vs. Total Orders

In the classical theory, it is typical to analyze extensions in which we start with a *total order* $\precsim$ on a set X and we want an extension $\precsim_E$ that is also a total order on the power set of X. However, sometimes this will not be possible. As a matter of fact, some combination of criteria imposed on the extension may oblige it to be a complete preorder instead of a total order, as shown in Proposition 1.

Proposition 1. *Let X be a finite set with at least three elements, endowed with a total order of* $\precsim$*. There is no extension* $\precsim_E$ *to a total order satisfying [SME] and [B].*

Proof. Let $X = \{a,b,c\}$ with $a \prec b \prec c$. It follows that $\{a\} \prec_E \{b\} \prec_E \{c\}$. Using [B], we have $\{a\} \prec_E \{a,b\} \prec_E \{b\} \prec_E \{b,c\} \prec_E \{c\}$. Now, by [SME], we would arrive at $\{a,c\} \prec_E \{a,b,c\}$, since $\{a\} \prec_E \{a,b\}$. Again, by [SME], it follows that $\{a,b,c\} \prec_E \{a,c\}$, because $\{b,c\} \prec_E \{c\}$. We arrive at a contradiction. Therefore, $\precsim_E$ cannot be a total order. □

Remark 7. *Under Proposition 1, we wonder if* $\precsim_E$ *could be a complete preorder. The answer is affirmative. An example is the extension given as follows:* $\{a\} \prec_E \{a,b\} \prec_E \{b\} \sim_E \{a,c\} \sim_E \{a,b,c\} \prec_E \{b,c\} \prec_E \{c\} \prec_E \emptyset$*.*

If X has at most two elements, then there exist extensions $\precsim_E$ *that are total orders and satisfy [G] and [SME]. For instance, if* $X = \{a,b\}$ *and* $a \prec b$*, we may take the extension* $\precsim_E$ *given by* $\emptyset \prec_E \{a\} \prec_E \{a,b\} \prec_E \{b\}$*.*

Definition 17. *Let X be a nonempty set endowed with a complete preorder* $\precsim$*. Let* $\precsim_E$ *be an extension of* $\precsim$ *to the power set of X. Then we say that* $\precsim_E$ *satisfies the Weak Monotonicity Criterion [WM] if for any nonempty subsets* $A, B \subseteq X$ *and any element* $x \in X \setminus (A \cup B)$ *it holds true that* $A \prec_E B \Rightarrow A \cup \{x\} \precsim_E B \cup \{x\}$*.*

Proposition 2. *The [SME] criterion implies [WM]. The converse is false.*

Proof. The implication follows directly from definitions. Now let $X = \{a,b,c\}$ and $a \prec b \prec c$. Consider the extension $\precsim_E$ given by $\emptyset \prec_E \{a\} \prec_E \{b\} \prec_E \{b\} \sim_E \{a,b\} \sim_E \{a,c\} \sim_E \{b,c\} \sim_E \{a,b,c\}$. It satisfies [WM] but not [SME]. □

Lemma 1. *Let X be a nonempty finite set endowed with a total order* $\precsim$*. Let* $\precsim_E$ *be a quasi-transitive extension satisfying [G] and [WM]. Then, for every nonempty subset A of X, it holds true that* $A \sim_E \{\min(A), \max(A)\}$*.*

Proof. The proof can be seen in [2], but we include it here for the sake of completeness, since it is decisive in what follows then. Assume that X has at least three elements, since otherwise the result becomes evident. So let $n \geq 3$, and $X = \{a_1, a_2 \ldots, a_n\}$ with $a_1 \prec a_2 \prec \ldots \prec a_n$. Iterating [G] and using the transitivity of $\prec_E$, it follows that $\{a_1\} \prec_E \{a_1, a_2\} \prec_E \ldots \prec_E \{a_1, \ldots a_{n-1}\}$. Using [WM], we get $\{a_1, a_n\} \precsim_E X$. Since, by [G] again, $\{a_2, \ldots a_n\} \prec_E \{a_3, \ldots, a_n\} \prec_E \ldots \prec_E \{a_n\}$, once more, by [WM], we arrive at $X \precsim_E \{a_1, a_n\}$. Therefore, $X \sim_E \{a_1, a_n\}$. □

4.2. On Kannai-Peleg Theorem

Kannai and Peleg, in a seminal paper published in 1984 (see [2]), proved an impossibility theorem where the Gärdenfors Principle [G], the Weak Monotonicity Criterion [WM], and the fact that X is finite but has at least six different elements were crucial. Now we pay attention to some variants of the original theorem.

Proposition 3 (Kannai-Peleg Theorem –1984–, see [2])**.** *Let X be a finite set endowed with a total order $\precsim$. If X has at least 6 elements, then there is no extension $\precsim_E$ that is also a complete preorder and satisfies both [G] and [WM].*

Proof. See the proof in the classical reference [2]. We omit it here. Instead, we will prove in Proposition 5 below a result whose proof follows, so-to-say, similar ideas. □

Proposition 4. *Let X be a finite set whose cardinality is at most five, endowed with a complete preorder $\precsim$. Then there exists an extension $\precsim_E$ that is also a complete preorder and satisfies [G] and [WM].*

Proof. There is no loss of generality in assuming that X has five elements, that is, $X = \{a_1, a_2, a_3, a_4, a_5\}$, and it is endowed with a total order $\precsim$ such that $a_1 \prec a_2 \prec a_3 \prec a_4 \prec a_5$. Consider now the extension $\precsim_E$ given by: $\emptyset \prec_E \{a_1\} \prec_E \{a_1, a_2\} \prec_E \{a_1, a_3\} \sim_E \{a_1, a_2, a_3\} \prec_E \{a_1, a_4\} \sim_E \{a_1, a_2, a_4\} \sim_E \{a_1, a_2, a_3, a_4\} \sim_E \{a_1, a_3, a_4\} \prec_E \{a_2\} \prec_E \{a_2, a_3\} \prec_E \{a_3\} \sim_E \{a_2, a_4\} \sim_E \{a_2, a_3, a_4\} \sim_E \{a_1, a_5\} \sim_E \{a_1, a_2, a_5\} \sim_E \{a_1, a_3, a_5\} \sim_E \{a_1, a_4, a_5\} \sim_E \{a_1, a_2, a_3, a_5\} \sim_E \{a_1, a_2, a_4, a_5\} \sim_E \{a_1, a_3, a_4, a_5\} \sim_E \{a_1, a_2, a_3, a_4, a_5\} \prec_E \{a_3, a_4\} \prec_E \{a_4\} \prec_E \{a_2, a_5\} \sim_E \{a_2, a_3, a_5\} \sim_E \{a_2, a_4, a_5\} \sim_E \{a_2, a_3, a_4, a_5\} \prec_E \{a_3, a_5\} \sim_E \{a_3, a_4, a_5\} \prec_E \{a_4, a_5\} \prec_E \{a_5\}$. □

Definition 18. *Let X be a nonempty set endowed with a complete preorder $\precsim$. Let $\precsim_E$ be an extension of $\precsim$ to the power set of X. Then we say that $\precsim_E$ satisfies the Independence Criterion [IND] if for any nonempty subsets $A, B \subseteq X$ and any element $x \in X \setminus (A \cup B)$ it holds true that $A \precsim_E B \Rightarrow A \cup \{x\} \precsim_E B \cup \{x\}$.*

Notice that the independence criterion [IND] implies [WM].

Proposition 5. *Let X be a finite set endowed with a total order $\precsim$. If X has at least five elements, there is no extension $\precsim_E$ being a complete preorder satisfying [G] and [IND].*

Proof. By Lemma 1, since [IND] implies [WM], given a nonempty subset $A \subseteq X$, we have that $A \sim_E \{\min(A), \max(A)\}$. There is no loss of generality in assuming now that $X = \{a_1, a_2, a_3, a_4, a_5\}$ and $a_1 \prec a_2 \prec a_3 \prec a_4 \prec a_5$. Assume that there is an extension $\precsim_E$ that satisfies [G] and [IND]. First, we prove that $\{a_2, a_4\} \precsim_E \{a_3\}$. To do so, we will assume that this is false, and this will lead us to a contradiction: In fact, if $\{a_3\} \prec_E \{a_2, a_4\}$, using [IND] as regards a_5, we get $\{a_3, a_5\} \precsim_E \{a_2, a_4, a_5\}$. By Lemma 1, this implies $\{a_3, a_5\} \sim_E \{a_3, a_4, a_5\}, \{a_2, a_4, a_5\} \sim_E \{a_2, a_5\} \sim_E \{a_2, a_3, a_4, a_5\}$, so that, in particular, we have $\{a_3, a_4, a_5\} \precsim_E \{a_2, a_3, a_4, a_5\}$. Let now $B = \{a_3, a_4, a_5\}$. Since $a_2 \prec a_3 \prec a_4 \prec a_5$, by [G] we get $B \cup \{a_2\} \prec_E B$ or equivalently $\{a_2, a_3, a_4, a_5\} \prec_E \{a_3, a_4, a_5\}$. This is a contradiction. Therefore $\{a_2, a_4\} \precsim_E \{a_3\}$.

At this stage, again proceeding by contradiction, we prove that [G] is incompatible with [IND]. In fact, using now [IND], we get $\{a_1, a_2, a_4\} \precsim_E \{a_1, a_3\}$. In a similar way to the argument above, we use again Lemma 1 and obtain $\{a_1, a_3\} \sim_E \{a_1, a_2, a_3\}, \{a_1, a_2, a_4\} \sim_E \{a_1, a_4\} \sim_E \{a_1, a_2, a_3, a_4\}$. Hence $\{a_1, a_2, a_3, a_4\} \precsim_E \{a_1, a_2, a_3\}$. Let now $C = \{a_1, a_2, a_3\}$. Since $a_3 \prec a_4$ and $a_3 = \max C$, by [G] we have that $C \prec_E C \cup \{a_4\}$, or equivalently $\{a_1, a_2, a_3\} \prec_E \{a_1, a_2, a_3, a_4\}$. This is a contradiction. □

Example 1. *The extension $\precsim_E$ introduced in the example that appears in the proof of Proposition 4 satisfies [WM]. However, it fails to satisfy [IND]. To see this, notice that it is $\{a_3\} \sim_E \{a_2, a_4\}$, so that in particular, $\{a_3\} \precsim_E \{a_2, a_4\}$. However, $\{a_3, a_5\} \precsim \{a_2, a_4, a_5\}$ does not hold.*

If in Proposition 5 above we substitute the condition of X having at least five elements, putting instead that X has at most four elements, we get compatibility.

Proposition 6. *Let X be a finite set whose cardinality is at most four, endowed with a total order of $\precsim$. Then there is an extension $\precsim_E$ that satisfies both [G] and [IND].*

Proof. Without loss of generality, let $X = \{a_1, a_2, a_3, a_4\}$ and $\precsim$ such that $a_1 \prec a_2 \prec a_3 \prec a_4$. Take the extension $\precsim_E$ given by $\emptyset \prec_E \{a_1\} \prec_E \{a_1, a_2\} \prec_E \{a_1, a_3\} \sim_E \{a_1, a_2, a_3\} \prec_E \{a_2\} \prec_E \{a_2, a_3\} \sim_E \{a_1, a_4\} \sim_E \{a_1, a_2, a_4\} \sim_E \{a_1, a_3, a_4\} \sim_E \{a_1, a_2, a_3, a_4\} \prec_E \{a_3\} \prec_E \{a_2, a_4\} \sim_E \{a_2, a_3, a_4\} \prec_E \{a_3, a_4\} \prec_E \{a_4\}$. □

If we substitute [IND] by [SME] in the statement of Proposition 5 above, not only does the result become true, but now, just with three elements, we get incompatibility, as already stated in Proposition 1.

Concerning compatibility of criteria, the following classical result that involves, as well as weak monotonicity [WM], the Kelly criterion [K] instead of the more demanding Gärdenfors property [G] was proved in [18].

Proposition 7. *Let X stand for a finite nonempty set. Let $\precsim$ be a total order defined on X. Then there exists an extension $\precsim_E$ of $\precsim$ to the power set of X that is a complete preorder and satisfies both [K] and [WM].*

Proof. See Proposition 2 in [18]. □

In the same spirit, we now prove the following result on the compatibility of the criteria.

Proposition 8. *Let X stand for a finite nonempty set. Let $\precsim$ be a total order defined on X. Then there exists an extension $\precsim_E$ of $\precsim$ to the power set of X that is a complete preorder and satisfies both [R] and [SME].*

Proof. Just notice that given $X = \{a_1, a_2 \ldots, a_n\}$ with $a_1 \prec a_2 \prec \ldots \prec a_n$, the lexicographic order (that considers each element of X as a letter and each subset of X as a word where its case letters are also ordered by means of $\prec$, and the words are ordered lexicographically, as in a dictionary) satisfies both [R] and [SME]. □

5. Scores: Definitions, Hierarchies, and Incompatibility Results

A substantial part of this Section 5 already appeared in [3]. We have decided to include some definitions and results here for the sake of completeness and the well-understanding of the ideas.

5.1. Some Background on Scores

The most commonly used scores (see, for example, [8,19–27]) are typical (in the sense of Definition 6), as they are defined on the family of all finite subsets of the unit interval. Some examples may be seen in [3]. From the second condition in Definition 6, a score can be understood as a "mean value" (see [28] for more details).

As a sample, we furnish now another example, not included in [3]. This defines a function $s\colon \mathcal{F}([0,1]) \to [0,1]$ that is a typical score: For each $E = \{e_1, \ldots, e_n\} \in \mathcal{F}([0,1])$, with $e_1 < \ldots < e_n$, and p a natural number, we define $s(E) = (e_1^p + \ldots + e_n^p)^{\frac{1}{p}}$.

To classify scores on HFEs, we may pay attention to the following aspects:

(i) Properties based on certain coherence features of the score (e.g., the addition of better elements should never decrease the score).

(ii) Properties based on the specific type of HFEs for which the score will be considered (e.g., finite HFEs, interval HFEs, etc.).

Remark 8. *The classification will not necessarily produce mutually exclusive sets.*

5.2. On Coherence Features of Scores

First, we introduce some definitions and results related to coherence.

Definition 19. *Suppose $\mathcal{G} \subseteq \mathcal{P}([0,1])$. A score s on $\mathcal{G}$ is said to be best-worst monotonic for elements [BWME] if for any $x, y \in [0,1]$ such that $x < y$, and additionally $\{x\}, \{x,y\}$ and $\{y\}$ belong to $\mathcal{G}$, it holds that $s(\{x\}) < s(\{x,y\}) < s(\{y\})$.*

Notice that this property is essentially the Barberà-Pattanaik property [BP] (see Definition 9) adapted for scores.

Definition 20. *Suppose that $\mathcal{G} \subseteq \Pi([0,1])$. We say that a score s on $\mathcal{G}$ is strongly monotonic with respect to unions [SMU] when for each $A, B, A \cup B \in \mathcal{G}$ and such that $a < b$ for every $a \in A$, $b \in B$, it holds true that: $s(A) < s(A \cup B) < s(B)$.*

[SMU] captures the following intuition: Adding better elements to a subset should increase its score, and removing worse elements should also increase its score.

Remark 9. *Notice that [SMU] implies [BWME]. However, the converse is not true in general. A counterexample appears in [3].*

Given a non-empty subset A of the unit interval $[0,1]$ and real numbers $\alpha, \beta > 0$, we define the set αA as $\alpha A = \{\alpha \cdot t : t \in A\}$. Also, we define the set $\beta + A$ as follows: $\beta + A = \{\beta + t : t \in A\}$. Depending on A and α, β, the resulting sets αA and/or $\beta + A$ may or may not be subsets of $[0,1]$.

Definition 21. *Suppose now that $\mathcal{G} \subseteq \Pi([0,1])$. We say that a score s on $\mathcal{G}$ is algebraically coherent with respect to a dilatation [ACD] when for each $A \in \mathcal{G}$ and $\alpha > 0$ such that αA also belongs to $\mathcal{G}$ it holds true that $s(\alpha A) = \alpha s(A)$.*

Similarly, we say that a score s on $\mathcal{G}$ is algebraically coherent with respect to a translation [ACT] when for each $A \in \mathcal{G}$ and $\beta > 0$ such that $A + \beta$ also belongs to $\mathcal{G}$ it holds true that $s(A + \beta) = s(A) + \beta$.

Remark 10. *Many typical scores were introduced in [3] satisfy both [ACD] and [ACT].*

Definition 22. *Suppose $\mathcal{G} \subseteq \mathcal{P}([0,1])$. We say that a score s on $\mathcal{G}$ satisfies translation invariance [TI] when for each $A \in \mathcal{G}$ such that $A + \varepsilon \in \mathcal{G}$ (with $\varepsilon > 0$), it holds true that $s(A) < s(A + \varepsilon)$.*

Definition 23. *Let $\mathcal{G} \subseteq \mathcal{P}([0,1])$. We say that a score s on $\mathcal{G}$ satisfies the (adapted (We say here "adapted" Gärdenfors property since this property was introduced by Gärdenfors [4] in 1976 just to deal with finite subsets, so that the original definition was stated making reference to maxima and minima instead of suprema and infima. (Remember Definition 8 above))) Gärdenfors property [G] if for every $A \in \Pi([0,1])$ and any element $x \notin A$ such that $A, A \cup \{x\} \in \mathcal{G}$ the following two conditions hold:*

$$[G1]\ x < \inf A \Rightarrow s(A \cup \{x\}) < s(A); \quad [G2]\ \sup A < x \Rightarrow s(A) < s(A \cup \{x\}).$$

Remark 11. *Among the classical scores (see [3]) defined for finite subsets of the unit interval (TFHE's), the minimum satisfies [G1], but not [G2]. Similarly, the maximum satisfies [G2] but not [G1].*

Definition 24. *Let $\mathcal{G} \subseteq \mathcal{P}([0,1])$. We say that a score s on $\mathcal{G}$ satisfies the weak monotonicity property [WM] if for every $A, B \in \Pi([0,1])$ and $x \notin A \cup B$, such that $A, B, A \cup \{x\}, B \cup \{x\} \in \mathcal{G}$, it holds true that $s(A) < s(B) \Rightarrow s(A \cup \{x\}) < s(B \cup \{x\})$.*

Remark 12. *Notice that [WM] is a particular case of Definition 17 above for extensions of orderings. It is not easy to find scores satisfying [WM], except maybe in situations dealing with special classes of finite subsets of the unit interval (see, e.g., [29,30]).*

A classical theorem by Kannai and Peleg [2] states this incompatibility result.

Proposition 9. *Let $\mathcal{G} \subseteq \mathcal{P}([0,1])$ such that there exists a subset $A \subseteq [0,1]$ whose cardinality is at least 6, such that A and all its subsets belong to $\mathcal{G}$. Then no score s on $\mathcal{G}$ satisfies both the adapted Gärdenfors property [G] and the weak monotonicity property [WM].*

Proof. See [2], p. 174. Observe also that this Proposition 9 is also a direct consequence of Proposition 3 for extensions of orderings. □

5.3. Scores Defined on Special Classes of Sets

All the results of this subsection appear in [3]. For this reason, we do not include their proofs here. Nevertheless, the ideas involved in these results may affect Section 6.2 of the next Section 6 when trying to induce impossibility results for extensions of orderings mimicking the results got here for scores on HFEs.

Bearing this in mind, we now pay attention to scores that sometimes are not defined on the whole set $\mathcal{P}([0,1])$ but instead act only on THFEs (namely, $\mathcal{F}([0,1])$) or on intervals $\mathcal{I}([0,1])$. In particular, we pay attention to scores defined on families $\mathcal{G}$ that include the set of intervals $\mathcal{I}([0,1])$.

This gives rise to new definitions and results concerning the compatibility of the newly introduced properties when imposed on those scores. These have been studied in depth in [3]. We keep here just a sample of these new definitions and results to motivate their use in Section 6 of the present paper when adapted to utility functions related to extensions of orderings.

Definition 25. *Suppose $\mathcal{I}([0,1]) \subseteq \mathcal{G} \subseteq \mathcal{P}([0,1])$. A score $s : \mathcal{G} \to [0,1]$ is extremes monotonic [EM] when it satisfies the following two conditions: [EM1]: $0 \leqslant b < b' \leqslant 1$ implies $s([a,b]) < s([a,b'])$ for each $a \in [0,b]$; [EM2]: $0 \leqslant a < a' \leqslant 1$ implies $s([a,b]) < s([a',b])$ for each $b \in [a',1]$.*

Moreover, s is strongly extreme monotonic [SEM] if for each $a, b \in [0,1]$ with $a < b$ it holds that $s([0,a)) < s([0,a]) < s([0,b))$.

Proposition 10. *Consider a score $s\colon \mathcal{G} \to [0,1]$ with $\mathcal{I}([0,1]) \subseteq \mathcal{G} \subseteq \mathcal{P}([0,1])$. Then s cannot satisfy the property [SEM] of strong extremes monotonicity.*

Moreover, s cannot satisfy both [SMU] and [EM1].

Proof. See Lemmas 1 and 2 in [3].

6. Comparisons and Analogies between the Theory of Extension of Orderings and the Theory of Scores on Hesitant Fuzzy Elements

6.1. From Extensions of Orderings to Scores

The theory of scores on HFEs can be considered a particular case of that of extensions of orderings from a set to its power set. In fact, if we consider the usual total order $\leq$ on the unit interval $[0,1]$ and a score s defined on $\Pi([0,1])$, the order $\leq$ is immediately extended to a complete preorder $\leq_E$ on $\Pi([0,1])$, but just declaring that $A \leq_E \Leftrightarrow s(A) \leq s(B)$, for any nonempty subsets A, B of $[0,1]$.

Even if we only consider the theory of extensions of orderings on finite sets, we can also adapt its results to particular cases of scores, namely scores defined on the family $\mathcal{F}([0,1])$

THFEs. Actually, to compare two nonempty finite sets of $[0,1]$, say $A = \{x_1, \ldots, x_n\}$ and $B = \{y_1, \ldots, y_k\}$, first we endow the finite set $A \cup B$ with the total order $\leq$ inherited from the usual order of $[0,1]$, and, through a score s defined on $\mathcal{F}([0,1])$, we extend it to a complete preorder $\leq$ on the nonempty parts of $A \cup B$ by declaring that $C \leq D \Leftrightarrow s(C) \leq s(D)$.

What is clear now is that the impossibility results arising from extensions of orderings from a set to its power set have a parallel result on scores: Proposition 3 on extensions induces Proposition 9 on scores.

Definition 26. *Let $\mathcal{G} \subseteq \mathcal{P}([0,1])$. We say that a score s on $\mathcal{G}$ satisfies the weak monotonicity property of second type [WM2] if for every $A, B \in \Pi([0,1])$ and $x \notin A \cup B$, such that $A, B, A \cup \{x\}, B \cup \{x\} \in \mathcal{G}$, it holds true that $s(A) < s(B) \Rightarrow s(A \cup \{x\}) \leq s(B \cup \{x\})$. Similarly, it satisfies the weak monotonicity property of the third type [WM3] if for every $A, B \in \Pi([0,1])$ and $x \notin A \cup B$, such that $A, B, A \cup \{x\}, B \cup \{x\} \in \mathcal{G}$, it holds true that $s(A) \leq s(B) \Rightarrow s(A \cup \{x\}) \leq s(B \cup \{x\})$.*

Proposition 3 for extensions of orderings also implies the following Proposition 11.

Proposition 11. *Let $\mathcal{G} \subseteq \mathcal{P}([0,1])$ such that there exists a subset $A \subseteq [0,1]$ whose cardinality is at least 6, such that A and all its subsets belong to $\mathcal{G}$. Then there is no score s on $\mathcal{G}$ that satisfies both the adapted Gärdenfors property [G] and the weak monotonicity property of the second type [WM2].*

And Proposition 5 immediately gives rise to the following result for scores.

Proposition 12. *Let $\mathcal{G} \subseteq \mathcal{P}([0,1])$ such that there exists a subset $A \subseteq [0,1]$ whose cardinality is at least 5, such that A and all its subsets belong to $\mathcal{G}$. Then there is no score s on $\mathcal{G}$ that satisfies both the adapted Gärdenfors property [G] and the weak monotonicity property of the third type [WM3].*

Needless to say, these cases are just a sample. That is, the key fact here is that impossibility results encountered in the framework of the extensions of orderings from a set to its power set (a.k.a, "ranking sets of objects", see [1]) immediately induce parallel results on the theory of scores on hesitant fuzzy elements.

A remarkable fact here is that, to the extent of what we know and perceive, a large part of these results of the impossibility of extensions of orderings turn out to be little known, if not ignored, by a substantial part of the researchers in fuzzy set theory, dealing with scores on hesitant fuzzy elements.

6.2. From Scores to Extensions of Orderings

The converse situation, that is, using some results on scores on hesitant fuzzy elements to get new (im)possibility theorems on extensions of total orders from a nonempty set to its power set, is not so direct. Since the extension of orderings is a more general setting, it may happen that some results on scores could not be translated into one for extensions of orderings, in the most general case (extensions that start from a total order defined on a nonempty set X, not necessarily finite).

In addition, the results on scores have $[0,1]$ as a starting point, that is, as a totally ordered set through the usual order $\leq$, on whose power set we want to define some score s accomplishing some properties or criteria. To translate this to some abstract set X endowed with a total order $\precsim$, we will need that X has some additional structure (e.g., topological) that could, in some way, remind us of $[0,1]$.

Let us now discuss how to adapt Definition 25, Proposition 10, and some other definitions and results introduced in [1] to the framework of extensions of orderings from a set to its power set.

In what follows, X will stand for a nonempty set, and we will also assume that it is endowed with a total order $\precsim$. We want to extend $\precsim$ to a complete preorder (i.e., now we will admit ties), say $\precsim_E$, defined on the power set of X.

Given $a, b \in X$ with $a \prec b$ we denote by $[a, b] = \{x \in X : a \precsim x \precsim b\}$; $(a, b) = \{x \in X : a \prec x \prec b\}$; $[a, b) = \{x \in X : a \precsim x \prec b\}$; $(a, b] = \{x \in X : a \prec x \precsim b\}$. Also, given $a \in X$, we denote $(a, \rightarrow) = \{x \in X : a \prec x\}$; $[a, \rightarrow) = \{x \in X : a \precsim x\}$; $(\leftarrow, a) = \{x \in X : x \prec a\}$; $(\leftarrow, a] = \{x \in X : x \precsim a\}$.

Definition 27. *An extension $\precsim_E$ of $\precsim$ to the power set of X satisfies the property of interval compatibility [IC] if given $a, b \in X$ with $a \prec b$, it holds true that $(a, b) \sim_E [a, b] \sim_E [a, b) \sim_E (a, b]$.*

Remark 13. *This property [IC] is very demanding. It forces the total order $\precsim$ on X to be dense-in-itself. That is, if $a \prec b$, there is some other element $c \in X$ such that $a \prec c \prec b$. The reason is that the extension $\precsim_E$ should accomplish, on the one hand, that $\{a\} \prec_E \{b\}$. But, on the other hand, by [IC] we have that $[a, b) = \{a\}$, whereas $(a, b] = \{b\}$, and [IC] then would imply that $\{a\} \sim_E \{b\}$, a contradiction. As a clear consequence X should be infinite.*

Definition 28. *An extension $\precsim_E$ of $\precsim$ to the power set of X is extreme monotonic [EM] when it satisfies the following two conditions:*

[EM1] If $b \prec c$, then $[a, b] \prec_E [a, c]$ holds for every $a \precsim b$ $(a, b, c \in X)$;

[EM2] If $b \prec c$, then $[b, a] \prec_E [c, a]$ holds for every $c \precsim a$ $(a, b, c \in X)$.

Moreover, it is strongly extreme monotonic [SEM] if for each $a, b \in X$ with $a \prec b \prec c$, it holds that $[a, b) \prec_E [a, b] \prec_E [a, c) \prec_E [a, c]$.

Remark 14. *This new property, [SEM], is also very demanding. If X has at least three elements, the total order $\precsim$ is almost "dense-in-itself" in the following sense: If $a \prec b \prec c$ there exists $d \in X$ with $b \prec d \prec c$. Otherwise, $[a, b] = [a, c)$, so that $[a, b] \prec_E [a, c)$ becomes impossible. In particular, X should be infinite, too.*

The results we can obtain now for extensions of orderings, after adapting some impossibility results for scores, have a strong relationship with the existence of a utility function. The reason is the following: If a complete preorder $\precsim$ on a nonempty set Z has a utility representation u, there is no loss of generality in considering that u takes values in $[0, 1]$. Consequently, impossibility results for scores, adapted now to criteria of extensions of orderings from a set to its power set (see, e.g., Proposition 13 below), tell us the following key fact Either the extension $\precsim_E$ cannot be a complete preorder, or if it is, then it does not admit a utility function. The first situation is typical for impossibility results on finite sets. The reason is that any complete preorder on a finite set always admits a utility representation. Moreover, if a set X is finite, its power set is also finite (see e.g., the first three chapters in [31]).

Proposition 13. *Let X be a set whose cardinality is at least that of the continuum. Suppose that X is endowed with a total order $\precsim$. Let $\precsim_E$ be a complete preorder on the power set of X that is an extension of $\precsim$. Then, if $\precsim_E$ satisfies [SEM], it does not admit a utility representation.*

Proof. This is analogous to the first part of Proposition 10 for scores. Suppose by contradiction that u is a utility function for $\precsim_E$. Take $x \in X$ with either $[x, \rightarrow)$ or $(\leftarrow, x]$ having the cardinality of the continuum. (We will assume the first possibility, without loss of generality). Now, observe that for any $x, y \in X$ with $x \prec y, z$, we have that $a = u([x, y)) < b = u([x, y]) < c = u([x, z)) < d = u([x, z])$. So there exists a rational number $q_y \in (a, b)$ as well as another rational number $q_z \in (c, d)$, and by construction $q_z > q_y$. But this leads to a contradiction since $\mathbb{Q}$ is countable. □

Proposition 14. *Let X be a set whose cardinality is at least that of the continuum. Suppose that X is endowed with a total order $\precsim$. Let $\precsim_E$ be a complete preorder on the power set of X that is an extension of $\precsim$. Then, if $\precsim_E$ satisfies [GBP], it does not admit a utility representation.*

Proof. This is analogous to the second part of Proposition 10 for scores. Suppose by contradiction that u is a utility function for $\precsim_E$. Take $x \in X$ with either $[x, \rightarrow)$ or $(\leftarrow, x]$ having the cardinality of the continuum. (We will assume the first possibility, without loss of generality). Now, observe that for any $x, y \in X$ with $x \prec y, z$, we have that $[x, y) \prec_E [x, y] \prec_E \{y\}$ by [GBP]:Moreover, for the same reason, $[x, y] \prec_E [x, z) \prec_E (y, z)$. As in Proposition 14, $a = u([x, y)) < b = u([x, y]) < c = u([x, z)) < d = u([x, z])$, and we finally arrive at a contradiction. □

7. Concluding Remarks and Lines for Future Research

The main objective of this paper has been to show analogies between two apparently disparate theories. We have established some parallelism between the (more general) theory of extensions of ordering from a set to its power set and the theory of defining scores on hesitant fuzzy elements. Thus, each possibility or impossibility result encountered in the theory of ranking sets of objects (i.e., extending total orders from a set to its power set) immediately generates a parallel result of the same kind (possibility or impossibility) concerning scores on HFEs. Moreover, some possibility or impossibility results on scores on HFEs can be adapted somehow to obtain parallel results on extensions of orderings.

As a line for future research, we suggest exploring in depth both theories, searching for (im)-possibility results that, being well known in one of the frameworks, have not been used or even commented on in the other. Therefore, we could complete (or, at least, enlarge) the panorama of possibility/impossibility theorems arising in both theories: extension of orderings from a set to its power set vs. scores on HFEs.

Author Contributions: Conceptualization, E.I., A.M. and M.S.S.; Methodology, E.I., A.M. and M.S.S.; Formal analysis, E.I., A.M. and M.S.S.; Investigation, E.I., A.M. and M.S.S.; Writing—original draft, E.I., A.M. and M.S.S. All authors have read and agreed to the published version of the manuscript.

Funding: This work was supported by the project of reference PID2022-136627NB-I00 from MCIN/AEI/10.13039/501100011033/FEDER, UE, and by ERDF A way of making Europe.

Data Availability Statement: The original contributions presented in the study are included in the article, further inquiries can be directed to the corresponding author.

Acknowledgments: Thanks are given to the editors and three anonymous referees for their valuable suggestions and comments.

Conflicts of Interest: The authors have no conflicts of interest to declare. All co-authors have seen and agreed with the contents of the paper, and there is no financial interest to report. We certify that this is original work, and it is not under review at any other publication.

References

1. Barberà, S.; Bossert, W.; Pattanaik, P. Ranking sets of Objects. In *Chapter 17 of Handbook of Utility Theory: Volume 2 Extensions*; Barberà, S., Hammond, P., Seidl, C., Eds.; Kluwer Academic Publishers: Dordrecht, The Netherlands, 2004; pp. 893–979.
2. Kannai, Y.; Peleg, B. A note on the extension of an order on a set to the power set. *J. Econ. Theory* **1984**, *32*, 172–175. [CrossRef]
3. Alcantud, J.C.R.; Campión, M.J.; Induráin, E.; Munárriz, A. Scores of hesitant fuzzy elements revisited: "Was sind und was sollen". *Inf. Sci.* **2023**, *648*, 119500. [CrossRef]
4. Gärdenfors, P. Manipulation of social choice functions. *J. Econ. Theory* **1976**, *13*, 217–228. [CrossRef]
5. Alcantud, J.C.R.; Giarlotta, A. Necessary and possible hesitant fuzzy sets: A novel model for group decision making. *Inf. Fusion* **2019**, *46*, 63–76. [CrossRef]
6. Zadeh, L.A. Fuzzy sets. *Inf. Control* **1965**, *8*, 338–353. [CrossRef]
7. Torra, V. Hesitant fuzzy sets. *Int. J. Intell. Syst.* **2010**, *25*, 529–539. [CrossRef]
8. Xia, M.; Xu, Z. Hesitant fuzzy information aggregation in decision making. *Int. J. Approx. Reason.* **2011**, *52*, 395–407. [CrossRef]
9. Bustince, H.; Barrenechea, E.; Pagola, M.; Fernández, J.; Xu, Z.; Bedregal, B.; Montero, J.; Hagras, H.; Herrera, F.; De Baets, B. A historical account of types of fuzzy sets and their relationships. *IEEE Trans. Fuzzy Syst.* **2016**, *24*, 179–194. [CrossRef]
10. Narukawa, Y.; Torra, V. An score index for hesitant fuzzy sets based on the Choquet integral. In Proceedings of the 2021 IEEE International Conference on Fuzzy Systems, Luxembourg, 11–14 July 2021; pp. 1–5.
11. Torra, V.; Narukawa, Y. On hesitant fuzzy sets and decision. In Proceedings of the 2009 IEEE International Conference on Fuzzy Systems, Jeju Island, Republic of Korea, 20–24 August 2009; pp. 1378–1382.
12. Mendel, J.M. Type-2 fuzzy sets: Some questions and answers. *IEEE Connect. Newsl. IEEE Neural Netw. Soc.* **2003**, *1*, 10–13.

13. Mendel, J.M.; John, R.I. Type-2 fuzzy sets made simple. *IEEE Trans. Fuzzy Syst.* **2002**, *10*, 117–127. [CrossRef]
14. Zadeh, L.A. The concept of a linguistic variable and its application to approximate reasoning. *Inf. Sci.* **1975**, *8*, 199–249. [CrossRef]
15. Bedregal, B.; Reiser, R.; Bustince, H.; López-Molina, C.; Torra, V. Aggregation functions for typical hesitant fuzzy elements and the action of automorphisms. *Inf. Sci.* **2014**, *255*, 82–99. [CrossRef]
16. De Miguel, J.R.; Goicoechea, M.I.; Induráin, E.; Olóriz, E. Criterios de extensión al conjunto potencia de ordenaciones sobre un conjunto finito. *Rev. Real Acad. Cienc. Exactas Físicas Nat. Madr.* **2000**, *94*, 83–92.
17. Kelly, J. Strategy proofness and social choice functions without singlevaluedness. *Econometrica* **1977**, *45*, 439–446. [CrossRef]
18. Barberà, S.; Pattanaik, P. Extending an order on a set to the power set: Some remarks on Kannai and Peleg's approach. *J. Econ. Theory* **1984**, *32*, 185–191. [CrossRef]
19. Chen, N.; Xu, Z.; Xia, M. Correlation coefficients of hesitant fuzzy sets and their applications to clustering analysis. *Appl. Math. Model.* **2013**, *37*, 2197–2211. [CrossRef]
20. Farhadinia, B. A novel method of ranking hesitant fuzzy values for multiple attribute decision-making problems. *Int. J. Intell. Syst.* **2013**, *28*, 752–767. [CrossRef]
21. Farhadinia, B. A series of score functions for hesitant fuzzy sets. *Inf. Sci.* **2014**, *277*, 102–110. [CrossRef]
22. Rodríguez, R.; Bedregal, B.; Bustince, H.; Dong, Y.; Farhadinia, B.; Kahraman, C.; Martínez, L.; Torra, V.; Xu, Y.; Xu, Z.; et al. A position and perspective analysis of hesitant fuzzy sets on information fusion in decision making. Towards high quality progress. *Inf. Fusion* **2016**, *29*, 89–97. [CrossRef]
23. Rodríguez, R.; Martínez, L.; Herrera, F. Hesitant fuzzy linguistic term sets for decision making. *IEEE Trans. Fuzzy Syst.* **2012**, *20*, 109–119. [CrossRef]
24. Rodríguez, R.; Martínez, L.; Torra, V.; Xu, Z.; Herrera, F. Hesitant fuzzy sets: State of the art and future directions. *Int. J. Intell. Syst.* **2014**, *29*, 495–524. [CrossRef]
25. Wang, B.; Liang, J.; Pang, J. Deviation degree: A perspective on score functions in hesitant fuzzy Sets. *Int. J. Fuzzy Syst.* **2019**, *21*, 2299–2317. [CrossRef]
26. Xu, Z. *Hesitant Fuzzy Sets Theory;* Volume 314 of Studies in Fuzziness and Soft Computing; Springer International Publishing: Berlin/Heidelberg, Germany, 2014.
27. Xu, Z.; Xia, M. Distance and similarity measures for hesitant fuzzy sets. *Inf. Sci.* **2011**, *181*, 2128–2138. [CrossRef]
28. Campión, M.J.; Candeal, J.C.; Catalán, R.G.; Giarlotta, A.; Greco, S.; Induráin, E.; Montero, J. An axiomatic approach to finite means. *Inf. Sci.* **2018**, *457*, 12–28. [CrossRef]
29. Bossert, W. Preference extension rules for ranking sets of alternatives with a fixed cardinality. *Theory Decis.* **1995**, *39*, 301–317. [CrossRef]
30. Maly, J.; Woltran, S. Ranking Specific Sets of Objects. In *Lecture Notes in Informatics (LNI), Gesellschaft für Informatik;* Springer: Berlin/Heidelberg, Germany, 2017; pp. 193–201.
31. Bridges, D.S.; Mehta, G.B. *Representation of Preference Orderings;* Springer: Berlin/Heidelberg, Germany, 1995.

axiom s

MDPI

Article

Modelling Up-and-Down Moves of Binomial Option Pricing with Intuitionistic Fuzzy Numbers

Jorge de Andrés-Sánchez

Social and Business Research Lab., Universitat Rovira i Virgili, Campus de Bellissens, 43204 Reus, Spain; jorge.deandres@urv.cat

Abstract: Since the early 21st century, within fuzzy mathematics, there has been a stream of research in the field of option pricing that introduces vagueness in the parameters governing the movement of the underlying asset price through fuzzy numbers (FNs). This approach is commonly known as fuzzy random option pricing (FROP). In discrete time, most contributions use the binomial groundwork with up-and-down moves proposed by Cox, Ross, and Rubinstein (CRR), which introduces epistemic uncertainty associated with volatility through FNs. Thus, the present work falls within this stream of literature and contributes to the literature in three ways. First, analytical developments allow for the introduction of uncertainty with intuitionistic fuzzy numbers (IFNs), which are a generalization of FNs. Therefore, we can introduce bipolar uncertainty in parameter modelling. Second, a methodology is proposed that allows for adjusting the volatility with which the option is valued through an IFN. This approach is based on the existing developments in the literature on adjusting statistical parameters with possibility distributions via historical data. Third, we introduce into the debate on fuzzy random binomial option pricing the analytical framework that should be used in modelling upwards and downwards moves. In this sense, binomial modelling is usually employed to value path-dependent options that cannot be directly evaluated with the Black–Scholes–Merton (BSM) model. Thus, one way to assess the suitability of binomial moves for valuing a particular option is to approximate the results of the BSM in a European option with the same characteristics as the option of interest. In this study, we compared the moves proposed by Renddleman and Bartter (RB) with CRR. We have observed that, depending on the moneyness degree of the option and, without a doubt, on options traded at the money, RB modelling offers greater convergence to BSM prices than does CRR modelling.

Keywords: intuitionistic fuzzy numbers; probability–possibility transformation; fuzzy binomial option pricing; zero-coupon bond options; binomial up-and-down modelling

MSC: 62A88; 91G20; 91G30

Citation: Andrés-Sánchez, J.d. Modelling Up-and-Down Moves of Binomial Option Pricing with Intuitionistic Fuzzy Numbers. *Axioms* **2024**, *13*, 503. https://doi.org/10.3390/axioms13080503

Academic Editors: Ta-Chung Chu and Wei-Chang Yeh

Received: 23 June 2024
Revised: 24 July 2024
Accepted: 25 July 2024
Published: 26 July 2024

1. Introduction

The Black–Scholes–Merton (BSM) model for valuing European options [1,2] has been one of the fundamental pillars of financial economics since the late 20th century [3]. The approach used to determine the BSM formula, which is based on the no-arbitrage argument, allows the valuation of not only options but also any asset containing some form of optionality, using a few parameters that are relatively easy to estimate because they do not depend on subjective risk perception. Thus, option pricing theory enables pricing not only for a great deal of derivative assets but also for some embedded rights, such as convertibility rights and early amortization in bonds or financial assets such as life insurance or mortgage loans [3]. It also allows for the valuation of companies [1] or investment projects using real options theory [3].

However, while the BSM philosophy allows for the valuation of a wide range of economic rights, continuous-time option valuation models do not allow for the evaluation of the majority of path-dependent options. This encompasses American options, several types of exotic options, or flexibilities associated with real options [4]. Thus, one of the main derivations of the BSM is the binomial approximation, also called the two-state model [4], where up-and-down moves are instantaneous, which is equivalent to the BSM formula [5]. Therefore, the use of a binomial methodology in valuing path-dependent options allows the application of the philosophy and assumptions underlying the BSM [5].

The so-called binomial approximation is not a single model but comprises a great deal of up-and-down binomial moves modelling. The most widely used and well-known method is the one proposed by Cox, Ross, and Rubinstein [6] (CRR hereafter), although there are many more approximations. In this regard, we can mention the ones simultaneously published by Rendleman and Bartter [7] (RB hereafter) or [8,9]. In fact, [5] identified up to 11 possible variants of the binomial method.

Conventional option valuation models assume that the parameters governing variations in underlying asset prices are crisp values. However, in practical situations, there is often imprecision and/or vagueness regarding their values. For example, the historical volatility of the underlying asset must be estimated through sample values; thus, a more comprehensive but also imprecise estimation requires at least the use of confidence interval estimations associated with a significance level [10]. In the case of real options, parameters such as the exercise price or even its date may be imprecisely estimated by the evaluator or manager [11,12]. Thus, at the beginning of the 21st century, a trend in fuzzy mathematics emerged, which we can label fuzzy random option pricing (FROP). A considerable number of studies have modelled uncertainty in valuation parameters through possibility distributions [13]. These works are based on conventional option valuation frameworks, such as the BSM or the binomial method, which introduce the vagueness of parameters governing movements of the underlying asset through fuzzy subsets [13]. In most cases, the type of fuzzy subset used is the type-1 fuzzy number, which is typically triangular or trapezoidal [12,13] and should be considered an epistemic fuzzy set [14].

FROP development spans both discrete and continuous periods. Over time, contributions within the BSM framework have been particularly numerous [10,11,15–21]. However, FROP has utilized possibility distributions to model the parameter uncertainty of other price variation models, such as multivariate Brownian geometric models [22] or Levy processes [23,24].

Discrete-time developments have mainly focused on extending the binomial model to price options for stocks [25–30] and real options [31–34]. In these papers, all binomial models shape the up-and-down moves with the analytical groundwork of the CRR without considering any of the numerous alternatives provided in the literature. However, there is no reason not to choose any other binomial model from those mentioned earlier, such as the RB model [31]. Likewise, to the best of our knowledge, modelling vagueness over the value of parameters governing the movement of the underlying asset price is performed through fuzzy numbers, which are typically linear [13]. An exception to this assertion is the intuitionistic triangular extension to CRR [35], which uses soft set parameters to model CRR. Notably, a significant number of FROP binomial extensions consider volatility to be the main source of uncertainty [22,26,36].

The reflections outlined in the preceding paragraphs lead to the development of the present work, in which we extend the two-state model of option valuation to estimate the parameters governing underlying assets quantified by intuitionistic fuzzy numbers (IFNs), with particular emphasis on volatility. This study introduces the following novelties to the FROP literature:

1. We model uncertainty via IFNs, which generalize FNs, in an option pricing context. Introducing parameter quantification with IFNs allows for the incorporation of bipolar information; that is, capturing values that can actually take the parameter, as well as those that are definitely not [37]. It should not be understood that IFNs introduce more uncertainty in parameter estimation but rather introduce new information [38].

Although intuitionistic fuzzy uncertainty has been considered in some studies [39–43], it is quite residual and absent in fuzzy binomial modelling.

2. We propose a methodology that allows for the adjustment of the volatility necessary to value the option as an IFN using the historical volatility approach [4] and the concept of coherent probability–possibility transformation [44]. This focus has been adopted to fit fuzzy number parameters in an FROP setting to price stock options [10], in a real options setting [45], and in the field of valuation of interest-sensitive instruments [46].
3. We contribute to FROP in a binomial setting by critically proposing the modelling of up-and-down moves in the valuation of the path-dependent option under assessment. We compared the commonly used fuzzy literature CRR with the alternative of Rendleman and Bartter [7]. In this sense, given that the use of the binomial model is justified by its convergence to the BSM, the evaluation of binomial models is carried out by comparing the proximity of their calculated price with the BSM in a European option with the same characteristics as those intended to be evaluated [47].

The paper is organized as follows. The following section presents the analytical foundations of fuzzy mathematics used in this paper and proposes an intuitionistic fuzzy estimate of option volatility on the basis of the concept of historical volatility. In the Section 3, intuitionistic expressions of the BSM and binomial option prices are developed. Fourth, we evaluated the modelling of binomial up-and-down moves with CRR and RB in an intuitionistic fuzzy setting. We assume that volatility is a unique uncertain parameter. To test up-and-down moves, we used historical data from the IBEX-35 Futures Index, which is the reference index for the most traded options on stocks in the Spanish derivatives financial market.

2. Intuitionistic Fuzzy Estimates of Statistical Parameters and Intuitionistic Fuzzy Number Arithmetic

2.1. Fuzzy Numbers, Intuitionistic Fuzzy Numbers, and Distance between Intuitionistic Fuzzy Numbers

Definition 1. *A fuzzy set in a universe of discourse X, $\widetilde{A}$, is $\widetilde{A} = \{\langle x, \mu_A(x)\rangle,\ x \in X\}$, where $\mu_A : X \longrightarrow [0,1]$ is the so-called membership function [48]. Conversely, $\widetilde{A}$ can be represented through level sets or α-cuts: A_α:$A_\alpha = \{x|\mu_A(x) \geq \alpha,\ \alpha \in (0,1]\}$.*

Definition 2. *A fuzzy number (FN), $\widetilde{A}$, is a fuzzy subset of a real line. It is normal* (i.e., $\exists x|\mu_A(x) = 1)$) *or convex* (i.e., $\forall x_1, x_2 \in \mathbb{R}, 0 \leq \lambda \leq 1,\ \mu_A(\lambda x_1 + (1-\lambda)x_2) \geq$ *$min(\mu_A(x_1), \mu_A(x_2)))$ [49]. Therefore, the level sets of $\widetilde{A}$ and A_α are confidence intervals:*

$$A_\alpha = \{x|\mu_A(x) \geq \alpha, \alpha \in (0,1]\} = [\underline{A}_\alpha, \overline{A}_\alpha], \tag{1}$$

where $\underline{A}_\alpha$ increases with α and where $\overline{A}_\alpha$ decreases with respect to α.

Remark 1. *A fuzzy number $\widetilde{A}$ is also known as a possibility distribution, and $\mu_A(x)$ is known as the possibility distribution function.*

Definition 3. *The intuitionistic fuzzy set (IFS) $\widetilde{A}^I$ in the universe of discourse X is $\widetilde{A}^I =$ $\{\langle x, \mu_A(x), v_A(x)\rangle,\ x \in X\}$, where $\mu_A : X \longrightarrow [0,1]$ is the membership value of x in $\widetilde{A}^I$ and where $v_A : X \longrightarrow [0,1]$ is the nonmembership value. The following relation holds: $0 \leq \mu_A(x) + v_A(x) \leq 1$ [50].*

Remark 2. *The degree of hesitancy of $\widetilde{A}^I$, $h_A(x)$, is $h_A(x) = 1 - \mu_A(x) - v_A(x)$. Note that an IFS generalizes the concept of an FS such that if $h_A(x) = 0\ \forall x$, $\widetilde{A}^I$ is a conventional FS $\widetilde{A}$.*

Definition 4. *An intuitionistic fuzzy number (IFN)* $\overset{\sim I}{A}$ *is an IFS defined on real numbers such that [51]:*

(i) is normal, $\exists x | \mu_A(x) = 1 \Rightarrow v_A(x) = h_A(x) = 0$.

(ii) $\mu_A(x)$ *is convex, which implies that* $\forall x_1, x_2 \in \mathbb{R}, 0 \leq \lambda \leq 1$, $\mu_A(\lambda x_1 + (1-\lambda)x_2) \geq min(\mu_A(x_1), \mu_A(x_2))$, *and* $v_A(x)$ *are concave; that is,* $\forall x_1, x_2 \in \mathbb{R}, 0 \leq \lambda \leq 1$, $v_A(\lambda x_1 + (1-\lambda)x_2) \leq max(v_A(x_1), v_A(x_2))$.

Remark 3. *An IFN* $\overset{\sim I}{A}$ *can be represented throughout its level sets or* $\langle \alpha, \beta \rangle$*-cuts,* $A_{\langle \alpha,\beta \rangle}$*, as:*

$$A_{\langle \alpha,\beta \rangle} = \langle A_\alpha = \left[\underline{A_\alpha}, \overline{A_\alpha}\right], A'_\beta = \left[\underline{A'_\beta}, \overline{A'_\beta}\right], 0 \leq \alpha + \beta \leq 1, \alpha, \beta \in (0,1) \rangle. \quad (2)$$

where $\underline{A_\alpha}$ *and* $\overline{A'_\beta}$ *increase with respect to* α *and* β*, respectively. Similarly,* $\overline{A_\alpha}$ *and* $\underline{A'_\beta}$ *decrease with respect to these arguments.*

Remark 4. *In an IFN,* $\mu_A(x)$ *is the lower possibility distribution function of* $\overset{\sim I}{A}$*, and* $\mu^*_A(x) = 1 - v_A(x)$ *is its upper distribution function [52]. Consequently, the* α*-cut representation of* $\overset{\sim *}{A}$*,* A^*_α *and the* β*-cut representation in (2),* A'_β*, accomplishes that for* $\beta = 1-\alpha$, $A'_{1-\alpha} = A^*_\alpha$*, i.e.,*

$$\left[\underline{A'_{1-\alpha}}, \overline{A'_{1-\alpha}}\right] = \left[\underline{A^*_\alpha}, \overline{A^*_\alpha}\right] \quad (3)$$

Thus, $\mu^*_A(x)$ and $\mu_A(x)$ can be interpreted as bipolar measurements of the reliability of A as x [38]. Thus, $\mu^*_A(x)$ quantifies for x the potential possibility and $\mu_A(x)$ its real possibility. Figure 1 shows the shape of an intuitionistic triangular fuzzy number.

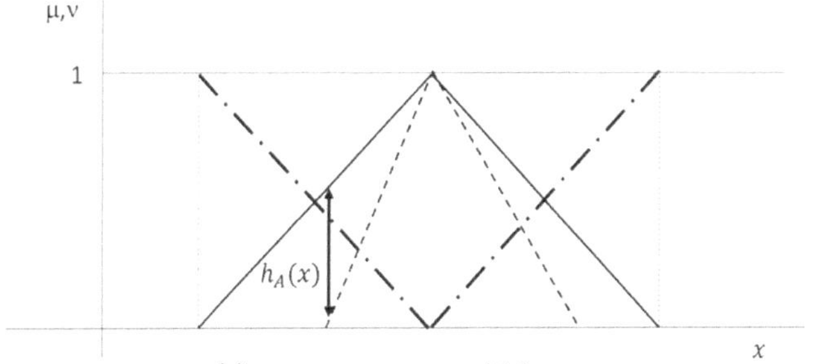

Figure 1. Shape of a triangular intuitionistic fuzzy number.

Definition 5. *The expected value of an intuitionistic fuzzy number,* $EV(\overset{\sim I}{A})$*, can be defined via (2) and (3) as follows [53]:*

$$EV(\overset{\sim I}{A}) = \frac{1}{2}\int_0^1 f(\alpha)\left(\underline{A_\alpha} + \overline{A_\alpha} + \underline{A'_{1-\alpha}} + \overline{A'_{1-\alpha}}\right)d\alpha. \quad (4)$$

where $f(\alpha)$ *is a function that satisfies* $f(0) = 0$ *and* $\int_0^1 f(\alpha)d\alpha = 0.5$*. Therefore, we consider in this paper* $f(\alpha) = \alpha$.

Definition 6. *Let there be two IFNs:* $\overset{\sim I}{A}$ *and* $\overset{\sim I}{B}$*. The distance between these IFNs is defined by (2) and (3) as follows [54]:*

$$D(\overset{\sim I}{A},\overset{\sim I}{B}) = \frac{1}{2}\int_0^1 f(\alpha)\left[\left(\underline{A_\alpha} - \underline{B_\alpha}\right)^2 + \left(\overline{A_\alpha} - \overline{B_\alpha}\right)^2 + \left(\underline{A'_{1-\alpha}} - \underline{B'_{1-\alpha}}\right)^2 + \left(\overline{A'_{1-\alpha}} - \overline{B'_{1-\alpha}}\right)^2\right]d\alpha. \quad (5)$$

where $f(\alpha)$ is the same function as in (4).

Note that (4) and (5) can be implemented by using any of the numerical approximations to integral calculations existing in the literature, such as Simpson's rule.

2.2. Fitting Statistical Parameters with Intuitionistic Fuzzy Numbers

Fuzzy numbers are frequently employed in fuzzy mathematics to represent the epistemic uncertainty of parameters and play an analogous role to random variables in probability theory [38]. Consequently, several studies have investigated the equivalence between fuzzy numbers and random variables with the aim of facilitating consistent ways to transform random variables into possibility distribution functions [44,55,56].

Definition 7. *From a random variable **A** and a family of confidence intervals A_α such that $P(x \in A_\alpha) \geq 1 - \alpha$, where $P(\cdot)$ is a probability measure, we can induce an equivalent FN $\tilde{A}$ whose α-cut representation $A_\alpha$is [55]:*

$$A_\alpha = \left[\underline{A_\alpha}, \overline{A_\alpha}\right] = \left[\left\{x \middle| P(\mathbf{A} \leq x) = \frac{\alpha}{2}\right\}, \left\{x \middle| P(\mathbf{A} \leq x) = 1 - \frac{\alpha}{2}\right\}\right]. \quad (6)$$

Thus, the membership function of the fuzzy number $\tilde{A}$ equivalent to **A** is as follows:

$$\mu_A(x) = sup\{\alpha | x \in A_\alpha\}. \quad (7)$$

This analytical connection between random variables and possibility distributions has led several authors to propose adjusting statistical parameters, such as the mean or variance, with fuzzy numbers that are built up by overlaying sample confidence intervals from the lowest to the highest level of significance [57–61]. Therefore, we define:

Definition 8. *Let (i) be a sample of independent and identically distributed random variables with an unknown parameter θ that allows its interval estimate to be obtained with a significance level $(1-\alpha)$%, $\left[\underline{\theta}(\alpha), \overline{\theta}(\alpha)\right]$. (ii) be a monotonic function $g(\gamma,\alpha) : (0,1] \to \left[\frac{\gamma}{2}, 0.5\right], \gamma \in (0,1)$. Then, from $\left[\underline{\theta}(\alpha), \overline{\theta}(\alpha)\right]$, we can induce a fuzzy number estimate for θ, ${}^{\gamma}\tilde{\theta}$ whose α-cuts are [59]:*

$${}^{\gamma}\theta_\alpha = \left[\underline{{}^{\gamma}\theta_\alpha}, \overline{{}^{\gamma}\theta_\alpha}\right] = \left[\underline{\theta}(2g(\gamma,\alpha)), \overline{\theta}(2g(\gamma,\alpha))\right]. \quad (8)$$

The function $g(\cdot)$ transforms the significance level of the probabilistic confidence interval to a possibilistic membership degree.

Remark 5. *According to [59], the parameter γ determines the width of the support of ${}^{\gamma}\tilde{\theta}$. The determination of γ to encompass all potential values considered in ${}^{\gamma}\tilde{\theta}$ can be interpreted through the application of the 95% and 99.7% rules, which are commonly employed in finance and business modelling [62]. A 95% value (implying a significance level of 5%) covers typical scenarios, incorporating those that are reasonable but not entirely extreme. Conversely, the utilization of significance levels γ close to 0, such as 0.3% (i.e., the 99.7% rule), assumes virtually all conceivable scenarios in support of ${}^{\gamma}\tilde{\theta}$.*

Remark 6. *In this study, the function $g(\gamma,\alpha)$ is defined as a linear function [59]:*

$$g(\gamma,\alpha) = \left(\frac{1}{2} - \frac{\gamma}{2}\right)\alpha + \frac{\gamma}{2}. \quad (9)$$

Definition 9. *Let be a sample with n observations and a sample variance $\hat{\sigma}^2$(standard deviation $\hat{\sigma}$). For variance σ^2, we can build a possibilistic estimate ${}^{\gamma}\tilde{\sigma}^2$ whose membership function is [59]:*

$$\mu_{\gamma\sigma^2}(x) = \begin{cases} \frac{2-\gamma}{1-\gamma} - \frac{2}{1-\gamma}F\left(\frac{(n-1)\hat{\sigma}^2}{x}\right) & \frac{(n-1)\hat{\sigma}^2}{\chi^2_{n-1;\frac{2-\gamma}{2}}} \le x \le \frac{(n-1)\hat{\sigma}^2}{M} \\ \frac{2}{1-\gamma}F\left(\frac{(n-1)\hat{\sigma}^2}{x}\right) - \frac{\gamma}{1-\gamma} & \frac{(n-1)\hat{\sigma}^2}{M} \le x \le \frac{(n-1)\hat{\sigma}^2}{\chi^2_{n-1;\frac{\gamma}{2}}}, \end{cases} \tag{10}$$

where $F(\cdot)$ is the distribution function of a Chi-squared distribution with $n-1$ degrees of freedom; $\chi^2_{n-1;(\cdot)}$ is its inverse for a probability level $(\cdot)$; and M is the median of the Chi-squared distribution. Therefore, the α-levels of ${}^{\gamma}\tilde{\sigma}^2$ and ${}^{\gamma}\sigma^2_{\alpha}$ are as follows:

$$ {}^{\gamma}\sigma^2_{\alpha} = \left[\underline{{}^{\gamma}\sigma^2_{\alpha}}, \overline{{}^{\gamma}\sigma^2_{\alpha}}\right] = \left[\frac{(n-1)\hat{\sigma}^2}{\chi^2_{n-1;\frac{2-\gamma}{2}-\left(\frac{1}{2}-\frac{\gamma}{2}\right)\alpha}}, \frac{(n-1)\hat{\sigma}^2}{\chi^2_{n-1;\left(\frac{1}{2}-\frac{\gamma}{2}\right)\alpha+\frac{\gamma}{2}}}\right]. \tag{11}$$

Remark 7. *Therefore, for the standard deviation, σ, the possibility distribution function estimate, ${}^{\gamma}\tilde{\sigma}$, can be obtained by performing $\mu_{\gamma\sigma}(x) = \mu_{\gamma\sigma^2}(x^2)$; thus, the α-cuts ${}^{\gamma}\sigma_{\alpha}$ are as follows:*

$$ {}^{\gamma}\sigma_{\alpha} = \left[\underline{{}^{\gamma}\sigma_{\alpha}}, \overline{{}^{\gamma}\sigma_{\alpha}}\right] = \left[\sqrt{\frac{(n-1)\hat{\sigma}^2}{\chi^2_{n-1;\frac{2-\gamma}{2}-\left(\frac{1}{2}-\frac{\gamma}{2}\right)\alpha}}}, \sqrt{\frac{(n-1)\hat{\sigma}^2}{\chi^2_{n-1;\left(\frac{1}{2}-\frac{\gamma}{2}\right)\alpha+\frac{\gamma}{2}}}}\right]. \tag{12}$$

Definition 10. *Let us suppose a sample of a random variable with an associated unknown parameter θ that allows us to obtain an interval estimate $\left[\underline{\theta}(\alpha), \overline{\theta}(\alpha)\right]$ with a significance level of $(1-\alpha)\%$. Therefore, we can adjust an intuitionistic estimate $\tilde{\theta}^I$ by fitting its lower distribution function* via *(8) and (9) γ, $({}^{\gamma}\tilde{\theta})$, that is, $\mu_{\theta}(x) = \mu_{\gamma\theta}(x)$, and its upper distribution function with $\gamma^* \le \gamma$ in (7), $({}^{\gamma^*}\tilde{\theta})$, that is, $\mu^*_{\theta}(x) = \mu_{\gamma^*\theta}(x)$. Therefore, for $\tilde{\theta}^I$, we can state that:*

$$\mu^*_{\theta}(x) = \mu_{\gamma^*\theta}(x) \text{ and so } v_{\theta}(x) = 1 - \mu_{\gamma^*\theta}(x). \tag{13}$$

Thus, $\theta_{\langle\alpha,\beta\rangle} = \langle\theta_{\alpha} = \left[\underline{\theta}_{\alpha}, \overline{\theta}_{\alpha}\right], \theta'_{\beta} = \left[\underline{\theta'}_{\beta}, \overline{\theta'}_{\beta}\right], 0 \le \alpha+\beta \le 1, \alpha, \beta \in (0,1)\rangle$, where:

$$\theta_{\alpha} = {}^{\gamma}\theta_{\alpha} = \left[\underline{{}^{\gamma}\theta_{\alpha}}, \overline{{}^{\gamma}\theta_{\alpha}}\right] = \left[\underline{\theta}(2h(\gamma,\alpha)), \overline{\theta}(2h(\gamma,\alpha))\right], \tag{14}$$

$$\theta'_{\beta} = {}^{\gamma^*}\theta_{1-\beta} = \left[\underline{{}^{\gamma^*}\theta_{1-\beta}}, \overline{{}^{\gamma^*}\theta_{1-\beta}}\right] = \left[\underline{\theta}(2h(\gamma^*,1-\beta)), \overline{\theta}(2h(\gamma^*,1-\beta))\right] \tag{15}$$

Remark 8. *Note that the proposed approach utilizes, on the one hand, the transformation of confidence intervals for statistical parameters into possibility distributions. On the other hand, an IFN can be delimited through two possibility distributions: a lower distribution, which gathers the values of the parameters of interest considered real according to the available evidence, and an upper distribution, which gathers the potential values of the parameters [38].*

Definition 11. *Let be a sample with n observations and sample variance $\hat{\sigma}^2$. From Definition 10, we can fit intuitionistic variance $\tilde{\sigma}^{I2}$ by adjusting ${}^{\gamma}\tilde{\sigma}^2$ and ${}^{\gamma^*}\tilde{\sigma}^2$. From (13), we can define a possibilistic estimate, $\tilde{\sigma}^{I2} = \{\langle x, \mu_{\sigma^2}(x), v_{\sigma^2}(x)\rangle,\ x \in X\}$, where:*

$$\mu_{\sigma^2}(x) = \mu_{\gamma\sigma^2}(x),\ \mu^*_{\sigma^2}(x) = \mu_{\gamma^*\sigma^2}(x) \text{ and } v_{\sigma^2}(x) = 1 - \mu^*_{\sigma^2}(x). \tag{16}$$

Therefore, the level set representation $\tilde{\sigma}^{I2}$ *can be denoted as* $\sigma^2_{\langle\alpha,\beta\rangle} = \langle \sigma^2_\alpha = \left[\underline{\sigma^2_\alpha}, \overline{\sigma^2_\alpha}\right], \sigma^{2'}_\beta = \left[\underline{\sigma^{2'}_\beta}, \overline{\sigma^{2'}_\beta}\right], 0 \le \alpha+\beta \le 1, \alpha,\beta \in (0,1)\rangle$, *where by using (11), (14), and (15):*

$$\sigma^2_\alpha = {}^{\gamma}\sigma^2_\alpha = \left[\underline{{}^{\gamma}\sigma^2_\alpha}, \overline{{}^{\gamma}\sigma^2_\alpha}\right] = \left[\frac{(n-1)\hat{\sigma}^2}{\chi^2_{n-1;\frac{2-\gamma}{2}-\left(\frac{1}{2}-\frac{\gamma}{2}\right)\alpha}}, \frac{(n-1)\hat{\sigma}^2}{\chi^2_{n-1;\left(\frac{1}{2}-\frac{\gamma}{2}\right)\alpha+\frac{\gamma}{2}}}\right]. \quad (17)$$

$$\sigma^{2'}_\beta = \left[\underline{{}^{\gamma^*}\sigma^2_{1-\beta}}, \overline{{}^{\gamma^*}\sigma^2_{1-\beta}}\right] = \left[\frac{(n-1)\hat{\sigma}^2}{\chi^2_{n-1;\frac{2-\gamma^*}{2}-\left(\frac{1}{2}-\frac{\gamma^*}{2}\right)(1-\beta)}}, \frac{(n-1)\hat{\sigma}^2}{\chi^2_{n-1;\left(\frac{1}{2}-\frac{\gamma^*}{2}\right)(1-\beta)+\frac{\gamma^*}{2}}}\right]. \quad (18)$$

Remark 9. *Therefore, for the standard deviation* σ, *we can fit an IFN* $\tilde{\sigma}^{I}$ *obtained by performing* $\mu_\sigma(x) = \mu_{\sigma^2}(x^2)$ *and* $\upsilon_\sigma(x) = \upsilon_{\sigma^2}(x^2)$. *Therefore, the* $\langle\alpha,\beta\rangle$*-cuts* $\sigma_{\langle\alpha,\beta\rangle} = \langle \sigma_\alpha = \left[\underline{\sigma_\alpha}, \overline{\sigma_\alpha}\right], \sigma'_\beta = \left[\underline{\sigma'_\beta}, \overline{\sigma'_\beta}\right], 0 \le \alpha+\beta \le 1, \alpha,\beta \in (0,1)\rangle$ *are:*

$$\sigma_\alpha = \left[\sqrt{\underline{\sigma^{2'}_\alpha}}, \sqrt{\overline{\sigma^{2'}_\alpha}}\right] = \left[\sqrt{\underline{{}^{\gamma}\sigma^2_\alpha}}, \sqrt{\overline{{}^{\gamma}\sigma^2_\alpha}}\right] \text{ and } \sigma'_\beta = \left[\sqrt{\underline{\sigma^{2'}_\beta}}, \sqrt{\overline{\sigma^{2'}_\beta}}\right] = \left[\sqrt{\underline{{}^{\gamma^*}\sigma^2_\alpha}}, \sqrt{\overline{{}^{\gamma^*}\sigma^2_\alpha}}\right]. \quad (19)$$

Numerical application 1. The empirical applications developed in this section are based on the daily average values of the IBEX35 Futures Index, which serves as the underlying asset for most liquid stock options in the Spanish derivative market. This numerical analysis uses daily data from 11 August 2022, to 27 January 2023, comprising 121 observations. Figure 2 illustrates the evolution of the index throughout the specified period and the corresponding logarithmic returns.

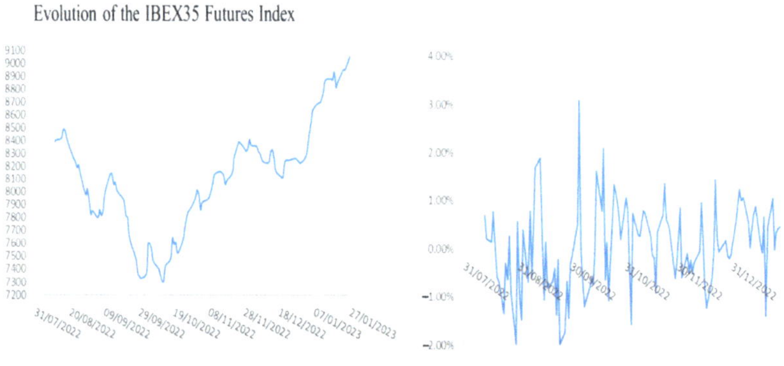

Figure 2. Evolution of IBEX 35 Futures from 4 November 2022, to 27 January 2023.

The volatility of the index requires a predefined time horizon that varies over time. In this example, we adjust an IFN for volatility over calculation horizons of $n = 60$ observations and $n = 120$ observations as of 27 January 2023. That is, we calculate the 60-day and 120-day volatility recorded on 27 January 2023.

The annualized 60-day volatility of the index is 9.806%. Thus, for a significance level of $\gamma = 5\%$ (i.e., with the 95% rule), we obtain a membership function for the annualized standard deviation ${}^{0.05}\tilde{\sigma}$, whose shape starts from (16) and Remark 5.

$$\mu_{{}^{0.05}\sigma}(x) = \begin{cases} \frac{1.95}{0.95} - \frac{2}{0.95}F\left(\frac{59\cdot 0.09806^2}{x^2}\right) & 0.08312 \le x \le 0.09806 \\ \frac{2}{0.95}F\left(\frac{59\cdot 0.09806^2}{x^2}\right) - \frac{0.05}{0.95} & 0.09806 \le x \le 0.11958 \end{cases}.$$

Therefore, the α-cuts (12) are:

$$^{0.05}\sigma_\alpha = \left[\underline{^{0.05}\sigma_\alpha}, \overline{^{0.05}\sigma^2_\alpha}\right] = \left[\sqrt{\frac{59 \cdot 0.09806^2}{\chi^2_{59;0.975-\frac{0.95}{2}\alpha}}}, \sqrt{\frac{59 \cdot 0.09806^2}{\chi^2_{59;\frac{0.95}{2}\alpha+0.025}}}\right].$$

We can construct an intuitionistic estimate for the volatility $\overset{\sim I}{\sigma}$ of the index in the period of interest on the basis of $^{0.05}\widetilde{\sigma}$ and adjust an upper possibility distribution whose support is set by the 99.7% rule in (16) and (19); that is, by stating γ^* = 0.3% and using $^{0.003}\widetilde{\sigma}$ to state the nonmembership function. Thus, $\mu_\sigma(x) = \mu_{^{0.05}\sigma}(x)$:

$$\mu_\sigma(x) = \begin{cases} \frac{1.95}{0.95} - \frac{2}{0.95}F\left(\frac{59 \cdot 0.09806^2}{x^2}\right) & 0.08312 \le x \le 0.09806 \\ \frac{2}{0.95}F\left(\frac{59 \cdot 0.09806^2}{x^2}\right) - \frac{0.05}{0.95} & 0.09806 \le x \le 0.11958 \end{cases}.$$

From (16) and (19), $\mu^*_\sigma(x) = \mu_{^{0.003}\sigma}(x)$:

$$\mu^*_\sigma(x) = \begin{cases} \frac{1.997}{0.997} - \frac{2}{0.997}F\left(\frac{59 \cdot 0.09806^2}{x^2}\right) & 0.07668 \le x \le 0.09806 \\ \frac{2}{0.997}F\left(\frac{59 \cdot 0.09806^2}{x^2}\right) - \frac{0.003}{0.997} & 0.09806 \le x \le 0.13338 \end{cases}.$$

In such a way, the nonmembership of $\overset{\sim I}{\sigma}$ is $\nu_\sigma(x) = 1 - \mu^*_\sigma(x)$:

$$\nu_\sigma(x) = \begin{cases} \frac{2}{0.997}F\left(\frac{59 \cdot 0.09806^2}{x^2}\right) - \frac{1}{0.997} & 0.07668 \le x \le 0.09806 \\ \frac{1}{0.997} - \frac{2}{0.997}F\left(\frac{59 \cdot 0.09806^2}{x^2}\right) & 0.09806 \le x \le 0.13338 \end{cases}$$

Similarly, $\sigma_{\langle\alpha,\beta\rangle} = \langle \sigma_\alpha = [\underline{\sigma_\alpha}, \overline{\sigma_\alpha}], \sigma'_\beta = [\underline{\sigma'_\beta}, \overline{\sigma'_\beta}], 0 \le \alpha+\beta \le 1, \alpha,\beta \in (0,1)\rangle$, where:

$$\sigma_\alpha = {^\gamma\sigma_\alpha} = \left[\underline{^\gamma\sigma_\alpha}, \overline{^\gamma\sigma_\alpha}\right] = \left[\sqrt{\frac{59 \cdot 0.09806^2}{\chi^2_{59;0.975-\frac{0.95}{2}\alpha}}}, \sqrt{\frac{59 \cdot 0.09806^2}{\chi^2_{59;\frac{0.95}{2}\alpha+0.025}}}\right],$$

and,

$$\sigma'_\beta = \left[\underline{^{\gamma^*}\sigma_{1-\beta}}, \overline{^{\gamma^*}\sigma_{1-\beta}}\right] = \left[\sqrt{\frac{59 \cdot 0.09806^2}{\chi^2_{59;0.9985-\frac{0.997}{2}(1-\beta)}}}, \sqrt{\frac{59 \cdot 0.09806^2}{\chi^2_{59;\frac{0.997}{2}(1-\beta)+0.0015}}}\right].$$

Figure 3 shows the shape of the possibilistic estimates of the annualized volatility for the last 60 days of the IBEX-35 Futures Index on 27 January 2023, $^{0.05}\widetilde{\sigma}$ and $^{0.003}\widetilde{\sigma}$, and the IFN developed above.

On the other hand, the annualized 120-day volatility of the IBEX-35 Futures index is 13.908%. Thus, analogous to the 60-day volatility, we can construct an intuitionistic estimate for the 120-day volatility $\overset{\sim I}{\sigma}$ of the index in the period of interest on the basis of $^{0.05}\widetilde{\sigma}$ and $^{0.003}\widetilde{\sigma}$. Thus, $\mu_\sigma(x) = \mu_{^{0.05}\sigma}(x)$:

$$\mu_{^{0.05}\sigma}(x) = \mu_\sigma(x) = \begin{cases} \frac{1.95}{0.95} - \frac{2}{0.95}F\left(\frac{119 \cdot 0.13908^2}{x^2}\right) & 0.12343 \le x \le 0.13908 \\ \frac{2}{0.95}F\left(\frac{119 \cdot 0.13908^2}{x^2}\right) - \frac{0.05}{0.95} & 0.13908 \le x \le 0.15931 \end{cases}.$$

From (16) and (19), $\mu^*_{\sigma}(x) = \mu_{0.003_{\sigma}}(x)$:

$$\mu^*_{\sigma}(x) = \begin{cases} \frac{1.997}{0.997} - \frac{2}{0.997}F\left(\frac{119 \cdot 0.13908^2}{x^2}\right) & 0.11635 \le x \le 0.13908 \\ \frac{2}{0.997}F\left(\frac{119 \cdot 0.13908^2}{x^2}\right) - \frac{0.003}{0.997} & 0.13908 \le x \le 0.17995 \end{cases}.$$

In such a way, the nonmembership of $\overset{\sim I}{\sigma}$ is $\nu_{\sigma}(x) = 1 - \mu^*_{\sigma}(x)$:

$$\nu_{\sigma}(x) = \begin{cases} \frac{2}{0.997}F\left(\frac{119 \cdot 0.13908^2}{x^2}\right) - \frac{1}{0.997} & 0.07668 \le x \le 0.09806 \\ \frac{1}{0.997} - \frac{2}{0.997}F\left(\frac{119 \cdot 0.13908^2}{x^2}\right) & 0.09806 \le x \le 0.13338 \end{cases}$$

Similarly, $\sigma_{\langle\alpha,\beta\rangle} = \langle \sigma_\alpha = [\underline{\sigma_\alpha}, \overline{\sigma_\alpha}], \sigma'_\beta = [\underline{\sigma'_\beta}, \overline{\sigma'_\beta}], 0 \le \alpha + \beta \le 1, \alpha, \beta \in (0,1) \rangle$, where:

$$\sigma_\alpha = {}^{\gamma}\sigma_\alpha = \left[\underline{{}^{\gamma}\sigma_\alpha}, \overline{{}^{\gamma}\sigma_\alpha}\right] = \left[\sqrt{\frac{119 \cdot 0.13908^2}{\chi^2_{119;0.975-\frac{0.95}{2}\alpha}}}, \sqrt{\frac{119 \cdot 0.13908^2}{\chi^2_{119;\frac{0.95}{2}\alpha+0.025}}}\right],$$

and,

$$\sigma'_\beta = \left[\underline{{}^{\gamma^*}\sigma_{1-\beta}}, \overline{{}^{\gamma^*}\sigma_{1-\beta}}\right] = \left[\sqrt{\frac{119 \cdot 0.13908^2}{\chi^2_{119;0.9985-\frac{0.997}{2}(1-\beta)}}}, \sqrt{\frac{119 \cdot 0.13908^2}{\chi^2_{119;\frac{0.997}{2}(1-\beta)+0.0015}}}\right].$$

Regarding the obtained results, the following points need to be clarified:

- To determine historical volatility, the desired time horizon (e.g., 30 days, 60 days) must be specified. The choice of horizon will determine the core of the IFN that quantifies volatility.
- The time horizon for volatility affects the breadth of the membership and nonmembership functions: a shorter time horizon implies fewer observations for calculating volatility and broader confidence intervals (17) and (18).
- The percentiles used to set the upper and lower possibility functions determine their breadth. The percentiles associated with lower probabilities result in narrower membership and nonmembership functions. For example, using 90% rules for the lower possibility function and 95% for the upper possibility function would result in narrower $\langle\alpha, \beta\rangle$-cuts.

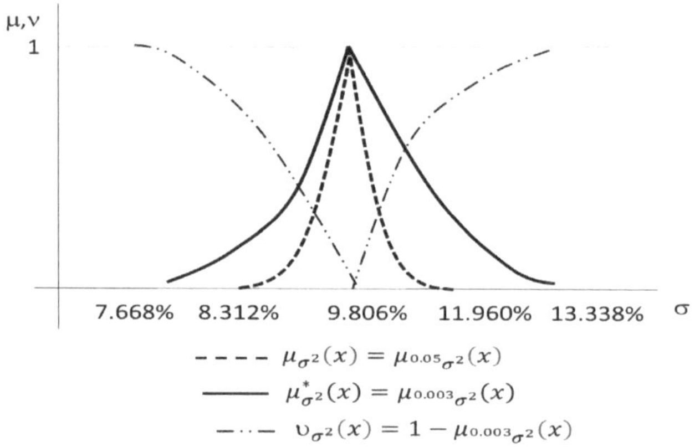

Figure 3. Fuzzy number and intuitionistic fuzzy number estimates of the 60-day volatility of the IBEX35 Futures in 27 January 2023.

2.3. Intuitionistic Fuzzy Number Arithmetic

We are now interested in evaluating continuous and differentiable functions $y = f(x_1, x_2, \ldots, x_n)$ such that the values of the arguments are estimated by IFNs $\overset{\sim I}{A}_{(i)}$, $i = 1, 2, \ldots, n$. Thus, $f(\cdot)$ generates an IFN $\overset{\sim I}{B}$, $\overset{\sim I}{B} = f\left(\overset{\sim I}{A}_{(1)}, \overset{\sim I}{A}_{(2)}, \ldots, \overset{\sim I}{A}_{(n)}\right)$. The membership and nonmembership functions of $\overset{\sim I}{B}$ can be obtained via the generalization of Zadeh's extension principle [63]:

$$\mu_B(y) = \max_{y=f(x_1,x_2,\ldots,x_n)} \min\left\{\mu_{A_{(1)}}(x_1), \mu_{A_{(2)}}(x_2), \ldots, \mu_{A_{(n)}}(x_n)\right\}, \tag{20}$$

$$v_B(y) = \min_{y=f(x_1,x_2,\ldots,x_n)} \max\left\{v_{A_{(1)}}(x_1), v_{A_{(2)}}(x_2), \ldots, v_{A_{(n)}}(x_n)\right\}. \tag{21}$$

Therefore, if $\overset{\sim I}{A}_{(i)}$, $i = 1, 2, \ldots, n$ are simply fuzzy numbers $\tilde{A}_{(i)}$, the result is also a fuzzy number $\tilde{B}$ whose shape only depends on $\mu_B(y)$ in (20).

The compatibility of Zadeh's extension with the α-cut arithmetic can also be extended to the evaluation of $f(\cdot)$. Thus, to obtain $B_{\langle\alpha,\beta\rangle}$ from $A_{(i)\,\langle\alpha,\beta\rangle}$, $i = 1, 2, \ldots, n$, we must implement:

$$B_{\langle\alpha,\beta\rangle} = f\left(A_{(1)\,\langle\alpha,\beta\rangle}, A_{(2)\,\langle\alpha,\beta\rangle}, \ldots, A_{(n)\,\langle\alpha,\beta\rangle}\right), \tag{22}$$

Thus, given that f is supposed to be continuous, $B_{\langle\alpha,\beta\rangle}$ can be represented as

$$B_{\langle\alpha,\beta\rangle} = \langle B_\alpha = \left[\underline{B_\alpha}, \overline{B_\alpha}\right], B'_\beta = \left[\underline{B'_\beta}, \overline{B'_\beta}\right], 0 \leq \alpha + \beta \leq 1, \alpha, \beta \in [0,1]\rangle,$$

where for $i = 1,2,\ldots,n$:

$$\underline{B_\alpha} = \inf\left\{y \middle| y = f(x_1, \ldots, x_n), x_i \in A_{(i)_\alpha}\right\}, \overline{B_\alpha} = \sup\left\{y \middle| y = f(x_1, \ldots, x_n), x_i \in A_{(i)_\alpha}\right\}, \tag{23}$$

$$\underline{B'_\beta} = \inf\left\{y \middle| y = f(x_1, \ldots, x_n), x_i \in A'_{(i)_\beta}\right\}, \overline{B'_\beta} = \sup\left\{y \middle| y = f(x_1, \ldots, x_n), x_i \in A'_{(i)_\beta}\right\}. \tag{24}$$

If $f(\cdot)$ increases with respect to x_i, $i = 1, 2, \ldots m$ and decreases in x_i, $i = m+1, m+2, \ldots, n$, $m \leq n$, we obtain [64]:

$$\underline{B_\alpha} = f\left(\underline{A_{(1)_\alpha}}, \underline{A_{(2)_\alpha}}, \ldots, \underline{A_{(m)_\alpha}}, \overline{A_{(m+1)_\alpha}}, \ldots, \overline{A_{(n)_\alpha}}\right), \tag{25}$$

$$\overline{B_\alpha} = f\left(\overline{A_{(1)_\alpha}}, \overline{A_{(2)_\alpha}}, \ldots, \overline{A_{(m)_\alpha}}, \underline{A_{(m+1)_\alpha}}, \ldots, \underline{A_{(n)_\alpha}}\right). \tag{26}$$

and thus, by induction, for $B'_\beta = \left[\underline{B'_\beta}, \overline{B'_\beta}\right]$:

$$\underline{B'_\beta} = f\left(\underline{A'_{(1)_\beta}}, \underline{A'_{(2)_\beta}}, \ldots, \underline{A'_{(m)_\beta}}, \overline{A'_{(m+1)_\beta}}, \ldots, \overline{A'_{(n)_\beta}}\right), \tag{27}$$

$$\overline{B'_\beta} = f\left(\overline{A'_{(1)_\beta}}, \overline{A'_{(2)_\beta}}, \ldots, \overline{A'_{(m)_\beta}}, \underline{A_{(m+1)_\beta}}, \ldots, \underline{A_{(n)_\beta}}\right). \tag{28}$$

In the case where we evaluate $f(\cdot)$ in the fuzzy numbers $\tilde{A}_{(i)}$, $i = 1,2,\ldots,n$, the α-cuts of the result, $\tilde{B}$, are simply $B_\alpha = f\left(A_{(1)_\alpha}, \ldots, A_{(n)_\alpha}\right)$, which can be fitted with (23), (25), and (26) when $f(\cdot)$ is a monotonic function.

In this regard, alternatives exist for directly computing (22)–(28) to reduce the computation time. For example, using piecewise linear approximations can significantly reduce computational overhead.

3. An Extension Black–Scholes–Merton and Binomial Option Pricing Model for the Use of Intuitionistic Fuzzy Parameters

3.1. Pricing European Options with the Black–Scholes–Merton Model and Intuitionistic Fuzzy Parameters

Let be a call European option for an asset with price S, strike price K, and volatility σ that can be exercised in T years. The BSM formula [1,2] for the price of a call option, $C(S,K,r,\sigma,T)$, for the free risk rate r is

$$C(S,K,r,\sigma,T) = S\Phi\left(\frac{ln\left(\frac{S}{K}\right) + \left(r + \frac{\sigma^2}{2}\right)T}{\sigma\sqrt{T}}\right) - e^{-rT}K\Phi\left(\frac{ln\left(\frac{S}{K}\right) + \left(r - \frac{\sigma^2}{2}\right)T}{\sigma\sqrt{T}}\right), \quad (29)$$

where $\Phi(\cdot)$ is the cumulative standard Gaussian function.

In the case of a European put option, the price is also a function of S, K, r, σ, and T such that

$$P(S,K,r,\sigma,T) = e^{-rT}K\Phi\left(-\frac{ln\left(\frac{S}{K}\right) + \left(r - \frac{\sigma^2}{2}\right)T}{\sigma\sqrt{T}}\right) - S\Phi\left(-\frac{ln\left(\frac{S}{K}\right) + \left(r + \frac{\sigma^2}{2}\right)T}{\sigma\sqrt{T}}\right). \quad (30)$$

Note that in the case of options on future contracts, Black [65] demonstrates that prices (29) and (30) are obtained by evaluating them at $r = 0$.

A mainstream research field within the FROP literature is located within the BSM framework. Within these applications, we outline the following settings:

1. Analytical aspects include computing prices (29)–(30) with possibility distributions [16,66,67].
2. The immunization measures derived from formulas (29)–(30), so-called 'the Greeks', are computed and analysed when data are given fuzzy numbers [10,68–70].
3. The most relevant parameter is computed, at least for options on financial assets traded in organized markets, which is volatility. Studies include both computing historical volatility [10] and calculating implied volatility [19–21,67,71].
4. A fuzzy BSM formula is used for corporate and real option pricing [11,15,72,73].

Obviously, the parameters that should be subject to fuzzification depend on the context in which optionality is embedded. In options for stocks traded in exchange markets, parameters that may be vague include the price of the underlying asset, the risk-free interest rate, and volatility. In contrast, exercise price and expiration are crisp parameters because they are standardized contracts [12]. However, in the valuation of real options, a common situation is that the exercise price or expiration date are parameters whose knowledge is not precise [11,12,72].

Without loss of generality, we assume that all the variables used to evaluate (29) and (30) S, K, r, σ, and T are given by IFNs $\overset{\sim I}{S}, \overset{\sim I}{K}, \overset{\sim I}{r}, \overset{\sim I}{\sigma}$, and $\overset{\sim I}{T}$, respectively. With the exception of $\overset{\sim I}{r}$, the remaining IFNs were defined strictly in $\Re^+$. Therefore, (29) induces an intuitionistic fuzzy price for a call option $\overset{\sim I}{C} = C(\overset{\sim I}{S}, \overset{\sim I}{K}, \overset{\sim I}{r}, \overset{\sim I}{\sigma}, \overset{\sim I}{T})$ whose level sets are obtained by evaluating $C_{\langle\alpha,\beta\rangle} = C(S_{\langle\alpha,\beta\rangle}, K_{\langle\alpha,\beta\rangle}, r_{\langle\alpha,\beta\rangle}, \sigma_{\langle\alpha,\beta\rangle}, T_{\langle\alpha,\beta\rangle})$. Therefore, by considering that $\frac{\partial C}{\partial S} \geq 0$, $\frac{\partial C}{\partial K} \leq 0$, $\frac{\partial C}{\partial r} \geq 0$, $\frac{\partial C}{\partial \sigma} \geq 0$ and $\frac{\partial C}{\partial T} \geq 0$ [3] and applying rules (25)–(28) in (29)

$$C_\alpha = \left[\underline{C_\alpha}, \overline{C_\alpha}\right] = \left[C\left(\underline{S_\alpha}, \overline{K_\alpha}, \underline{r_\alpha}, \underline{\sigma_\alpha}, \underline{T_\alpha}\right), C\left(\overline{S_\alpha}, \underline{K_\alpha}, \overline{r_\alpha}, \overline{\sigma_\alpha}, \overline{T_\alpha}\right)\right], \quad (31)$$

and

$$C'_{\beta} = \left[\underline{C'_{\beta}}, \overline{C'_{\beta}}\right] = \left[C\left(\underline{S'_{\beta}}, \overline{K'_{\beta}}, \underline{r'_{\beta}}, \underline{\sigma'_{\beta}}, \underline{T'_{\beta}}\right), C\left(\overline{S'_{\beta}}, \underline{K'_{\beta}}, \overline{r'_{\beta}}, \overline{\sigma'_{\beta}}, \overline{T'_{\beta}}\right)\right]. \quad (32)$$

Similarly, in the case of put options, we can induce an intuitionistic fuzzy price $\overset{\sim I}{P} = P(\overset{\sim I}{S}, \overset{\sim I}{K}, \overset{\sim I}{r}, \overset{\sim I}{\sigma}, \overset{\sim I}{T})$ whose level sets are obtained by evaluating $P_{\langle\alpha,\beta\rangle} = P(S_{\langle\alpha,\beta\rangle}, K_{\langle\alpha,\beta\rangle}, r_{\langle\alpha,\beta\rangle}, \sigma_{\langle\alpha,\beta\rangle}, T_{\langle\alpha,\beta\rangle})$. To do so, we must apply rules (22)–(28) in (30) by considering $\frac{\partial P}{\partial S} \leq 0$, $\frac{\partial P}{\partial K} \geq 0$, $\frac{\partial P}{\partial r} \leq 0$, and $\frac{\partial C}{\partial \sigma} \geq 0$ [3]. Therefore, we obtain the following for $P_{\alpha} = \left[\underline{P_{\alpha}}, \overline{P_{\alpha}}\right]$:

$$\underline{P_{\alpha}} = \text{minimum } P\left(\overline{S_{\alpha}}, \underline{K_{\alpha}}, \overline{r_{\alpha}}, \underline{\sigma_{\alpha}}, T\right), \text{subject to : } \underline{T_{\alpha}} \leq T \leq \overline{T_{\alpha}}, \quad (33)$$

$$\overline{P_{\alpha}} = \text{maximum } P\left(\underline{S_{\alpha}}, \overline{K_{\alpha}}, \underline{r_{\alpha}}, \overline{\sigma_{\alpha}}, T\right), \text{subject to : } \underline{T_{\alpha}} \leq T \leq \overline{T_{\alpha}}, \quad (34)$$

and $P'_{\beta} = \left[\underline{P'_{\beta}}, \overline{P'_{\beta}}\right]$ considering:

$$\underline{P'_{\beta}} = \text{minimum } P\left(\overline{S'_{\beta}}, \underline{K'_{\beta}}, \overline{r'_{\beta}}, \underline{\sigma'_{\beta}}, T\right), \text{subject to : } \underline{T'_{\beta}} \leq T \leq \overline{T'_{\beta}}, \quad (35)$$

$$\overline{P'_{\beta}} = \text{maximum } P\left(\underline{S'_{\beta}}, \overline{K'_{\beta}}, \underline{r'_{\beta}}, \overline{\sigma'_{\beta}}, T\right), \text{subject to : } \underline{T'_{\beta}} \leq T \leq \overline{T'_{\beta}}. \quad (36)$$

3.2. *Pricing European Options with a Binomial Model and Intuitionistic Fuzzy Parameters*

The origin of the binomial option pricing model can be traced back to the nearly simultaneous publication of Cox et al. [6] and Rendleman and Bartter [7]. In the binomial framework, variation in stock prices occurs in discrete time, and there are only two possible moves, which are growth (up) and decline (down). In both CRR and RB, as well as in their multiple variants, movements of the prices of the underlying asset are multiplicative, and both these movements and the risk-neutral probabilities depend on the volatility of the underlying asset and the risk-free rate [5,9].

Thus, we symbolize the upwards growth rate of the underlying asset as $u(r,\sigma) > 1$ and the downwards rate as $d(r,\sigma)$, where $0 < d(r,\sigma) < 1$. Similarly, we denote the risk-neutral probability of the upwards movement as $\pi_u(r,\sigma)$ and the associated probability of the downwards movement as $\pi_d(r,\sigma) = 1 - \pi_u(r,\sigma)$. Table 1 shows the models analysed in our work. The first is the most commonly used method in practice and is unanimously considered in the fuzzy literature; we call this the CRR [6]. The alternative tested in this paper is [7], which we call the RB. Note that both cases can be unified into a general formulation of up-and-down moves, $u(a,r,\sigma) = e^{a\cdot\sigma\sqrt{h}}$ and $d(b,r,\sigma) = e^{-b\cdot\sigma\sqrt{h}}$. Therefore, while in the case of CRR $a = b = 1$, for the RB, it can be checked that $a = \left(r - \sigma^2/2\right)h + \sigma\sqrt{h}$ and $b = -\left(r - \sigma^2/2\right)h + \sigma\sqrt{h}$.

Table 1. Models of binomial up-and-down moves used in this paper.

	CRR [6]	**RB [7]**
$u(r,\sigma)$	$e^{\sigma\sqrt{h}}$	$e^{(r-\sigma^2/2)h+\sigma\sqrt{h}}$
$d(r,\sigma)$	$e^{-\sigma\sqrt{h}}$	$e^{(r-\sigma^2/2)h-\sigma\sqrt{h}}$
$p_u(r,\sigma)$	$\frac{e^{rh}-e^{-\sigma\sqrt{h}}}{e^{\sigma\sqrt{h}}-e^{-\sigma\sqrt{h}}}$	$\frac{e^{\sigma^2 h/2}-e^{-\sigma\sqrt{h}}}{e^{\sigma\sqrt{h}}-e^{-\sigma\sqrt{h}}}$
$p_d(r,\sigma)$	$\frac{e^{\sigma\sqrt{h}}-e^{rh}}{e^{\sigma\sqrt{h}}-e^{-\sigma\sqrt{h}}}$	$\frac{e^{\sigma\sqrt{h}}-e^{\sigma^2 h/2}}{e^{\sigma\sqrt{h}}-e^{-\sigma\sqrt{h}}}$

In any binomial model, maturity T is divided into n periods of duration h years such that $T = n\cdot h$. Then, the price of a European call option is

$$C_b(S,K,r,\sigma,n,h) = e^{-r\cdot n\cdot h}\sum_{j=0}^{n}\binom{n}{j}\pi_u(r,\sigma)^j\pi_d(r,\sigma)^{n-j}\max\{Su(r,\sigma)^j d(r,\sigma)^{n-j}-K,0\} \tag{37}$$

For a put option,

$$P_b(S,K,r,\sigma,n,h) = e^{-r\cdot n\cdot h}\sum_{j=0}^{n}\binom{n}{j}\pi_u(r,\sigma)^j\pi_d(r,\sigma)^{n-j}\max\{K-Su(r,\sigma)^j d(r,\sigma)^{n-j},0\} \tag{38}$$

As $h \to 0$, Formulas (37)–(38) tend towards the price obtained with the BSM [5]. The primary utility of using binomial lattices lies in their applicability to options where the use of the BSM formula is challenging, such as American options, exotic options, or different flexibilities linked to real options [4].

In the FROP literature, the binomial model is the most widely used discrete-time option valuation method. In all the studies, modelling of the up-and-down moves and their probabilities was performed using the CRR model, with volatility being the parameter commonly considered fuzzy. Furthermore, in all contributions, uncertainty is introduced epistemically through fuzzy numbers, which are typically triangular or trapezoidal [12,13].

Next, we extend (18) and (19), assuming that all the parameters, except those associated with maturity, are IFNs $\overset{\sim I}{S}, \overset{\sim I}{K}, \overset{\sim I}{r}, \overset{\sim I}{\sigma}$. With the exception of $\overset{\sim I}{r}$, these IFNs are defined in $\Re^+$ The hypothesis that n and h are crisp parameters is a unanimous assumption in the FROP literature. Therefore, (37) induces an intuitionistic fuzzy price of call options $\overset{\sim I}{C_b} = C_b\left(\overset{\sim I}{S}, \overset{\sim I}{K}, \overset{\sim I}{r}, \overset{\sim I}{\sigma}, n, h\right)$ whose level sets are obtained by evaluating $C_{b_{\langle\alpha,\beta\rangle}} = C_b\left(S_{\langle\alpha,\beta\rangle}, K_{\langle\alpha,\beta\rangle}, r_{\langle\alpha,\beta\rangle}, \sigma_{\langle\alpha,\beta\rangle}, n, h\right)$ and by applying rules (25)–(28) in (37)

$$C_{b_\alpha} = \left[\underline{C_{b_\alpha}}, \overline{C_{b_\alpha}}\right] = \left[C_b\left(\underline{S_\alpha}, \overline{K_\alpha}, \underline{r_\alpha}, \underline{\sigma_\alpha}, n, h\right), C_b\left(\overline{S_\alpha}, \underline{K_\alpha}, \overline{r_\alpha}, \overline{\sigma_\alpha}, n, h\right)\right], \tag{39}$$

and

$$C'_{b_\beta} = \left[\underline{C'_{b_\beta}}, \overline{C'_{b_\beta}}\right] = \left[C_b\left(\underline{S'_\beta}, \overline{K'_\beta}, \underline{r'_\beta}, \underline{\sigma'_\beta}, n, h\right), C_b\left(\overline{S'_\beta}, \underline{K'_\beta}, \overline{r'_\beta}, \overline{\sigma'_\beta}, n, h\right)\right] \tag{40}$$

For the case of an intuitionistic fuzzy price of a put option $\overset{\sim I}{P_b} = P_b\left(\overset{\sim I}{S}, \overset{\sim I}{K}, \overset{\sim I}{r}, \overset{\sim I}{\sigma}, n, h\right)$, the level sets are obtained by evaluating $P_{b_{\langle\alpha,\beta\rangle}} = P_b\left(S_{\langle\alpha,\beta\rangle}, K_{\langle\alpha,\beta\rangle}, r_{\langle\alpha,\beta\rangle}, \sigma_{\langle\alpha,\beta\rangle}, n, h\right)$ analogous to the case of call options. Therefore, we obtain P_{b_α} by applying (25)–(28) in (38)

$$P_{b_\alpha} = \left[\underline{P_{b_\alpha}}, \overline{P_{b_\alpha}}\right] = \left[P_b\left(\overline{S_\alpha}, \underline{K_\alpha}, \overline{r_\alpha}, \underline{\sigma_\alpha}, n, h\right), P_b\left(\underline{S_\alpha}, \overline{K_\alpha}, \underline{r_\alpha}, \overline{\sigma_\alpha}, n, h\right)\right], \tag{41}$$

and

$$P'_{b_\beta} = \left[\underline{P'_{b_\beta}}, \overline{P'_{b_\beta}}\right] = \left[P\left(\overline{S'_\beta}, \underline{K'_\beta}, \overline{r'_\beta}, \underline{\sigma'_\beta}, n, h\right), P\left(\underline{S'_\beta}, \overline{K'_\beta}, \underline{r'_\beta}, \overline{\sigma'_\beta}, n, h\right)\right]. \tag{42}$$

Numerical application 2. In this section, we evaluate European call options for IBEX-35 futures via the intuitionistic versions of the BSM and binomial models developed in this section. Within the binomial models, we considered both CRR and RB for modelling moves. The only intuitionistic fuzzy parameter is the 60-day volatility ($\overset{\sim I}{\sigma}$), which was adjusted in the numerical application on 27 January 2023, in numerical application 1. Additionally, we consider the intuitionist estimate of the 60-day volatility obtained from the S&P 500 index on 4 November 2020.

In the application, we assume that the value of the underlying asset is $S = 1$ and $r = 0\%$ because we are pricing futures on options [65]. The maturity of all the options was $T = 1$ year. We consider three possible strike prices with three different degrees of moneyness as follows: $K = 0.9$ (in the money), $K = 1$ (in the money), and $K = 1.1$ (out of the money).

For the binomial models, we considered the following eight different jump frequencies: annual ($h = 1, n = 1$), semiannual ($h = \frac{1}{2}, n = 2$), quarterly ($h = \frac{1}{4}, n = 4$), monthly ($h = \frac{1}{12}$, $n = 12$), biweekly ($h = \frac{1}{24}, n = 24$), weekly ($h = \frac{1}{48}, n = 48$), daily ($h = \frac{1}{252}, n = 252$), and every 12 h $h = \frac{1}{504}, n = 504$).

Therefore, to calculate the prices of call options with the BSM formula, we use (29), (31), and (32) to implement $\overset{\sim I}{C} = C\left(1, K, 0, \overset{\sim I}{\sigma}, 1\right)$.

Thus, the level sets of the BSM option price in the case of the 60-day volatility obtained for the IBEX-35 of Futures obtained in numerical application 1 are $C_{\langle\alpha,\beta\rangle} = C(1, K, 0, \sigma_{\langle\alpha,\beta\rangle}, 1)$

$$C_\alpha = \left[\underline{C_\alpha}, \overline{C_\alpha}\right] = \left[C\left(1, K, 0, \sqrt{\frac{59 \cdot 0.09806^2}{\chi^2_{59;0.975-\frac{0.95}{2}\alpha}}}, 1\right), C\left(1, K, 0, \sqrt{\frac{59 \cdot 0.09806^2}{\chi^2_{59;\frac{0.95}{2}\alpha+0.025}}}, 1\right)\right],$$

and

$$C'_\beta = \left[\underline{C'_\beta}, \overline{C'_\beta}\right] = \left[\left(1, K, 0, \sqrt{\frac{59 \cdot 0.09806^2}{\chi^2_{59;0.9985-\frac{0.997}{2}(1-\beta)}}}, 1\right), C\left(1, K, 0, \sqrt{\frac{59 \cdot 0.09806^2}{\chi^2_{59;\frac{0.997}{2}(1-\beta)+0.0015}}}, 1\right)\right].$$

On the other hand, if call option prices are calculated using the binomial formula, they are evaluated via an evaluation of (37) with intuitionistic volatility: $\overset{\sim I}{C_b} = C_b\left(1, K, 0, \overset{\sim I}{\sigma}, n, h\right)$. The level sets are calculated using (39)–(40) as follows: $C_{b_{\langle\alpha,\beta\rangle}} = C_b\left(1, K, 0, \sigma_{\langle\alpha,\beta\rangle}, n, h\right)$, where

$$C_{b_\alpha} = \left[\underline{C_{b_\alpha}}, \overline{C_{b_\alpha}}\right] = \left[C_b\left(1, K, 0, \sqrt{\frac{59 \cdot 0.09806^2}{\chi^2_{59;0.975-\frac{0.95}{2}\alpha}}}, n, h\right), C_b\left(1, K, 0, \sqrt{\frac{59 \cdot 0.09806^2}{\chi^2_{59;\frac{0.95}{2}\alpha+0.025}}}, n, h\right)\right],$$

and

$$C'_{b_\beta} = \left[\underline{C'_{b_\beta}}, \overline{C'_{b_\beta}}\right] = \left[C_b\left(1, K, 0, \sqrt{\frac{59 \cdot 0.09806^2}{\chi^2_{59;0.9985-\frac{0.997}{2}(1-\beta)}}}, n, h\right), C_b\left(1, K, 0, \sqrt{\frac{59 \cdot 0.09806^2}{\chi^2_{59;\frac{0.997}{2}(1-\beta)+0.0015}}}, n, h\right)\right]$$

Obviously, the final form of the above $\langle\alpha, \beta\rangle$-cuts will depend on how we model the moves of the underlying asset, which, in this work, are those listed in Table 1. Table 2 shows the results of the expected values of the European call prices, which we calculate via (4); that is, $EV(\overset{\sim I}{C})$ in the case of prices calculated with the BSM and $EV(\overset{\sim I}{C_b})$ with those calculated with the binomial model.

When evaluating a particular option that cannot be valued with the BSM formula, it seems reasonable to choose a binomial model that yields prices of the European option of the same type (call or put) and with the closest strike prices, maturities, and of the same type as the BSM formula [47]. Therefore, Table 2 also shows the distance between the value of the price calculated with the BSM and with the binomial model, which is expressed in relation to the expected value of the BSM price; that is, $\frac{D(\overset{\sim I}{C}, \overset{\sim I}{C_b})}{EV(\overset{\sim I}{C})}$. It involves using (4) and (5). This ratio indicates the error caused by the binomial approximation of the BSM formula. The calculations of (4) and (5) were performed by computing the integrals using Simpson's rule.

Table 2. A comparison of European option prices (with $T = 1$ and $r = 0\%$) on futures over IBEX-35 obtained using the Black–Scholes–Merton model and binomial models such as CRR and RB via the intuitionistic fuzzy variance in numerical example 1.

$K = 900$							
h	n	**BSM (a)**	**CRR (a)**	**CRR (b)**	**RB (a)**	**RB (b)**	**(c)**
1	1	107.83	101.87	8.401%	102.97	7.417%	RB
1/2	2	107.83	109.36	2.179%	109.83	2.662%	CRR
1/4	4	107.83	106.83	1.660%	107.07	1.563%	RB
1/12	12	107.83	107.83	0.395%	107.94	0.374%	RB
1/24	24	107.83	107.84	0.165%	107.84	0.172%	CRR
1/48	48	107.83	107.81	0.085%	107.83	0.089%	CRR
1/252	252	107.83	107.83	0.018%	107.83	0.017%	RB
1/504	504	107.83	107.83	0.009%	107.83	0.009%	CRR
$K = 1000$							
h	n	**BSM (a)**	**CRR (a)**	**CRR (b)**	**RB (a)**	**RB (b)**	**(c)**
1	1	40.85	51.18	34.909%	51.04	34.461%	RB
1/2	2	40.85	36.20	15.712%	37.48	11.500%	RB
1/4	4	40.85	38.40	8.288%	39.34	5.181%	RB
1/12	12	40.85	40.01	2.844%	40.55	1.056%	RB
1/24	24	40.85	40.43	1.431%	40.80	0.221%	RB
1/48	48	40.85	40.64	0.717%	40.89	0.124%	RB
1/252	252	40.85	40.81	0.137%	40.89	0.124%	RB
1/504	504	40.85	40.83	0.068%	40.87	0.064%	RB
$K = 1100$							
h	n	**BSM (a)**	**CRR (a)**	**CRR (b)**	**RB (a)**	**RB (b)**	**(c)**
1	1	10.36	4.61	90.444%	2.97	109.045%	CRR
1/2	2	10.36	12.98	35.153%	12.48	28.995%	RB
1/4	4	10.36	9.48	19.089%	9.06	22.685%	CRR
1/12	12	10.36	10.60	4.641%	10.49	4.304%	RB
1/24	24	10.36	10.36	2.127%	10.38	2.040%	RB
1/48	48	10.36	10.38	1.097%	10.35	1.081%	RB
1/252	252	10.36	10.36	0.208%	10.36	0.219%	CRR
1/504	504	10.36	10.36	0.108%	10.36	0.106%	RB

Notes: i. Prices are expressed in notions of 1000 monetary units. ii. (a) stands for the expected value of the option; (b) stands for the ratio $\frac{D(\overset{\sim I}{C}, \overset{\sim I}{C_b})}{EV(\overset{\sim I}{C})}$; and (c) stands for the up-and-down binomial moves model that provides the nearest price to the BSM price.

We can draw the following conclusions from the results presented in Table 2:

- As expected, both the CRR and the RB converge to the BSM since $D(\overset{\sim I}{C}, \overset{\sim I}{C_b}) \to 0$ when $h \to 0$.
- For the in-the-money options, the number of times that the CRR and RB are the closest to the BSM is the same (50%). The RB provides the closest price when there is only one move, and the CRR provides the closest price when the number of moves is the maximum ($n = 504$).
- For out-of-the-money options, the RB tends to provide better approximations to the BSM. However, when the movement frequency is annual, the CRR is better, but when $h = \frac{1}{504}$, the price closest to the BSM comes from the RB.
- For at-the-money options, the best model is the RB model, regardless of the movement frequency.

- The price equations are monotonic functions of volatility, so the extremes of C_α and C'_β are easily programmable even in a spreadsheet. On the other hand, the calculation of the integrals to obtain $EV(\overset{\sim I}{C})$ and $D(\overset{\sim I}{C}, \overset{\sim I}{C_b})$ has been carried out via Simpson's rule with the discretization of the $\langle\alpha, \beta\rangle$-cuts, which in our case have been obtained as α = 0, 0.25, 0.5, 0.75, 1, and $\beta = 1 - \alpha$. Thus, these calculations are also easily implemented with a spreadsheet.

The 60-day volatility of the S&P 500 index on 4 November 2020, was 17.462%. Thus, from (16)–(19), the $\langle 1, 0\rangle$-cut of the intuitionistic volatility is $\sigma_{\langle 1,0\rangle} = \langle \sigma_1 = [0.17462, 0.17462], \sigma'_0 = [0.17462, 0.17462] \rangle$. Likewise, the $\langle 0, 1\rangle$-cut of the volatility is $\sigma_{\langle 0,1\rangle} = \langle \sigma_0 = [0.14801, 0.21298], \sigma'_1 = [0.13656, 0.23753] \rangle$. Therefore, the results of the approximation to BSM prices for the same call options analysed in Table 2 are presented in Table 3.

Table 3. A comparison of European option prices (with T = 1 and $r = 0\%$) on the S&P 500 obtained via the Black–Scholes–Merton model and binomial models such as the CRR and RB using the intuitionistic fuzzy volatility on 4 November 2020.

				K = 900			
h	n	**BSM (a)**	**CRR (a)**	**CRR (b)**	**RB (a)**	**RB (b)**	**(c)**
1	1	130.33	136.40	6.425%	140.19	10.295%	CRR
1/2	2	130.33	136.08	6.208%	134.95	5.181%	RB
1/4	4	130.33	133.57	3.426%	133.64	3.576%	CRR
1/12	12	130.33	129.81	0.855%	130.38	0.692%	RB
1/24	24	130.33	130.69	0.470%	130.40	0.365%	RB
1/48	48	130.33	130.35	0.177%	130.43	0.209%	CRR
1/252	252	130.33	130.34	0.036%	130.35	0.037%	CRR
1/504	504	130.33	130.34	0.018%	130.34	0.019%	CRR
				K = 1000			
h	n	**BSM (a)**	**CRR (a)**	**CRR (b)**	**RB (a)**	**RB (b)**	**(c)**
1	1	72.67	90.95	34.749%	90.16	33.333%	RB
1/2	2	72.67	64.41	15.713%	68.31	8.482%	RB
1/4	4	72.67	68.31	8.289%	71.15	3.035%	RB
1/12	12	72.67	71.18	2.844%	72.76	0.247%	RB
1/24	24	72.67	71.92	1.431%	72.96	0.527%	RB
1/48	48	72.67	72.30	0.717%	72.94	0.502%	RB
1/252	252	72.67	72.60	0.137%	72.71	0.086%	RB
1/504	504	72.67	72.64	0.068%	72.66	0.042%	RB
				K = 1100			
h	n	**BSM (a)**	**CRR (a)**	**CRR (b)**	**RB (a)**	**RB (b)**	**(c)**
1	1	36.58	45.50	33.176%	40.14	13.987%	RB
1/2	2	36.58	42.52	23.032%	43.30	25.578%	CRR
1/4	4	36.58	40.40	14.503%	39.90	12.453%	RB
1/12	12	36.58	36.17	3.145%	35.75	3.612%	CRR
1/24	24	36.58	36.78	1.553%	37.07	1.972%	CRR
1/48	48	36.58	36.62	0.787%	36.52	0.721%	RB
1/252	252	36.58	36.60	0.161%	36.59	0.162%	CRR
1/504	504	36.58	36.59	0.075%	36.58	0.074%	RB

Notes: i. Prices are expressed in notions of 1000 monetary units. ii. (a) stands for the expected value of the option; (b) stands for the ratio $\frac{D(\overset{\sim I}{C}, \overset{\sim I}{C_b})}{EV(\overset{\sim I}{C})}$; and (c) stands for the up-and-down binomial moves model that provides the nearest price to the BSM price.

Tables 2 and 3 both show that for options trading in the money and out of the money, whether one binomial approximation is better than the other depends on the frequency of the moves. However, for at-the-money options, the RB modelling of the moves is better than CRR regardless of their frequency.

Therefore, when deciding how to model up-and-down moves in a situation with fuzzy volatility, which is the most common assumption in models developing binomial FROP, it is worth considering Rendleman and Bartter modelling [7] as a more accurate alternative than the more common CRR modelling. In the following section, we develop a deeper analysis of the convergence of the RB and CRR to the BSM.

4. Assessment of the Convergence of Two Alternative Binomial Moves Modelling to Black–Scholes–Merton Prices with Intuitionistic Fuzzy Volatility

4.1. Materials and Methods

In this section, we compare the ability of the two modelling approaches of binomial moves (CRR and RB) to approximate the price obtained with the BSM for a European call option on the IBEX 35 futures. We assume a notional value of 1 monetary unit, T = 1 year, and strike prices of K = 0.9 (in the money), K = 1 (at the money), and K = 1.1 (out of the money). Because these are options for futures, we consider r = 0 [65]. Thus, as in numerical application 1, the only intuitionistic fuzzy parameter is volatility.

To evaluate the scenarios of fuzzy intuitionistic volatility, we considered the evolution of the IBEX 35 futures index from 27 April 2011, to 27 January 2023. Therefore, we used 3020 observations. For this index, we determined volatility, measured as the annualized standard deviation of logarithmic returns in a temporal window of 60 days. Figure 4 shows the evolution of the 60-day historical volatility during the entire analysed period. We observe that the average volatility is 17.21%, with a minimum of 8.71% and a maximum of 48.74%.

The empirical analysis was conducted using the following steps:

Step 1: We identified five scenarios of low volatility, five of medium volatility, and five of high volatility. The low-volatility scenarios are the 1st, 5th, 19th, 20th, and 30th percentiles of historical volatility. The medium volatility scenarios are determined from percentiles 40, 45, 50, 55, and 60 of the calculated standard deviations. The high-volatility scenarios are identified with 70th, 80th, 90th, 95th, and 99th percentiles. Table 4 shows the <0,1>-cut, <0.5,0.5>-cut, and <1,0>-cut of the volatility scenarios considered in this empirical application.

Step 2: For these volatility scenarios, we fit an intuitionistic estimation using Equations (16)–(19). We used 95 and 99.7 rules to adjust the membership and non-membership functions, respectively. Thus, with γ = 0.05, we obtain $\mu_\sigma(x) = \mu_{0.05_\sigma}(x)$, and with γ^* = 0.003, $\mu^*_\sigma(x) = \mu_{0.003_\sigma}(x)$ and so $v_\sigma(x) = 1 - \mu^*_\sigma(x)$.

Step 3: We determined the prices of the evaluated European call options (for K = 0.9, 1, 1.1, and T = 1) for all evaluated volatility scenarios. To calculate the binomial prices, we used periodicities $h = \{1/504, 1/252, 1/48, 1/24, 1/12, 1/4, 1/2, 1\}$.

Step 4: In all valuations, we determined the distance (5) between the value obtained with the BSM and the tested binomial models; that is, $D(\overset{\sim I}{C}, \overset{\sim I}{C_b})$. Comparing the distances of the prices obtained with CRR and RB in a specific option, volatility scenario, and move frequency with respect to the benchmark, the BSM, allows us to establish which model converges better to the BSM.

Step 5: We conducted three analyses of the convergence of the binomial IFN to the intuitionistic BSM as follows:

1. We analysed the level of convergence for each degree of moneyness (in the money, out of the money, and at the money) separately, considering all volatility scenarios and moving frequencies together.
2. We analysed the convergence levels by differentiating the degree of moneyness and movement frequency by considering conjointly all volatility scenarios. Within move frequencies, we differentiated between 'low' frequencies (monthly, quarterly, semi-annual, and annual) and 'high' frequencies (every 12 h, daily, weekly, and every half month).

3. We analysed the convergence levels by differentiating the moneyness degree and volatility scenarios without differentiating move periodicity. Within the volatility scenarios, we differentiated low-volatility, medium-volatility, and high-volatility scenarios, as indicated in Table 4.

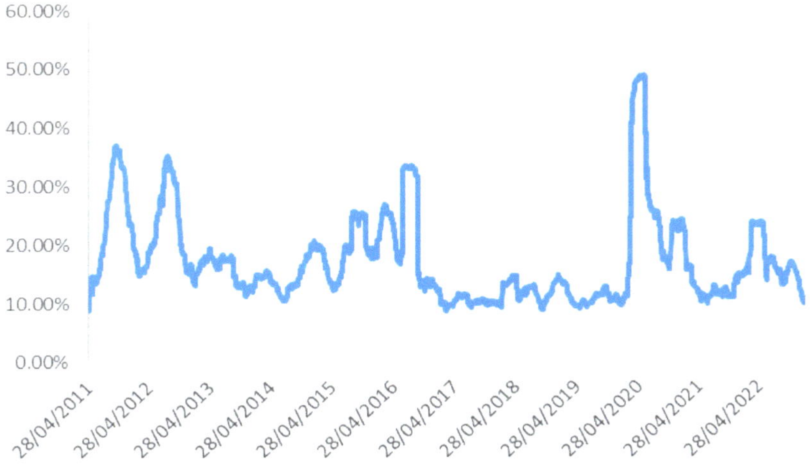

Figure 4. Evolution of the 60-day volatility of the IBEX35 Futures Index from 27 April 2011, to 27 January 2023.

In all three analyses, we calculated the proportion ρ, in which RB modelling was closer than CRR was to the BSM. Thus, if both methods provide approximations of equal quality, the proportion ρ in which the RB improves the CRR should be 0.5. If the RB is generally better than the CRR, $\rho > 0.5$, and if the CRR is better than the RB, then $\rho < 0.5$. In all the cases, we evaluated the null hypothesis about the proportion $\rho = 0.5$, with the test statistic $z = \frac{\rho - 0.5}{\sqrt{\frac{0.5(1-0.5)}{N}}}$, which tends to follow a standard normal distribution. where N is the number of simulations embedded in the statistical test.

Table 4. <1,0>-cut, <0.5, 0.5>-cut, and <0, 1>-cut of historical intuitionistic fuzzy volatilities of futures on the IBEX-35 linked to the 60-day windows are considered in this empirical analysis.

			$\sigma_1 = \sigma'_0$	$\sigma_{0.5}$		$\sigma'_{0.5}$		σ_0		σ'_1	
	Percentile	$\hat{\sigma}$	$\sigma_1 = \sigma'_0$	$\underline{\sigma_{0.5}}$	$\overline{\sigma_{0.5}}$	$\underline{\sigma'_{0.5}}$	$\overline{\sigma'_{0.5}}$	$\underline{\sigma_0}$	$\overline{\sigma_0}$	$\underline{\sigma'_1}$	$\overline{\sigma'_1}$
Low volatility	1%	9.32%	9.37%	8.61%	10.34%	8.30%	11.00%	7.90%	11.37%	7.29%	12.68%
	5%	9.88%	9.94%	9.13%	10.96%	8.80%	11.66%	8.37%	12.05%	7.73%	13.44%
	10%	10.28%	10.34%	9.50%	11.41%	9.16%	12.13%	8.71%	12.54%	8.04%	13.99%
	20%	11.27%	11.34%	10.42%	12.51%	10.05%	13.31%	9.56%	13.75%	8.82%	15.34%
	30%	12.55%	12.62%	11.59%	13.92%	11.18%	14.81%	10.63%	15.30%	9.81%	17.07%
Medium volatility	40%	13.44%	13.51%	12.41%	14.91%	11.97%	15.86%	11.39%	16.39%	10.51%	18.28%
	45%	14.04%	14.12%	12.97%	15.58%	12.51%	16.57%	11.90%	17.12%	10.98%	19.09%
	50%	14.78%	14.86%	13.65%	16.40%	13.17%	17.44%	12.53%	18.03%	11.56%	20.10%
	55%	15.66%	15.75%	14.46%	17.38%	13.95%	18.48%	13.27%	19.10%	12.24%	21.30%
	60%	16.53%	16.62%	15.27%	18.35%	14.73%	19.51%	14.01%	20.16%	12.93%	22.48%
High volatility	70%	18.30%	18.41%	16.91%	20.31%	16.31%	21.60%	15.51%	22.32%	14.31%	24.90%
	80%	23.15%	23.28%	21.39%	25.70%	20.63%	27.32%	19.62%	28.24%	18.11%	31.49%
	90%	26.71%	26.86%	24.67%	29.64%	23.80%	31.52%	22.64%	32.58%	20.89%	36.33%
	95%	33.25%	33.43%	30.71%	36.90%	29.62%	39.23%	28.18%	40.55%	26.00%	45.22%
	99%	47.87%	48.15%	44.23%	53.13%	42.66%	56.50%	40.58%	58.39%	37.44%	65.12%

4.2. Results

The results in Table 5 suggest that while in the in-the-money options, the CRR model converges more to the BSM than to the RB (44.17% = $\rho < 50\%$), this closer proximity in prices is not significant, because the p value (p) is 0.212. In contrast, if the option is out of the money, the better performance of RB (ρ = 60%) is significant (p = 0.029).

Table 5. Convergence of the CRR and RB binomial models to the BSM on the moneyness degree.

Moneyness Degree	Strike Price	ρ	z	p Value
In the money	$K = 0.9$	44.17%	−1.278	0.201
At the money	$K = 1$	100.00%	10.954	<0.001
Out of the money	$K = 1.1$	60.00%	2.191	0.029

Note: The number of observations for every strike price is N = 120.

Table 6 shows that the different comparative performances of the intuitionistic binomial models tested to approximate the BSM depend on the periodicity of the moves. Thus, the slightly greater performance of RB over CRR in the in-the-money and out-of-the-money options when the frequency is low (periodicities $h \geq 1/12$) is not significant. In contrast, better CRR convergence is observed for the in-the-money options when the frequency is high (periodicities lower than one month), which is significant (ρ = 36.67%, p = 0.039). Greater convergence of the RB to the BSM at these frequencies is also noted for the out-of-the-money options, which is also significant (ρ = 63.33%, p = 0.0389).

Table 7 shows that the volatility scenario also determines the significance at which the RB or CRR converges more to the BSM. Thus, within low- and high-volatility scenarios and in the in-the-money and out-of-the-money options, we do not observe that either of the two binomial modelling approaches significantly provides greater convergence to the BSM than the other approaches do. In contrast, in the 'medium' volatility scenarios and in-the-money options, CRR provides greater convergence to the BSM than RB does (ρ = 32.5%, p = 0.027). Conversely, in the out-of-the-money options and the same 'medium' volatility scenarios, the RB generates prices closer to the BSM than the CRR does (ρ = 80%, p < 0.001).

Table 6. Convergence of the CRR and RB binomial models to the BSM based on the moneyness degree and the periodicity of the moves.

	Low Frequency (h = 1, 1/2, 1/4, 1/12)				High Frequency (h = 1/24, 1/48, 1/252, 1/504)		
Moneyness Degree	**Strike Price**	**ρ**	**z**	**p Value**	**ρ**	**z**	**p Value**
In the money	$K = 0.9$	53.33%	0.516	0.606	36.67%	−2.066	0.039
At the money	$K = 1$	100%	7.746	<0.001	100.00%	7.746	<0.001
Out of the money	$K = 1.1$	55.00%	0.775	0.439	63.33%	2.066	0.039

Note: The number of observations for every strike price and move periodicity was N = 60.

Table 7. Convergence of the CRR and RB binomial models to the BSM based on the moneyness degree and the periodicity of moves.

		Low Volatility			Medium Volatility			High Volatility		
Moneyness Degree	**Strike Price**	**ρ**	**z**	**p Value**	**ρ**	**z**	**p Value**	**ρ**	**z**	**p Value**
In the money	$K = 0.9$	50%	0.000	1.000	32.5%	−2.214	0.027	50%	0	1.000
At the money	$K = 1$	100%	6.325	<0.001	100%	6.325	<0.001	100%	6.324	<0.001
Out of the money	$K = 1.1$	45%	−0.632	0.527	80%	3.795	<0.001	55%	0.632	0.527

Note: The number of observations for each strike price and volatility scenario is N = 40.

Tables 5–7 also show that in the at-the-money options, the RB binomial model better approximates BSM prices than does the CRR in all the simulations performed, regardless of the periodicity of the moves and the evaluated volatility scenario.

5. Conclusions and Further Research

Fuzzy random option pricing (FROP) is a branch of fuzzy mathematics that models the uncertainty of the parameters necessary for valuing options through fuzzy subsets [12]. A common approach in FROP involves assuming stochastic variation in the underlying asset price, where the uncertainty in the parameters governing these fluctuations is introduced through type-1 fuzzy numbers [13]. In FROP, possibility distributions must be interpreted as epistemic fuzzy sets [14].

Among the continuous-time frameworks considered by FROP, the most developed is the Black–Scholes–Merton (BSM) model [10,11,15–20,71]. Similarly, the most commonly used analytic groundwork for option pricing in discrete time in FROP is the binomial model [25–27,29–34]. In most fuzzy literature, the forms used to model uncertainty as parameters such as volatility are fuzzy numbers, usually with a linear shape [13], and the modelling of Cox, Ross, and Rubinstein moves [6], denoted as CRRs.

The contributions of this study are located in the FROP with a binomial framework, highlighting three aspects. First, the modelling of the parameters governing the valuation of options is carried out through intuitionistic fuzzy numbers (IFNs), which are a generalization of fuzzy numbers. Their use allows the introduction of bipolar uncertainty of parameters into option pricing. Thus, in financial analysis, they not only offer information about the possible values that variables may introduce but also information about those variables that are certainly not worth considering [37]. We are aware that IFNs increase computational complexity with respect to the use of simple type-1 fuzzy numbers. Therefore, the use of a Gaussian quadrature or other numerical integration techniques that require fewer evaluation points while maintaining accuracy and implement processing for the computation of level sets and integrals should be considered. Modern computational environments with multicore processors can be utilized to speed up these computations.

Second, we propose a methodology that allows for the adjustment of the volatility of the underlying asset, which is the key parameter in option valuation and typically involves the most uncertainty, with an IFN. The proposed methodology combines the concept of historical volatility [4] and the interpretation of the α-level sets of possibility distributions of the variable as the narrowest interval containing those values with a probability of occurrence of 1-α [55]. This approach has already been used in the field of FROP to adjust the volatility of the underlying asset through type-1 fuzzy numbers [10,45,46] by applying the contributions of possibilistic adjustments of statistical parameters [57–61].

The proposed method requires adjusting two possibility distributions. On the one hand, a lower possibility distribution, whose support can be aligned with the 95% rule, assesses the current possibility levels. On the other hand, an upper possibility distribution, encompassing a broader range of values and adjustable, for instance, using the 99.7% rule, measures the potential possibility of the parameter of interest taking specific values. The IFN resulting from applying the proposed methodology allows for the parametric representation of all possible outcomes for the parameter of interest. This includes both highly probable results (which form the core of the membership function and align with the 95% rule) and those that are almost certainly not considered for the parameter (i.e., values that are not part of the 99.7% rule selection and have a nonmembership level of 1).

Third, although the literature on option pricing has proposed a wide variety of methodologies for modelling up-and-down moves [5,8,9], developments in FROP have been limited to the most well-known method developed in [6], which we refer to as CRR. In this study, the suitability of using an alternative, which we refer to as RB [7], is analysed. Notably, the main justification for using the BSM is its ability to use the conceptual BSM framework in path-dependent options, where the direct application of the BSM does not apply [4]. To assess binomial approximations to the BSM in the valuation of path-dependent options, we compared their convergence to the value of the BSM in a European option with the same characteristics (strike price, expiration date, volatility, and initial price of the underlying asset). In this study, we evaluated the CRR and RB models on a call option on IBEX35 futures with a maturity of one year that can be traded in the money, at the money, and out of the money.

We observe that the moneyness of the option is relevant for determining the convergence of the proposed intuitionistic fuzzy binomial move models. While the CRR might perform better for options traded in the money, the RB tends to perform better for out-of-the-money options. Additionally, for options traded at the money, the RB always approximated BSM prices better than the CRR did.

The periodicity of moves is a relevant second-order factor. We observe that the best convergence of the CRR in-the-money options and that of the RB out-of-the-money options is significant when the periodicity of the movements is less than one month. However, in periods equal to or greater than monthly, it cannot be established whether any of the methods approximate these moneyness degrees better than the alternatives do. Similarly, we find that the volatility of the option determines which particular model of the binomial moves is better than that of the alternative. We observe that both the superiority of the CRR in the in-the-money options and the superiority of the RB in the out-of-the-money options are significant only in the central volatility scenarios.

The results presented in this study have various theoretical and practical implications. Studies introducing bipolar uncertainty through IFNs are scarce in financial economics, and option principles are nonexistent in fuzzy binomial option pricing. In this context, we outline contributions in the fields of FROP [36,39–43], capital budgeting [51,74,75], and risk evaluation [76,77].

In the contributions of FROP with a binomial framework, the modelling of up-and-down moves was carried out with the formulation of CRR without any critical consideration. We highlight the need to consider alternative models that may converge better to the BSM. In this work, we found that the contemporary modelling by Rendleman and Bartter [7] often converges better to BSM prices than do CRRs in the out-of-the-money options and always in the at-the-money options. The analysis presented in this work can be expanded by introducing alternative up-and-down moves, such as those in [8,9].

The findings of this paper can be useful for practitioners because although volatility or prices are assumed to be intuitionistic fuzzy subsets, their interpretation is very intuitive and does not require knowledge of fuzzy logic. This covers both the construction of the possibility distributions of volatility and the calculation of option prices. The construction of IFNs to measure bipolar uncertainty is based on common quantitative concepts in financial pricing, such as historical volatility [4] and the 95–99.7% rule [62]. Thus, the <1,0>-cut of the variable of interest is a singleton that indicates its possible value with maximum reliability. This is comparable to the result offered by the BSM model or the evaluated binomial models using crisp parameters. The <0,1>-cut set was delimited by extreme values. Thus, the 0-cut of the membership function indicates moderately extreme scenarios, comparable to those generated via the 95% rule. The 1-cut of the nonmembership function indicates extremely extreme scenarios, similar to those generated using the 99.7% rule. Using intermediate levels of membership and nonmembership allows for the structuring of several scenarios linked to prefixed membership and nonmembership levels of interest to the decision maker.

The developments presented in this paper open up several lines for future research, both within option price modelling and in more general areas of quantitative finance. This analysis can undoubtedly be extended to more binomial moves than the two considered in the work and can be applied not only to stock market data but also in the field of real option valuation. In any case, our work has shown that to value a specific non-European option with a proposed binomial model, alternative binomial moves to the CRR should be considered. To determine the optimum moves, it may be particularly useful to assume that these can take the general form $u(a) = e^{a \cdot \sigma \sqrt{h}}$ in the case of up and $d(b) = e^{-b \cdot \sigma \sqrt{h}}$ for down. Thus, in this problem, the decision variables are a and b, which must minimize the distance between the binomial and BSM prices of an equivalent European option that we actually want to value (same maturity, strike price, etc.).

The modelling of the parameters that determine price behavior proposed in this work can be generalized to option valuation models where the stochastic process governing the movement of the underlying asset's price is more complex than a univariate geometric Brownian motion. The uncertainty of other parameters that govern price movements, such as the correlation between the stochastic components in multivariate movements or the parameter quantifying the speed of return to the equilibrium value in mean-reverting processes, can be captured in the intuitional estimation through a probability–possibility transformation criterion, as presented in this work. In this context, [46] quantified the mean reversion parameter as a type-1 fuzzy number on the basis of its estimation through probabilistic intervals with a time series model and applied the criterion of Equation (8).

The intuitionistic fit of volatility presented in this work is based on the concept of historical volatility associated with the prices of the underlying asset [4], which has also been used in the FROP literature [10,45,46]. An alternative procedure could be to adjust intuitionistic fuzzy volatility from observed implicit volatilities. The modelling of implied volatility has been the subject of extensive debate in quantitative finance [78]. In the FROP literature, implicit volatility has been adjusted with fuzzy numbers, either with fuzzy regression [19,20] or with coherent transformations of the empirical distribution functions to possibility distributions [21,71]. These contributions can serve as an analytical basis for adjusting empirical volatilities through IFNs, either by using intuitionistic regression instruments or considering that IFNs can be modelled as bivariate distribution functions.

In financial modelling, methods aimed at making the best possible point predictions, such as neural networks and many machine learning algorithms, are certainly useful [79]. However, we also understand the importance of being able to parametrize predictions that take into account the variability of the parameter of interest, which will be the subject of future analysis. In this work, we have demonstrated that intuitionistic fuzzy modelling can be reliable.

Funding: This research received no external funding.

Data Availability Statement: The raw data supporting the conclusions of this article will be made available by the authors on request.

Conflicts of Interest: The author declares no conflicts of interest.

References

1. Black, F.; Scholes, M. The pricing of options and corporate liabilities. *J. Political Econ.* **1973**, *81*, 637–654. Available online: http://www.jstor.org/stable/1831029 (accessed on 13 October 2023). [CrossRef]
2. Merton, R.C. Theory of rational option pricing. *Bell J. Econ. Manag. Sci.* **1973**, *4*, 141–183. [CrossRef]
3. Hull, J.C. *Options Futures and Other Derivatives*; Pearson Education: New Delhi, India, 2008.
4. Van der Hoek, J.; Elliott, R.J. *Binomial Models in Finance*; Springer: New York, NY, USA, 2006.
5. Chance, D.M. A Synthesis of Binomial Option Pricing Models for Lognormally Distributed Assets. *J. Appl. Financ. (Former. Financ. Pract. Educ.)* **2008**, *18*, 38–56. [CrossRef]
6. Cox, J.; Ross, S.; Rubinstein, M. Option Pricing: A Simplified Approach. *J. Financ. Econ.* **1979**, *7*, 229–263. [CrossRef]
7. Rendleman, R.J., Jr.; Bartter, B.J. Two state option pricing. *J. Financ.* **1979**, *34*, 1092–1110.
8. Trigeorgis, L. A log-transformed binomial numerical analysis method for valuing complex multioption investments. *J. Financ. Quant. Anal.* **1991**, *26*, 309–326. [CrossRef]
9. Jabbour, G.M.; Kramin, M.V.; Young, S.D. Two-state option pricing: Binomial models revisited. *J. Futures Mark. Futures Options Other Deriv. Prod.* **2001**, *21*, 987–1001. [CrossRef]
10. Chrysafis, K.A.; Papadopoulos, B.K. On theoretical pricing of options with fuzzy estimators. *J. Comput. Appl. Math.* **2009**, *223*, 552–566. [CrossRef]
11. Carlsson, C.; Fuller, R. A fuzzy approach to real option valuation. *Fuzzy Sets Syst.* **2003**, *139*, 297–312. [CrossRef]
12. Andrés-Sánchez, J. A systematic review of the interactions of fuzzy set theory and option pricing. *Expert Syst. Appl.* **2023**, *223*, 119868. [CrossRef]
13. Muzzioli, S.; De Baets, B. Fuzzy approaches to option price modelling. *IEEE Trans. Fuzzy Syst.* **2016**, *25*, 392–401. [CrossRef]

14. Romaniuk, M.; Hryniewicz, O. Interval-based, nonparametric approach for resampling of fuzzy numbers. *Soft Comput.* **2019**, *23*, 5883–5903. [CrossRef]
15. Zmeskal, Z. Application of the fuzzy-stochastic methodology to appraising the firm value as a European call option. *Eur. J. Oper. Res.* **2001**, *135*, 303–310. [CrossRef]
16. Wu, H.C. Pricing European options based on the fuzzy pattern of Black-Scholes formula. *Comput. Oper. Res.* **2004**, *31*, 1069–1081. [CrossRef]
17. Guerra, M.L.; Sorini, L.; Stefanini, L. Option price sensitivities through fuzzy numbers. *Comput. Math. Appl.* **2011**, *61*, 515–526. [CrossRef]
18. Kim, Y.; Lee, E.B. Optimal Investment Timing with Investment Propensity Using Fuzzy Real Options Valuation. *Int. J. Fuzzy Syst.* **2018**, *20*, 1888–1900. [CrossRef]
19. Muzzioli, S.; Gambarelli, L.; De Baets, B. Indices for Financial Market Volatility Obtained Through Fuzzy Regression. *Int. J. Inf. Technol. Decis. Mak.* **2018**, *17*, 1659–1691. [CrossRef]
20. Muzzioli, S.; Gambarelli, L.; De Baets, B. Option implied moments obtained through fuzzy regression. *Fuzzy Optim. Decis. Mak.* **2020**, *19*, 211–238. [CrossRef]
21. Capotorti, A.; Figà-Talamanca, G. SMART-or and SMART-and fuzzy average operators: A generalized proposal. *Fuzzy Sets Syst.* **2020**, *395*, 1–20. [CrossRef]
22. Anzilli, L.; Villani, G. Cooperative R&D investment decisions: A fuzzy real option approach. *Fuzzy Sets Syst.* **2023**, *458*, 143–164. [CrossRef]
23. Zhang, H.; Watada, J. Fuzzy Levy-GJR-GARCH American option pricing model based on an infinite pure jump process. *IEICE Trans. Inf. Syst.* **2018**, *E101D*, 1843–1859. [CrossRef]
24. Nowak, P.; Pawłowski, M. Application of the Esscher Transform to Pricing Forward Contracts on Energy Markets in a Fuzzy Environment. *Entropy* **2023**, *25*, 527. [CrossRef] [PubMed]
25. Yoshida, Y. A discrete-time model of American put option in an uncertain environment. *Eur. J. Oper. Res.* **2003**, *151*, 153–166. [CrossRef]
26. Muzzioli, S.; Torricelli, C. A multiperiod binomial model for pricing options in a vague world. *J. Econ. Dyn. Control* **2004**, *28*, 861–887. [CrossRef]
27. Lee, C.F.; Tzeng, G.-H.; Wang, S.-Y. A fuzzy set approach for generalized CRR model: An empirical analysis of S&P 500 index options. *Rev. Quant. Financ. Account.* **2005**, *25*, 255–275. [CrossRef]
28. Wang, G.X.; Wang, Y.Y.; Tang, J.M. Fuzzy Option Pricing Based on Fuzzy Number Binary Tree Model. *IEEE Trans. Fuzzy Syst.* **2022**, *30*, 3548–3558. [CrossRef]
29. Meenakshi, K.; Kennedy, F.C. A study of european fuzzy put option buyers model on future contracts involving general trapezoidal fuzzy numbers. *Glob. Stoch. Anal.* **2021**, *8*, 47–59. [CrossRef]
30. Meenakshi, K.; Kennedy, F.C. On some properties of American fuzzy put option model on fuzzy future contracts involving general linear octagonal fuzzy numbers. *Adv. Appl. Math. Sci.* **2021**, *21*, 331–342. [CrossRef]
31. Zmeskal, Z. Generalized soft binomial American real option pricing model (fuzzy-stochastic approach). *Eur. J. Oper. Res.* **2010**, *207*, 1096–1103. [CrossRef]
32. Ho, S.H.; Liao, S.H. A fuzzy real option approach for investment. *Expert Syst. Appl.* **2011**, *38*, 15296–15302. [CrossRef]
33. Anzilli, L.; Facchinetti, G.; Pirotti, T. Pricing of minimum guarantees in life insurance contracts with fuzzy volatility. *Inf. Sci.* **2018**, *460*, 578–593. [CrossRef]
34. Zhang, X.Y.; Yin, J.B. Assessment of investment decisions in bulk shipping through fuzzy real options analysis. *Marit. Econ. Logist.* **2023**, *25*, 122–139. [CrossRef]
35. Zhang, M.J.; Qin, X.Z.; Nan, J.X. Binomial tree model of the European option pricing based on the triangular intuitionistic fuzzy numbers. *Syst. Eng. Theory Pract.* **2013**, *33*, 34–40. [CrossRef]
36. Zmeskal, Z.; Dluhosova, D.; Gurny, P.; Kresta, A. Generalized soft multimode real options model (fuzzy-stochastic approach). *Expert Syst. Appl.* **2022**, *192*, 116388. [CrossRef]
37. Dubois, D.; Prade, H. An overview of the asymmetric bipolar representation of positive and negative information in possibility theory. *Fuzzy Sets Syst.* **2009**, *160*, 1355–1366. [CrossRef]
38. Dubois, D.; Prade, H. Gradualness, uncertainty and bipolarity: Making sense of fuzzy sets. *Fuzzy Sets Syst.* **2012**, *192*, 3–24. [CrossRef]
39. Wu, L.; Liu, J.F.; Wang, J.T.; Zhuang, Y.M. Pricing for a basket of LCDS under fuzzy environments. *SpringerPlus* **2016**, *5*, 1747. [CrossRef]
40. Wu, L.; Zhuang, Y.M.; Li, W. A New Default Intensity Model with Fuzziness and Hesitation. *Int. J. Comput. Intell. Syst.* **2016**, *9*, 340–350. [CrossRef]
41. Wu, L.; Mei, X.B.; Sun, J.G. A New Default Probability Calculation Formula an Its Application under Uncertain Environments. *Discret. Dyn. Nat. Soc.* **2018**, *2018*, 3481863. [CrossRef]
42. Ersen, H.Y.; Tas, O.; Kahraman, C. Intuitionistic fuzzy real-options theory and its application to solar energy investment projects. *Eng. Econ.* **2018**, *29*, 140–150. [CrossRef]
43. Ersen, H.Y.; Tas, O.; Ugurlu, U. Solar Energy Investment Valuation with Intuitionistic Fuzzy Trinomial Lattice Real Option Model. *IEEE Trans. Eng. Manag.* **2023**, *70*, 2584–2593. [CrossRef]

44. Dubois, D.; Folloy, L.; Mauris, G.; Prade, H. Probability–possibility transformations, triangular fuzzy sets, and probabilistic inequalities. *Reliab. Comput.* **2004**, *10*, 273–297. [CrossRef]
45. Chrysafis, K.A.; Papadopoulos, B.K. Decision Making for Project Appraisal in Uncertain Environments: A Fuzzy-Possibilistic Approach of the Expanded NPV Method. *Symmetry* **2021**, *13*, 27. [CrossRef]
46. Andrés-Sánchez, J. A Fuzzy-Random Extension of Jamshidian's Bond Option Pricing Model and Compatible One-Factor Term Structure Models. *Axioms* **2023**, *12*, 668. [CrossRef]
47. Hull, J.; White, A. The use of the control variate technique in option pricing. *J. Financ. Quant. Anal.* **1988**, *23*, 237–251. [CrossRef]
48. Zadeh, L.A. Fuzzy Sets. *Inf. Control* **1965**, *8*, 338–353. [CrossRef]
49. Dubois, D.; Prade, H. Fuzzy numbers: An overview. In *Readings in Fuzzy Sets and Intelligent Systems*; Dubois, D., Prade, H., Yager, R.R., Eds.; Elsevier: Amsterdam, The Netherlands, 1993; pp. 112–148. [CrossRef]
50. Atanassov, K. Intuitionistic fuzzy sets. *Fuzzy Sets Syst.* **1986**, *20*, 87–96. [CrossRef]
51. Kahraman, C.; Onar, Ç.; Öztayşi, B. Engineering economic analyses using intuitionistic and hesitant fuzzy sets. *J. Intell. Fuzzy Syst.* **2015**, *29*, 1151–1168. [CrossRef]
52. Mitchell, H.B. Ranking-intuitionistic fuzzy numbers. *Int. J. Uncertain. Fuzziness Knowl.-Based Syst.* **2004**, *12*, 377–386. [CrossRef]
53. Arefi, M.; Taheri, S.M. Least-Squares Regression Based on Atanassov's Intuitionistic Fuzzy Inputs–Outputs and Atanassov's Intuitionistic Fuzzy Parameters. *IEEE Trans. Fuzzy Syst.* **2015**, *23*, 1142–1154. [CrossRef]
54. Mohan, S.; Kannusamy, A.P.; Samiappan, V. A new approach for ranking of intuitionistic fuzzy numbers. *J. Fuzzy Ext. Appl.* **2020**, *1*, 15–26.
55. Couso, I.; Montes, S.; Gil, P. The necessity of the strong α-cuts of a fuzzy set. *Int. J. Uncertain. Fuzziness Knowl. Based Syst.* **2001**, *9*, 249–262. [CrossRef]
56. Mauris, G.; Lasserre, V.; Foulloy, L. A fuzzy approach for the expression of uncertainty in measurement. *Measurement* **2001**, *29*, 165–177. [CrossRef]
57. Buckley, J.J. Fuzzy statistics: Hypothesis testing. *Soft Comput.* **2005**, *9*, 512–518. [CrossRef]
58. Falsafain, A.; Taheri, S.M. On Buckley's approach to fuzzy estimation. *Soft Comput.* **2011**, *15*, 345–349. [CrossRef]
59. Sfiris, D.S.; Papadopoulos, B.K. Nonasymptotic fuzzy estimators based on confidence intervals. *Inf. Sci.* **2014**, *279*, 446–459. [CrossRef]
60. Adjenughwure, K.; Papadopoulos, B. Fuzzy-statistical prediction intervals from crisp regression models. *Evol. Syst.* **2020**, *11*, 201–213. [CrossRef]
61. Al-Kandari, M.; Adjenughwure, K.; Papadopoulos, K. A Fuzzy-Statistical Tolerance Interval from Residuals of Crisp Linear Regression Models. *Mathematics* **2020**, *8*, 1422. [CrossRef]
62. Alostad, H.; Davulcu, H. Directional prediction of stock prices using breaking news on Twitter. *Web Intell.* **2017**, *15*, 1–17. [CrossRef]
63. Parvathi, R.; Malathi, C.; Akram, M.; Atanassov, K. Intuitionistic fuzzy linear regression analysis. *Fuzzy Optim. Decis. Mak.* **2013**, *12*, 215–229. [CrossRef]
64. Buckley, J.J.; Qu, Y. On using α-cuts to evaluate fuzzy equations. *Fuzzy Sets Syst.* **1990**, *38*, 309–312. [CrossRef]
65. Black, F. The pricing of commodity contracts. *J. Financ. Econ.* **1976**, *3*, 167–179. [CrossRef]
66. Thiagarajah, K.; Appadoo, S.S.; Thavaneswaran, A. Option valuation model with adaptive fuzzy numbers. *Comput. Math. Appl.* **2007**, *53*, 831–841. [CrossRef]
67. Andrés-Sánchez, J. Pricing European Options with Triangular Fuzzy Parameters: Assessing Alternative Triangular Approximations in the Spanish Stock Option Market. *Int. J. Fuzzy Syst.* **2018**, *20*, 1624–1643. [CrossRef]
68. Guerra, M.L.; Sorini, L.; Stefanini, L. Value Function Computation in Fuzzy Models by Differential Evolution. *Int. J. Fuzzy Syst.* **2017**, *19*, 1025–1031. [CrossRef]
69. Li, H.; Ware, A.; Di, L.; Yuan, G.; Swishchuk, A.; Yuan, S. The application of nonlinear fuzzy parameters PDE method in pricing and hedging European options. *Fuzzy Sets Syst.* **2018**, *331*, 14–25. [CrossRef]
70. Chen, H.M.; Hu, C.F.; Yeh, W.C. Option pricing and the Greeks under Gaussian fuzzy environments. *Soft Comput.* **2019**, *23*, 13351–13374. [CrossRef]
71. Capotorti, A.; Figa-Talamanca, G. On an implicit assessment of fuzzy volatility in the Black and Scholes environment. *Fuzzy Sets Syst.* **2013**, *223*, 59–71. [CrossRef]
72. Collan, M.; Carlsson, C.; Majlender, P. Fuzzy Black and Scholes real options pricing. *J. Decis. Syst.* **2003**, *12*, 391–416. [CrossRef]
73. Tolga, A.Ç.; Kahraman, C.; Demircan, M.L. A Comparative Fuzzy Real Options Valuation Model using Trinomial Lattice and Black-Scholes Approaches: A Call Center Application. *J. Mult.-Valued Log. Soft Comput.* **2010**, *16*, 135.
74. Boltürk, E.; Kahraman, C. Interval-valued and circular intuitionistic fuzzy present worth analyses. *Informatica* **2022**, *33*, 693–711. [CrossRef]
75. Haktanır, E.; Kahraman, C. Intuitionistic fuzzy risk adjusted discount rate and certainty equivalent methods for risky projects. *Int. J. Prod. Econ.* **2023**, *257*, 108757. [CrossRef]
76. Uzhga-Rebrov, O.; Grabusts, P. Methodology for Environmental Risk Analysis Based on Intuitionistic Fuzzy Values. *Risks* **2023**, *11*, 88. [CrossRef]
77. Andrés-Sánchez, J.D. Pricing Life Contingencies Linked to Impaired Life Expectancies Using Intuitionistic Fuzzy Parameters. *Risks* **2024**, *12*, 29. [CrossRef]

78. Di Persio, L.; Vettori, S. Markov Switching Model Analysis of Implied Volatility for Market Indexes with Applications to S&P 500 and DAX. *J. Math.* **2014**, *2014*, 753852. [CrossRef]
79. Di Persio, L.; Honchar, O. Multitask machine learning for financial forecasting. *Int. J. Circuits Syst. Signal Process.* **2018**, *12*, 444–451.

axioms

Article

Fuzzy Testing Model Built on Confidence Interval of Process Capability Index C_{PMK}

Wei Lo [1], Tsun-Hung Huang [2], Kuen-Suan Chen [2,3,4,*], Chun-Min Yu [2,*] and Chun-Ming Yang [5]

1 School of Business Administration, Guangxi University of Finance and Economics, Nanning 530007, China; 2018120010@gxufe.edu.cn
2 Department of Industrial Engineering and Management, National Chin-Yi University of Technology, Taichung 411030, Taiwan; toby@ncut.edu.tw
3 Department of Business Administration, Chaoyang University of Technology, Taichung 413310, Taiwan
4 Department of Business Administration, Asia University, Taichung 413305, Taiwan
5 School of Economics and Management, Dongguan University of Technology, Dongguan 523808, China; 2020812@dgut.edu.cn
* Correspondence: kschen@ncut.edu.tw (K.-S.C.); march@ncut.edu.tw (C.-M.Y.)

Abstract: A variety of process capability indices are applied to the quantitative measurement of the potential and performance of processes in manufacturing. As it is easy to understand the formulae of these indices, this method is easy to apply. Furthermore, a process capability index is frequently utilized by a manufacturer to gauge the quality of a process. This index can be utilized by not only an internal process engineer to assess the quality of the process but also as a communication tool for an external sales department. When the manufacturing process deviates from the target value T, the process capability index C_{PMK} can be quickly detected, which is conducive to the promotion of smart manufacturing. Therefore, this study applied the index C_{PMK} as an evaluation tool for process quality. As noted by some studies, process capability indices have unknown parameters and therefore must be estimated from sample data. Additionally, numerous studies have addressed that it is essential for companies to establish a rapid response mechanism, as they wish to make decisions quickly when using a small sample size. Considering the small sample size, this study proposed a 100 $(1-\alpha)$% confidence interval for the process capability index C_{PMK} based on suggestions from previous studies. Subsequently, this study built a fuzzy testing model on the 100 $(1-\alpha)$% confidence interval for the process capability index C_{PMK}. This fuzzy testing model can help enterprises make decisions rapidly with a small sample size, meeting their expectation of having a rapid response mechanism.

Keywords: process capability indices; unknown parameters; confidence interval; fuzzy testing model; mathematical programming method

MSC: 62C05; 62C86

Citation: Lo, W.; Huang, T.-H.; Chen, K.-S.; Yu, C.-M.; Yang, C.-M. Fuzzy Testing Model Built on Confidence Interval of Process Capability Index C_{PMK}. *Axioms* **2024**, *13*, 379. https://doi.org/10.3390/axioms13060379

Academic Editors: Ta-Chung Chu, Wei-Chang Yeh and Salvatore Sessa

Received: 14 April 2024
Revised: 30 May 2024
Accepted: 2 June 2024
Published: 4 June 2024

1. Introduction

A number of process capability indices are unitless, measuring items produced in processes to determine whether they can achieve the quality level required by the product designer [1–3]. Indeed, process capability indices are common tools utilized by companies to gauge process quality. They can be offered to internal process engineers to assess process quality as well as viewed as communication tools for sales departments in external companies [4–7]. The two most widely used capability indices, C_P and C_{PK}, as Kane [8] suggested, are displayed below:

$$C_P = \frac{USL - LSL}{6\sigma} = \frac{d}{3\sigma} \quad (1)$$

and
$$C_{PK} = Min\left\{\frac{USL-\mu}{3\sigma}, \frac{\mu-LSL}{3\sigma}\right\} = \frac{d-|\mu-M|}{3\sigma}. \quad (2)$$

In the above equations, USL denotes the Upper Specification Limit, LSL denotes the Lower Specification Limit, μ refers to the process mean, σ refers to the process standard deviation, and $M = (LSL + USL)/2$ refers to the midpoint of the specification interval (LSL, USL). Boyles [9] noted that indices C_P and C_{PK} are based on yields and independent of the target T. Accordingly, they may fail to explain process centering, which refers to the capability of gathering data around the target. To tackle this problem, Chan, Cheng, and Spiring [10] came up with the Taguchi capability index C_{PM}, as presented below:

$$C_{PM} = \frac{USL-LSL}{6\sqrt{\sigma^2+(\mu-T)^2}} = \frac{d}{3\sqrt{\sigma^2+(\mu-T)^2}}, \quad (3)$$

where T represents the target value. $(\mu-T)^2+\sigma^2 = E(X-T)^2$ refers to the expected loss function of Taguchi. Considering the Taguchi capability index C_{PM}, Pearn, Kotz, and Johnson [11] took the following example with $T = \{3(USL) + (LSL)\}/4$ and $\sigma = d/3$. Process A, with $\mu_A = T - d/2$ = m, and process B, with $\mu_B = T + d/2 = USL$, both yielded the same result $-$ $C_{PM} = 0.555$. Nonetheless, the expected non-conforming proportions were approximately 0.27% and 50%, respectively. We can tell, in this case, that the Taguchi capability index C_{PM} measures process capability inconsistently. To overcome the problem, the process capability index C_{PMK} proposed by Choi and Owen [12] is employed to handle the processes with asymmetric tolerances. For symmetrical tolerances, the index C_{PMK} is expressed as follows:

$$C_{PMK} = \frac{d-|\mu-T|}{3\sqrt{\sigma^2+(\mu-T)^2}}. \quad (4)$$

As noted by Vännman [13], the rankings of indices C_P, C_{PK}, C_{PM}, and C_{PMK} are in the following order: (1) C_{PMK}, (2) C_{PM}, (3) C_{PK}, and (4) C_P, based on their sensitivity to the departure of the process mean μ from the target value T. These four process capability indices are favorable for processes with quality characteristics of the nominal-the-better (*NTB*) type. Among them, C_{PMK} integrates the numerator of index C_{PK} with the denominator of index C_{PM}. The deviation can thus be detected quickly as the manufacturing process deviates from the target value T, which helps promote smart manufacturing. Therefore, this paper utilizes the process capability index C_{PMK} as an evaluation tool for process quality. As noted by some studies, process capability indices have unknown parameters and therefore must be estimated from sample data [14]. In addition, as highlighted by many studies, companies typically seek a rapid response mechanism, enabling them to make decisions quickly while utilizing a small sample size [15,16]. If decisions, however, are made based on a small number of samples, there will be a risk of misjudgment due to sampling error. Given the case of small sample size, this paper follows some suggestions from previous studies and derives a 100 $(1-\alpha)$% confidence interval for the process capability index C_{PMK}. Next, building upon the 100 $(1-\alpha)$% confidence interval for the process capability index C_{PMK}, this paper develops a fuzzy testing model. This model, on the basis of the confidence interval, helps enterprises make quick decisions with a small sample size, fulfilling their need for a rapid response mechanism.

As noted by various studies, machine tools made in Taiwan won first place worldwide in terms of output value and sales volume. They are mainly sold to emerging markets in Southeast Asia and Eastern Europe. The central region of Taiwan is an industrial center for precision machinery and machine tools. It combines machine tool parts factories, aerospace, and medical industries, and connects parts processing and maintenance industries, forming a large cluster of machine tools and machinery industries [17,18]. Additionally, several studies have indicated that the high clustering effect of the machine tool industry in Taiwan has enabled central Taiwan to develop a robust industry chain for machine tools;

therefore, Taiwan plays a vital role in the world machine tool industry [19]. In view of this, we demonstrate how to implement the proposed fuzzy evaluation model using an axis produced by a machining factory in the central region of Taiwan.

In this paper, we organize the remaining sections as follows. In the Section 2, we demonstrate how to derive the Maximum Likelihood Estimator (*MLE*) as well as 100 $(1-\alpha)\%$ confidence regions for the process mean and the process standard deviation, respectively. This study utilizes the process capability index C_{PMK} as the object function and adopts the 100 $(1-\alpha)\%$ confidence regions for the process mean and regions for the process standard deviation as the feasible solution areas. Subsequently, we apply mathematical programming to find a 100 $(1-\alpha)\%$ confidence interval for the process capability index C_{PMK}. In the Section 3, we develop a fuzzy testing model using the 100 $(1-\alpha)\%$ confidence interval of the process capability index C_{PMK} to measure the process quality, so as to learn whether it reaches the required quality level. In this model, we first derive a triangular fuzzy number and then obtain its membership function. Next, based on fuzzy testing rules, we can determine whether the process quality satisfies the requirement, which can serve as a reference for other industries. As mentioned before, central Taiwan is an industrial center for machine tools. Therefore, in the Section 4, an axis manufactured by a machining factory in the central region of Taiwan is used as an empirical example to illustrate how to apply the proposed fuzzy testing model. In the Section 5, conclusions are presented.

2. Confidence Interval for Process Capability Index C_{PMK}

A random variable, denoted with X, has a normal distribution with the mean (μ) and the standard deviation (σ). Let $(X_1, X_2, \ldots, Xn)$ be a random sample received from a normal process. Then the Maximum Likelihood Estimators (*MLEs*) of the process mean (μ) and the process standard deviation (σ) are written in Equation (5) and Equation (6), respectively:

$$\mu^* = \frac{1}{n}\sum_{i=1}^{n} X_i, \tag{5}$$

and

$$\sigma^* = \sqrt{\frac{1}{n}\sum_{i=1}^{n}(X_i - \overline{X})^2}. \tag{6}$$

Furthermore, the estimator of the process capability index C_{PMK} is denoted by

$$C^*_{PMK} = \frac{d - |\mu^* - T|}{3\sqrt{\sigma^{*2} + (\mu^* - T)^2}}. \tag{7}$$

Let random variables be $Z = \sqrt{n}(\mu^* - \mu)/\sigma$ and $K = n\sigma^{*2}/\sigma^2$. As normality is assumed, μ^* and σ^{*2} are mutually independent, and so are random variables Z and K [20]. The random variable Z is denoted as $Z \sim N(0,1)$, following a normal distribution, while the random variable K, denoted with χ^2_{n-1}, represents a chi-squared distribution including $n-1$ degrees of freedom. Therefore, we have

$$p\left(-Z_{0.5-\sqrt{1-\alpha}/2} \leq \sqrt{n}(\mu^* - \mu)/\sigma \leq Z_{0.5-\sqrt{1-\alpha}/2}\right) = \sqrt{1-\alpha} \tag{8}$$

and

$$p\left(\chi^2_{0.5-\sqrt{1-\alpha}/2;n-1} \leq n\sigma^{*2}/\sigma^2 \leq \chi^2_{0.5+\sqrt{1-\alpha}/2;n-1}\right) = \sqrt{1-\alpha}, \tag{9}$$

where $Z_{0.5-\sqrt{1-\alpha}/2}$ is the upper $0.5-\sqrt{1-\alpha}/2$ quintile of the standard normal distribution, $\chi^2_{0.5-\sqrt{1-\alpha}/2;n-1}$ is the lower $0.5-\sqrt{1-\alpha}/2$ quintile of χ^2_{n-1}, and $\chi^2_{0.5+\sqrt{1-\alpha}/2;n-1}$ is the lower $0.5+\sqrt{1-\alpha}/2$ quintile of χ^2_{n-1}. Thus, we can further obtain $p(A) = 1-\alpha$, where

$$A = p\left\{\mu^* - e \times \sigma \leq \mu \leq \overline{X} + e \times \sigma, \sqrt{\frac{n}{\chi^2_{0.5+\sqrt{1-\alpha}/2;n-1}}}\sigma^* \leq \sigma \leq \sqrt{\frac{n}{\chi^2_{0.5-\sqrt{1-\alpha}/2;n-1}}}\sigma^*\right\}. \quad (10)$$

where we have $e = Z_{0.5-\sqrt{1-\alpha}/2}/\sqrt{n}$. In the random sample $(X_1, X_2, \ldots, X_n)$, the observed values are written as $(x_1, x_2, \ldots, x_n)$. Let the observed values of μ^* and σ^* be μ_0^* and σ_0^*, expressed as follows:

$$\mu_0^* = \frac{1}{n}\sum_{i=1}^{n} x_i \quad (11)$$

and

$$\sigma_0^* = \sqrt{\frac{1}{n}\sum_{i=1}^{n}(x_i - \mu_0^*)^2}. \quad (12)$$

Then the observed value for the estimator C_{PMK}^* is denoted by

$$C_{PMK0}^* = \frac{d - |\mu_0^* - T|}{3\sqrt{\sigma_0^{*2} + (\mu_0^* - T)^2}}. \quad (13)$$

Furthermore, the 100 $(1-\alpha)$% confidence region of (μ, σ), denoted with $CR(\mu, \sigma)$, is written as follows:

$$CR(\mu, \sigma) = \{(\mu, \sigma) | \mu_0^* - e \times \sigma \leq \mu \leq \mu_0^* + e \times \sigma,\ \sigma_L \leq \sigma \leq \sigma_U\}, \quad (14)$$

where we have $\sigma_L = \sigma_0^*\sqrt{n/\chi^2_{0.5+\sqrt{1-\alpha}/2;n-1}}$ and $\sigma_U = \sigma_0^*\sqrt{n/\chi^2_{0.5-\sqrt{1-\alpha}/2;n-1}}$. Chen [14] thinks that since the index C_{PMK} is a function of (μ, σ), then the probability of (μ, σ) belonging to $CR(\mu, \sigma)$ is as high as $1-\alpha$. Thus, in this paper, the process capability index C_{PMK} was employed as an object function while the $1-\alpha$ confidence region of (μ, σ) was used as a feasible solution area. Therefore, when $p\{(\mu, \sigma) \in CR(\mu, \sigma)\} \geq 1-\alpha$, then we have $p\{LC_{PMK} \leq C_{PMK} \leq UC_{PMK}\} \geq 1-\alpha$. Accordingly, the upper confidence limit for the process capability index C_{PMK} is defined in the model of mathematical programming as

$$\begin{cases} UC_{PMK} = Max\ C_{PMK}(\mu, \sigma) \\ subject\ to \\ \mu_0^* - e \times \sigma \leq \mu \leq \mu_0^* + e \times \sigma \\ \sigma_L \leq \sigma \leq \sigma_U \end{cases}. \quad (15)$$

For any process standard deviation, when σ is bigger than or equal to σ_L $(\sigma \geq \sigma_L)$, then we have $C_{PMK}(\mu, \sigma) \leq C_{PMK}(\mu, \sigma_L)$. Therefore, the mathematical programming model of Equation (15) can be rewritten as below:

$$\begin{cases} UC_{PMK} = Max\ C_{PMK}(\mu, \sigma_L) \\ subject\ to \\ \mu_0^* - e \times \sigma_L \leq \mu \leq \mu_0^* + e \times \sigma_L \end{cases}. \quad (16)$$

Similarly, for any process standard deviation, when σ is smaller than or equal to σ_U $(\sigma \leq \sigma_U)$, then we have $C_{PMK}(\mu, \sigma)\ C_{PMK}(\mu, \sigma_U)$. Therefore, the lower confidence limit for the process capability index C_{PMK} in the mathematical programming model is displayed as follows:

$$\begin{cases} LC_{PMK} = Min\ C_{PMK}(\mu, \sigma_U) \\ subject\ to \\ \mu_0^* - e \times \sigma_U \leq \mu \leq \mu_0^* + e \times \sigma_U \end{cases}. \quad (17)$$

Based on the above, this article proposes a process to explain how to use a mathematical programming method to solve the 100 $(1-\alpha)$% upper confidence limit and lower confidence limit of the index C_{PMK}. Then, a confidence interval-based fuzzy testing method was developed using Chen's method [14]. The development process of this fuzzy test is as follows:

Step 1: Expressing the index as a function of μ mean and standard deviation σ is as follows:

$$C_{PMK}(\mu,\sigma) = \frac{d-|\mu - T|}{3\sqrt{\sigma^2 + (\mu - T)^2}}. \tag{18}$$

Step 2: Derive the 100 $(1 - \alpha)$% confidence region of (μ,σ) as shown in Equation (14).

Step 3: Taking $C_{PMK}(\mu,\sigma)$ as the objective function and $CR(\mu,\sigma)$ as the feasible solution area, the maximum value UC_{PMK} and the minimum value LC_{PMK} are respectively obtained as shown in Equations (16) and (17).

Step 4: According to the confidence interval of index C_{PMK}, the resemble triangular fuzzy number and its membership function are constructed as shown in Equation (38). Then, develop a confidence interval-based fuzzy test method.

Next, we derive the 100 $(1 - \alpha)$% confidence interval for the process capability index C_{PMK} based on Case 1$\mu_0^* - e \times \sigma_U \leq T \leq \mu_0^* + e \times \sigma_U$, Case 2 $T < \mu_0^* - e \times \sigma_U$, and Case 3 $\mu_0^* + e \times \sigma_U < T$, respectively, as follows:

Case 1: $\mu_0^* - e \times \sigma_U \leq T \leq \mu_0^* + e \times \sigma_U$

In this case, we find $\mu = T$ and process capability index $C_{PMK} = d/(3\sigma)$. According to Equations (16) and (17), the 100 $(1 - \alpha)$% confidence interval for the process capability index C_{PMK} is $[LC_{PMK}, UC_{PMK}]$, where

$$LC_{PMK} = \frac{d}{3\sigma_U} = C_{PMK0}^* \times \sqrt{\frac{\chi^2_{0.5-\sqrt{1-\alpha}/2;n-1}}{n}}; \tag{19}$$

$$UC_{PMK} = \frac{d}{3\sigma_L} = C_{PMK0}^* \times \sqrt{\frac{\chi^2_{0.5+\sqrt{1-\alpha}/2;n-1}}{n}}. \tag{20}$$

Case 2: $T < \mu_0^* - e \times \sigma_U$

In this case, for any $\mu \leq \mu_0^* + e \times \sigma_U$, we have $C_{PMK}(\mu,\sigma_U) \geq C_{PMK}(\mu_0^* + e \times \sigma_U, \sigma_U)$. Based on Equation (17), the lower confidence limit for the process capability index C_{PMK} is depicted below:

$$LC_{PMK} = \frac{d - (\mu_0^* + e \times \sigma_U - T)}{3\sqrt{\sigma_U^2 + (\mu_0^* + e \times \sigma_U - T)^2}}. \tag{21}$$

Similarly, for any $\mu_0^* - e \times \sigma_L \leq \mu$, we have $C_{PMK}(\mu,\sigma_L) \leq C_{PMK}(\mu_0^* - e \times \sigma_L, \sigma_L)$. Based on Equation (16), the upper confidence limit for the process capability index C_{PMK} is shown as follows:

$$UC_{PMK} = \frac{d - (\mu_0^* - e \times \sigma_L - T)}{3\sqrt{\sigma_L^2 + (\mu_0^* - e \times \sigma_L - T)^2}}. \tag{22}$$

Case 3: $\mu_0^* + e \times \sigma_U < T$

In this case, for any $\mu_0^* - e \times \sigma_U \leq \mu$, we have $C_{PMK}(\mu,\sigma_U) \geq C_{PMK}(\mu_0^* - e \times \sigma_U, \sigma_U)$. Based on Equation (17), the lower confidence limit for the process capability index C_{PMK} is displayed below:

$$LC_{PMK} = \frac{d - (T - (\mu_0^* - e \times \sigma_U))}{3\sqrt{\sigma_U^2 + (T - (\mu_0^* - e \times \sigma_U))^2}}. \tag{23}$$

Similarly, for any $\mu \leq \mu_0^* + e \times \sigma_L$, we have $C_{PMK}(\mu,\sigma_L) \leq C_{PMK}(\mu_0^* + e \times \sigma_L, \sigma_L)$. Based on Equation (16), the upper confidence limit of the process capability index C_{PMK} can be depicted as follows:

$$UC_{PMK} = \frac{d - (T - (\mu_0^* + e \times \sigma_L))}{3\sqrt{\sigma_L^2 + (T - (\mu_0^* + e \times \sigma_L))^2}}. \tag{24}$$

Based on the above three cases, this paper builds a method for fuzzy testing upon the confidence interval for the process capability index C_{PMK}.

3. Fuzzy Testing Model Based on Confidence Interval of Process Capability Index C_{PMK}

As mentioned above, in pursuit of a rapid response mechanism, companies usually operate with a small sample size. Following several suggestions from previous studies, in this paper, we constructed a fuzzy testing method on the basis of the confidence interval for the process capability index C_{PMK}, given a small sample size. Pearn and Chen [21] defined the levels required by the process capability indices in the following table.

To identify whether the process capability index C_{PMK} is greater than or equal to C, the null hypothesis, denoted with H_0, and the alternative hypothesis, denoted with H_1, for fuzzy testing are stated below:

H_0: $C_{PMK} \geq C$ (indicating the process capability has achieved the desired level);

H_1: $C_{PMK} < C$ (indicating the process capability has not achieved the desired level).

Customers or process engineers can propose the required value C corresponding to the process capability index C_{PMK} with reference to Table 1. Based on the statistical testing rules mentioned above and Chen's method [14], this paper builds the fuzzy testing model upon the observed values for the estimator and the 100 $(1-\alpha)$% confidence interval of the process capability index C_{PMK}. According to Chen and Lin [22], the α—cuts of the triangular fuzzy number $\widetilde{C}_{PMK}$ can be written as follows:

$$\widetilde{C}_{PMK}[\alpha] = \begin{cases} [C_{PMK1}(\alpha), C_{PMK2}(\alpha)], \ for\ 0.01 \leq \alpha \leq 1 \\ [C_{PMK1}(0.01), C_{PMK2}(0.01)], \ for\ 0 \leq \alpha \leq 0.01 \end{cases}. \tag{25}$$

Table 1. The levels required by the process capability indices.

Required Level	Capability Index Value
Inadequate	$C_{PMK} < 1.00$
Capable	$1.00 \leq C_{PMK} < 1.33$
Satisfactory	$1.33 \leq C_{PMK} < 1.50$
Excellent	$1.50 \leq C_{PMK} < 2.00$
Superb	$2.00 \leq C_{PMK}$

As mentioned earlier, this paper derived the 100 $(1-\alpha)$% confidence interval for the process capability index C_{PMK} from Case 1 $\mu_0^* - e \times \sigma_U \leq T \leq \mu_0^* + e \times \sigma_U$, Case 2 $T < \mu_0^* - e \times \sigma_U$, and Case 3 $\mu_0^* + e \times \sigma_U < T$. According to these three cases, C_{PMK1} (α) and C_{PMK2} (α) can be depicted separately as follows.

Case 1: $\mu_0^* - e \times \sigma_U \leq T \leq \mu_0^* + e \times \sigma_U$

In this case, the process capability index C_{PMK} is represented as

$$C_{PMK} = \frac{d}{3\sigma}. \tag{26}$$

The observed values of the estimator C_{PMK}^* is represented as

$$C_{PMK0}^* = \frac{d}{3\sigma_0^*}. \tag{27}$$

Based on the above equations, Equations (19) and (20), $C_{PMK1}(\alpha)$ and $C_{PMK2}(\alpha)$ are expressed as follows:

$$C_{PMK1}(\alpha) = C^*_{PMK0} \times \sqrt{\frac{\chi^2_{0.5-\sqrt{1-\alpha}/2;n-1}}{n}}, \tag{28}$$

$$C_{PMK2}(\alpha) = C^*_{PMK0} \times \sqrt{\frac{\chi^2_{0.5+\sqrt{1-\alpha}/2;n-1}}{n}}. \tag{29}$$

Case 2: $T < \mu_0^* - e \times \sigma_U$

The process capability index C_{PMK} is represented as

$$C_{PMK} = \frac{d-(\mu - T)}{3\sqrt{\sigma^2 + (\mu - T)^2}}. \tag{30}$$

The observed value of the estimator C^*_{PMK} is represented as

$$C^*_{PMK0} = \frac{d-(\mu_0^* - T)}{3\sqrt{\sigma_0^{*2} + (\mu_0^* - T)^2}}. \tag{31}$$

Based on the above equations, Equations (21) and (22), $C_{PMK1}(\alpha)$, and $C_{PMK2}(\alpha)$ are expressed as follows:

$$e = Z_{0.5-\sqrt{1-\alpha}/2}/\sqrt{n},\ \sigma_L = \sigma_0^* \sqrt{n/\chi^2_{0.5+\sqrt{1-\alpha}/2;n-1}},\ \text{and}\ \sigma_U = \frac{n\sigma_0^{*2}}{\chi^2_{0.5-\sqrt{1-\alpha}/2;n-1}}$$

$$C_{PMK1}(\alpha) = \frac{d - \left(\mu_0^* + \sigma_0^* \sqrt{\frac{Z_{0.5-\sqrt{1-\alpha}/2}}{\chi^2_{0.5-\sqrt{1-\alpha}/2;n-1}}} - T\right)}{3\sqrt{\frac{n\sigma_0^{*2}}{\chi^2_{0.5-\sqrt{1-\alpha}/2;n-1}} + \left(\mu_0^* + \sigma_0^* \sqrt{\frac{Z_{0.5-\sqrt{1-\alpha}/2}}{\chi^2_{0.5-\sqrt{1-\alpha}/2;n-1}}} - T\right)^2}} \tag{32}$$

$$C_{PMK2}(\alpha) = \frac{d - \left(\mu_0^* - \sigma_0^* \sqrt{\frac{Z_{0.5-\sqrt{1-\alpha}/2}}{\chi^2_{0.5+\sqrt{1-\alpha}/2;n-1}}} - T\right)}{3\sqrt{\frac{n\sigma_0^{*2}}{\chi^2_{0.5+\sqrt{1-\alpha}/2;n-1}} + \left(\mu_0^* - \sigma_0^* \sqrt{\frac{Z_{0.5-\sqrt{1-\alpha}/2}}{\chi^2_{0.5+\sqrt{1-\alpha}/2;n-1}}} - T\right)^2}} \tag{33}$$

Case 3: $\mu_0^* + e \times \sigma_U < T$

In this case, the process capability index C_{PMK} is defined as:

$$C_{PMK} = \frac{d-(\mu - T)}{3\sqrt{\sigma^2 + (\mu - T)^2}}. \tag{34}$$

The observed values of the estimator C^*_{PMK} is defined as:

$$C^*_{PMK0} = \frac{d-(\mu_0^* - T)}{3\sqrt{\sigma_0^{*2} + (\mu_0^* - T)^2}}. \tag{35}$$

Based on the above equations, Equations (23) and (24), $C_{PMK1}(\alpha)$ and $C_{PMK2}(\alpha)$ are derived as follows:

$$C_{PMK1}(\alpha) = \frac{d - \left(T - \left(\mu_0^* - \sigma_0^* \sqrt{\frac{Z_{0.5-\sqrt{1-\alpha}/2}}{\chi^2_{0.5-\sqrt{1-\alpha}/2;n-1}}}\right)\right)}{3\sqrt{\frac{n\sigma_0^{*2}}{\chi^2_{0.5-\sqrt{1-\alpha}/2;n-1}} + \left(T - \left(\mu_0^* - \sigma_0^* \sqrt{\frac{Z_{0.5-\sqrt{1-\alpha}/2}}{\chi^2_{0.5-\sqrt{1-\alpha}/2;n-1}}}\right)\right)^2}} \tag{36}$$

and

$$C_{PMK2}(\alpha) = \frac{d - \left(T - \left(\mu_0^* + \sigma_0^* \sqrt{\frac{Z_{0.5-\sqrt{1-\alpha}/2}}{\chi^2_{0.5+\sqrt{1-\alpha}/2;n-1}}}\right)\right)}{3\sqrt{\frac{n\sigma_0^{*2}}{\chi^2_{0.5+\sqrt{1-\alpha}/2;n-1}} + \left(T - \left(\mu_0^* + \sigma_0^* \sqrt{\frac{Z_{0.5-\sqrt{1-\alpha}/2}}{\chi^2_{0.5+\sqrt{1-\alpha}/2;n-1}}}\right)\right)^2}}. \tag{37}$$

Therefore, we have the resemble triangular fuzzy number, $\widetilde{C}_{PMK} = \Delta\,(K_L, K_M, K_R)$, where $K_L = C_{PMK1}(0.01)$, $K_M = C_{PMK1}(1) = C_{PMK2}(1)$ and $K_R = C_{PMK2}(0.01)$. Its membership function is expressed as follows:

$$h(x) = \begin{cases} 0 & if\ x < K_L \\ \alpha_1 & if\ K_L \le x < K_M \\ 1 & if\ x = K_M \\ \alpha_2 & if\ K_M < x \le K_R \\ 0 & if\ x > K_R \end{cases}, \tag{38}$$

where α_1 is determined by $C_{PMK1}(\alpha_1) = x$ and α_2 is determined by $C_{PMK2}(\alpha_2) = x$. Before proposing the fuzzy testing method for the process capability index C_{PMK}, we reviewed the following statistical testing rules:

(1) When the upper confidence limit of the process capability index C_{PMK} exceeds or equals C ($UC_{PMK} \ge C$), do not reject H_0 and conclude that $C_{PMK} \ge C$.
(2) When the upper confidence limit of the process capability index C_{PMK} is smaller than C ($UC_{PMK} < C$), reject H_0 and conclude that $C_{PMK} < C$.

Then, this paper developed a fuzzy testing method, considering the confidence interval for the process capability index C_{PMK} based on statistical testing rules. According to Equation (36), for the fuzzy number $\widetilde{C}_{PMK}$, its membership function $h(x)$ is presented with the vertical line $x = C$ in Figure 1.

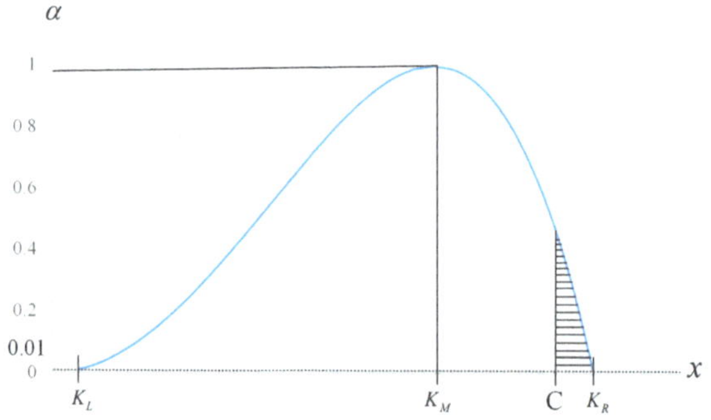

Figure 1. Membership function $h(x)$ with vertical line $x = C$.

Based on Chen and Lin [22], let A_T denote the area in the diagram of membership function $h(x)$, and let A_R denote the area in the same graph but to the right of membership function $h(x)$ from the vertical line $x = C$, such that

$$A_T = \{(x, \alpha) | C_{PMK1}(\alpha) \le x \le C_{PMK2}(\alpha)\}. \tag{39}$$

and

$$A_R = \{(x, \alpha) | C \le x \le C_{PMK2}(\alpha)\}. \tag{40}$$

According to Yu et al. [23] and based on Equations (39) and (40), we let $d_T = K_R - K_L$ and $d_R = K_R - C$. Then, we have $d_R/d_T = (K_R - C)/(K_R - K_L)$. Also, we denote Case 1 as $\mu_0^* - e \times \sigma_U \le T \le \mu_0^* + e \times \sigma_U$, Case 2 as $T < \mu_0^* - e \times \sigma_U$, and Case 3 as $\mu_0^* + e \times \sigma_U < T$. More detailed explanations are listed below:

Case 1: $\mu_0^* - e \times \sigma_U \le T \le \mu_0^* + e \times \sigma_U$

$$d_R/d_T = \frac{K_R - C}{2(K_R - K_M)}, \tag{41}$$

where

$$K_R = C_{PMK0}^* \times \sqrt{\frac{\chi^2_{0.9975;n-1}}{n}} \tag{42}$$

and

$$K_M = C_{PMK0}^* \times \sqrt{\frac{\chi^2_{0.5;n-1}}{n}}. \tag{43}$$

Case 2: $T < \mu_0^* - e \times \sigma_U$

$$d_R/d_T = \frac{K_R - C}{2(K_R - K_M)}, \tag{44}$$

where

$$K_R = \frac{d - \left(\mu_0^* - \sigma_0^* \sqrt{\frac{Z_{0.0025}}{\chi^2_{0.9975;n-1}}} - T\right)}{3\sqrt{\frac{n\sigma_0^{*2}}{\chi^2_{0.9975;n-1}} + \left(\mu_0^* - \sigma_0^* \sqrt{\frac{Z_{0.0025}}{\chi^2_{0.9975;n-1}}} - T\right)^2}} \tag{45}$$

and

$$K_M = \frac{d - \left(\mu_0^* + \sigma_0^* \sqrt{\frac{Z_{0.0025}}{\chi^2_{0.5;n-1}}} - T\right)}{3\sqrt{\frac{n\sigma_0^{*2}}{\chi^2_{0.5;n-1}} + \left(\mu_0^* + \sigma_0^* \sqrt{\frac{Z_{0.0025}}{\chi^2_{0.5;n-1}}} - T\right)^2}}. \tag{46}$$

Case 3: $\mu_0^* + e \times \sigma_U < T$

$$d_R/d_T = \frac{K_R - C}{2(K_R - K_M)}, \tag{47}$$

where

$$K_R = \frac{d - \left(T - \left(\mu_0^* + \sigma_0^* \sqrt{\frac{Z_{0.0025}}{\chi^2_{0.9975;n-1}}}\right)\right)}{3\sqrt{\frac{n\sigma_0^{*2}}{\chi^2_{0.9975;n-1}} + \left(T - \left(\mu_0^* + \sigma_0^* \sqrt{\frac{Z_{0.0025}}{\chi^2_{0.9975;n-1}}}\right)\right)^2}} \tag{48}$$

and

$$K_M = \frac{d - \left(T - \left(\mu_0^* - \sigma_0^* \sqrt{\frac{Z_{0.0025}}{\chi^2_{0.5;n-1}}}\right)\right)}{3\sqrt{\frac{n\sigma_0^{*2}}{\chi^2_{0.5;n-1}} + \left(T - \left(\mu_0^* - \sigma_0^* \sqrt{\frac{Z_{0.0025}}{\chi^2_{0.5;n-1}}}\right)\right)^2}}. \tag{49}$$

Based on Yu et al. [23], we let $0 < \phi \le 0.5$, then the fuzzy testing rules of the process capability index C_{PMK} are shown as follows:

(1) If $d_R/d_T \le \phi$, then H_0 is rejected, and we can conclude that $C_{PMK} < C$.
(2) If $d_R/d_T > \phi$, then H_0 is not rejected, and we can conclude that $C_{PMK} \ge C$.

4. An Application Example

It is known that the central region of Taiwan is a machine tool center. Therefore, this section of this paper demonstrates how to apply the proposed fuzzy testing model through an empirical example involving the axis machined by a manufacturer in central Taiwan. The fuzzy testing model built on the confidence interval of the process capability index C_{PMK} is an effective approach for deciding whether the process capability is acceptable or requires improvement. The target value T of the axis machined by the factory is 1.80 mm (T = 1.80), and the tolerance is 0.05 mm. Accordingly, the lower specification limit (LSL) is 1.75 mm (LSL = 1.75) and the upper specification limit (USL) is 1.85 mm (USL = 1.85). Thus, $T = (USL + LSL)/2$ and $d = (USL - LSL)/2$. According to customer requirements, the process engineer sets the required value for the process capability index C_{PMK} as 1.00. Aiming to gauge whether the value of the process capability index C_{PMK} exceeds or equals 1.00, the null hypothesis (H_0) and the alternative hypothesis (H_1) for fuzzy testing are listed as follows:

H_0: $C_{PMK} \geq 1.00$ (showing that the process capability is sufficient);

H_1: $C_{PMK} < 1.00$ (showing that the process capability is insufficient).

As mentioned earlier, in pursuit of a quick response mechanism, companies often opt for a small sample size. Let ($x_1, x_2, \ldots, x_{16}$) be the observed values for a random sample ($X_1, X_2, \ldots, X_{16}$). Then the observed values of μ^* and σ^* are μ_0^* and σ_0^*, respectively, as shown below:

$$\mu_0^* = \frac{1}{16}\sum_{i=1}^{16} x_i = 1.083 \tag{50}$$

and

$$\sigma_0^* = \sqrt{\frac{1}{16}\sum_{i=1}^{16}(x_i - \mu_0^*)^2} = 0.022. \tag{51}$$

Thus,

$$\mu_0^* - e \times \sigma_U = 1.083 - 0.702 \times 0.044 = 1.793$$

and

$$\mu_0^* + e \times \sigma_U = 1.083 + 0.702 \times 0.044 = 1.834.$$

The target value T belongs to the interval (1.793, 1.834). Thus, the observed value of the estimator C_{PMK}^* is calculated as follows:

$$C_{PMK0}^* = \frac{d}{3\sigma_0^*} = 0.758. \tag{52}$$

Based on Equation (50), we obtain the following values of K_R and K_M:

$$K_R = C_{PMK0}^* \times \sqrt{\frac{\chi^2_{0.9975;n-1}}{n}} = 0.758 \times \sqrt{\frac{34.950}{16}} = 1.120 \tag{53}$$

and

$$K_M = C_{PMK0}^* \times \sqrt{\frac{\chi^2_{0.5;n-1}}{n}} = 0.758 \times \sqrt{\frac{14.399}{16}} = 0.717. \tag{54}$$

Thus, the value of d_R/d_T is calculated as follows:

$$d_R/d_T = \frac{K_R - C}{2(K_R - K_M)} = \frac{1.120 - 1.00}{2 \times (1.120 - 0.717)} = 0.15 \tag{55}$$

Based on Yu et al. [23], we let $0 < \phi \leq 0.5$ and reviewed the fuzzy testing rules of the process capability index C_{PMK}. We obtained the following result:

(1) If $d_R/d_T \leq \phi$, then H_0 is rejected, and we can conclude that $C_{PMK} < C$.

The process engineer, drawing from past professional experience, analyzed and set the value of ϕ to 0.2; that is, ϕ = 0.2. According to the above fuzzy testing rule, since d_R/d_T is less than 0.2, the null hypothesis H_0 is rejected, and the conclusion $C_{PMK} < 1.00$ is drawn. In fact, the observed value of the estimator C^*_{PMK} is 0.758 ($C^*_{PMK0} = 0.758$), the upper confidence limit of the process capability index C_{PMK} is 1.120 ($UC_{PMK} = 1.120$) with $\alpha = 0.01$. If the result of the statistical inference shows that $C_{PMK} \geq 1.00$, it indicates that the proposed fuzzy testing model in this paper demonstrates greater practicality compared to the conventional statistical testing model.

5. Conclusions

Various process capability indices are applied to the quantitative measurement of the potential and performance of a process in the manufacturing industry. Not only can an internal process engineer use them to assess process quality, but an external sales department can also utilize them as a communication tool. The process capability index C_{PMK} can quickly detect process deviations from the target value, which is conducive to the promotion of smart manufacturing. Therefore, in this paper, we utilized the process capability index as a tool to evaluate process quality. Process capability indices, as noted by some studies, have unknown parameters and therefore must be estimated from sample data. In addition, as highlighted by many studies, companies typically pursue a rapid response mechanism, so they need to make decisions using a small sample size. Consequently, this study, based on some suggestions from previous studies for the case of small sample size, proposed the process capability index C_{PMK} with a 100 $(1 - \alpha)$% confidence interval. In the normal process condition where the sample mean and the sample variation are mutually independent, this study derived the 100 $(1 - \alpha)$% confidence region of (μ, σ). Then, this study adopted the process capability index C_{PMK} as an object function as well as the 100 $(1 - \alpha)$% confidence region of (μ, σ) as a feasible solution area, aiming to acquire the 100 $(1 - \alpha)$% confidence interval of the process capability index C_{PMK}. Immediately afterward, the 100 $(1 - \alpha)$% confidence interval of the process capability index C_{PMK} was utilized to establish a fuzzy testing model to evaluate process quality and see if it can achieve the required quality level. In this model, we first derived the triangular fuzzy number $\widetilde{C}_{PMK}$ and then obtained its membership function $h(x)$. According to the membership function $h(x)$, this study established fuzzy testing rules. Through these rules, we can tell if the process quality attains the required level, which can serve as a reference for other industries. As mentioned earlier, central Taiwan is an industrial center for machine tools. Accordingly, this study illustrated the use of the proposed fuzzy testing model with an example of the axis machined by a factory located in the central region of Taiwan. It is evident from this example that the proposed fuzzy testing model can exhibit greater practicality compared to the conventional statistical testing model.

Author Contributions: Conceptualization, W.L. and K.-S.C.; methodology, W.L., K.-S.C. and C.-M.Y. (Chun-Min Yu); software, T.-H.H.; validation, T.-H.H. and C.-M.Y. (Chun-Ming Yang); formal analysis, W.L., K.-S.C. and C.-M.Y. (Chun-Min Yu); resources, W.L.; data curation, C.-M.Y. (Chun-Ming Yang); writing—original draft preparation, W.L., T.-H.H., K.-S.C., C.-M.Y. (Chun-Min Yu) and C.-M.Y. (Chun-Ming Yang); writing—review and editing, W.L., K.-S.C. and C.-M.Y. (Chun-Min Yu); visualization, C.-M.Y. (Chun-Min Yu); supervision, K.-S.C.; project administration, W.L.; funding acquisition, W.L. All authors have read and agreed to the published version of the manuscript.

Funding: This research was partially funded by the Guangxi Philosophy and Social Sciences Research Project under grant No. 23AGL001 and the National Natural Science Foundation of China under grant No. 72361002.

Data Availability Statement: The data that support the findings of this study are available from the corresponding author upon reasonable request.

Conflicts of Interest: The authors declare no conflicts of interest.

References

1. Anderson, N.C.; Kovach, J.V. Reducing welding defects in turnaround projects: A lean six sigma case study. *Qual. Eng.* **2014**, *26*, 168–181. [CrossRef]
2. Gijo, E.V.; Scaria, J. Process improvement through Six Sigma with Beta correction: A case study of manufacturing company. *Int. J. Adv. Manuf. Technol.* **2014**, *71*, 717–730. [CrossRef]
3. Shafer, S.M.; Moeller, S.B. The effects of Six Sigma on corporate performance: An empirical investigation. *J. Oper. Manag.* **2012**, *30*, 521–532. [CrossRef]
4. Liao, M.Y.; Pearn, W.L. Modified weighted standard deviation index for adequately interpreting a supplier's lognormal process capability. *Proc. Inst. Mech. Eng. Part B-J. Eng. Manuf.* **2019**, *233*, 999–1008. [CrossRef]
5. Li, W.; Liu, G. Dynamic failure mode analysis approach based on an improved Taguchi process capability index. *Reliab. Eng. Syst. Saf.* **2022**, *218*, 108152. [CrossRef]
6. Shu, M.H.; Wang, T.C.; Hsu, B.M. Generalized quick-switch sampling systems indexed by Taguchi capability with record traceability. *Comput. Ind. Eng.* **2022**, *172*, 108577. [CrossRef]
7. Sanchez-Marquez, R.; Jabaloyes Vivas, J. Building a cpk control chart—A novel and practical method for practitioners. *Comput. Ind. Eng.* **2021**, *158*, 107428. [CrossRef]
8. Kane, V.E. Process capability indices. *J. Qual. Technol.* **1986**, *18*, 41–52. [CrossRef]
9. Boyles, R.A. The Taguchi Capability Index. *J. Qual. Technol.* **1991**, *23*, 17–26. [CrossRef]
10. Chan, L.K.; Cheng, S.W.; Spiring, F.A. A New Measure of Process Capability: C_{pm}. *J. Qual. Technol.* **1988**, *20*, 162–175. [CrossRef]
11. Pearn, W.L.; Kotz, S.; Johnson, N.L. Distributional and Inferential Properties of Process Capability Indices. *J. Qual. Technol.* **1992**, *24*, 216–231. [CrossRef]
12. Choi, B.C.; Owen, D.B. A study of a New Capability Index. *Commun. Stat.-Theory Methods* **1990**, *19*, 1231–1245. [CrossRef]
13. Vännman, K. A unified approach to capability indices. *Stat. Sin.* **1995**, *5*, 805–820.
14. Chen, K.S. Fuzzy testing decision-making model for intelligent manufacturing process with Taguchi capability index. *J. Intell. Fuzzy Syst.* **2020**, *38*, 2129–2139. [CrossRef]
15. Wu, C.H.; Hsu, Y.C.; Pearn, W.L. An improved measure of quality loss for notching processes. *Qual. Reliab. Eng. Int.* **2021**, *37*, 108–122. [CrossRef]
16. Arif, O.H.; Aslam, M.; Jun, C.H. Acceptance sampling plan for multiple manufacturing lines using EWMA process capability index. *J. Adv. Mech. Des. Syst. Manuf.* **2017**, *11*, JAMDSM0004. [CrossRef]
17. Chien, C.F.; Hong, T.Y.; Guo, H.Z. An empirical study for smart production for TFT-LCD to empower Industry 3.5. *J. Chin. Inst. Eng.* **2017**, *40*, 552–561. [CrossRef]
18. Wu, M.F.; Chen, H.Y.; Chang, T.C.; Wu, C.F. Quality evaluation of internal cylindrical grinding process with multiple quality characteristics for gear products. *Int. J. Prod. Res.* **2019**, *57*, 6687–6701. [CrossRef]
19. Lo, W.; Yang, C.M.; Lai, K.K.; Li, S.Y.; Chen, C.H. Developing a novel fuzzy evaluation model by one-sided specification capability indices. *Mathematics* **2021**, *9*, 1076. [CrossRef]
20. Chen, K.S.; Yu, C.M.; Huang, M.L. Fuzzy selection model for quality-based IC packaging process outsourcers. *IEEE Trans. Semicond. Manuf.* **2022**, *35*, 102–109. [CrossRef]
21. Pearn, W.L.; Chen, K.S. One-sided capability indices C_{pu} and C_{pl}: Decision making with sample information. *Int. J. Qual. Reliab. Manag.* **2002**, *19*, 221–245. [CrossRef]
22. Chen, H.Y.; Lin, K.P. Fuzzy supplier selection model based on lifetime performance index. *Expert Syst. Appl.* **2022**, *208*, 118135. [CrossRef]
23. Yu, C.M.; Lai, K.K.; Chen, K.S.; Chang, T.C. Process-quality evaluation for wire bonding with multiple gold wires. *IEEE Access* **2020**, *8*, 106075–106082. [CrossRef]

axioms

MDPI

Article

Fuzzy Decision-Making and Resource Management Model of Performance Evaluation Indices

Kuen-Suan Chen [1,2,3], Tsung-Hua Hsieh [4,5], Chia-Pao Chang [1,*], Kai-Chao Yao [4,*] and Tsun-Hung Huang [1]

1 Department of Industrial Engineering and Management, National Chin-Yi University of Technology, Taichung 411030, Taiwan; kschen@ncut.edu.tw (K.-S.C.); toby@ncut.edu.tw (T.-H.H.)
2 Department of Business Administration, Chaoyang University of Technology, Taichung 413310, Taiwan
3 Department of Business Administration, Asia University, Taichung 413305, Taiwan
4 Department of Industrial Education and Technology, National Changhua University of Education, Changhua 500208, Taiwan; dhheish@cyut.edu.tw
5 Department of Leisure Service Management, Chaoyang University of Technology, Taichung 413310, Taiwan
* Correspondence: chiapc@ncut.edu.tw (C.-P.C.); kcyao@cc.ncue.edu.tw (K.-C.Y.)

Abstract: The Performance Evaluation Matrix (PEM) is an excellent decision-making tool for assessment and resource management. Satisfaction Index and Importance Index are two important evaluation indicators of construction and PEM. Managers can decide whether the service item needs to be improved based on the Satisfaction Index of the service item. When resources are limited, managers can determine the priority of improving the service item based on the Importance Index. In order to avoid the risk of misjudgment caused by sample errors and meet the needs of enterprises' rapid decision-making, this study proposed a fuzzy test built on the confidence intervals of the above two key indicators to decide whether essential service items should be improved and determine the priority of improvement. Since the fuzzy test was relatively complex, this study further came up with fuzzy evaluation values and fuzzy evaluation critical values of service items following fuzzy testing rules. Besides, evaluation rules were established to facilitate industrial applications. This approach can be completed with any common word processing software, so it is relatively convenient in application and easy to manage. Finally, an application example was presented in this paper to explain the applicability of the proposed approach.

Keywords: performance evaluation matrix; satisfaction index; importance index; fuzzy evaluation critical values; service operating system

MSC: 62C05; 62C86

Citation: Chen, K.-S.; Hsieh, T.-H.; Chang, C.-P.; Yao, K.-C.; Huang, T.-H. Fuzzy Decision-Making and Resource Management Model of Performance Evaluation Indices. *Axioms* **2024**, *13*, 198. https://doi.org/10.3390/axioms13030198

Academic Editor: Oscar Castillo

Received: 6 February 2024
Revised: 11 March 2024
Accepted: 14 March 2024
Published: 15 March 2024

1. Introduction

The Performance Evaluation Matrix (PEM) is an outstanding evaluation and improvement tool for various service operating systems [1–4]. Many papers have been devoted to conducting research into the PEM, aiming to evaluate the performance of various service operating systems and determine whether they have reached the required level [5–7]. The PEM method, mainly based on the service items provided by the service operating systems and then designed into questionnaire scales, can be employed to investigate customers' or users' satisfaction and importance for each service item as well as to set the Satisfaction Index and the Importance Index [8–10].

Additionally, a few studies have used confidence intervals of indicators to create evaluation coordinate points of the Satisfaction Index and the Importance Index for each service item [11,12]. Observing where the evaluation coordinate points of all service items are located in the service quality zones of PEM can help determine which service item needs improvement or whether resource transfer is required [13,14]. The service item that falls into the service quality improvement zone has high customer importance and low customer satisfaction, so it needs improvement. The service item that falls into the service quality maintenance zone has equal customer importance and satisfaction, so it needs maintenance. The service item that falls into the resource transfer zone has low customer importance and high customer satisfaction, showing that customers are fully satisfied with the service item, but its importance is not high; therefore, this item must be reviewed, and a resource transfer must be considered to increase the overall satisfaction of the entire service operating system [15,16].

Yu et al. [17] have indicated that the above-mentioned method of performance evaluation may fail to identify improvement points due to customers' different cultures and mindsets. With the spirit of continuous improvement and total quality management, PEM is divided into four quadrants by the average values of the Satisfaction Index and the Importance Index. Service items in quadrants 2 and 4, in principle, are those whose values are lower than the average as well as the items which require improvement. It seems that the evaluation is directly conducted by statistics calculated from the sample data, but the evaluation method has taken sampling errors into account. To solve the problem of sampling errors, some studies have also made statistical inferences through the confidence intervals of the above two indices [11,12]. However, considering cost and effectiveness, Chen and Yu [18] have suggested that the number of samples is usually not too large in practice for making decisions quickly and accurately in a short time, thereby affecting the accuracy of statistical inferences. Obviously, sampling errors, assessment accuracy, and limited resources are issues that need to be considered and solved in the development of PEM. Aiming to solve the problem concerning the maintenance of evaluation accuracy in the case of small samples, this study, based on Chen and Yu [18] and the confidence interval proposed by some studies [11,12,19], develops a complete fuzzy testing method to evaluate which service item needs improvement. Meanwhile, when resources are limited, this method helps decide which service item should be a top priority for improvement. Next, following the fuzzy testing rules, this study derives the fuzzy decision-making value for satisfaction improvement and the fuzzy decision-making value for improvement priority of importance, so as to facilitate managers' decision-making [18,19].

In the PEM method, first, based on the service system which needs to be evaluated, we need to design a corresponding questionnaire and corresponding questions which are called service items by this study. In order not to lose generality, this study, like other studies, assumes that the number of service items provided by the service system is k; then k questions are designed to conduct a survey targeted at learners about satisfaction and importance for k service items [20].

Lin et al. [19] let random variable X_h represent the hth service item of satisfaction, then random variable X_h is distributed as a Beta distribution with the first parameter α_h and the second parameter β_h, denoted by $X_h \sim Beta(\alpha_h, \beta_h), h = 1, 2, \ldots, k$. Furthermore, let random variable Y_h indicate the hth service item of importance, then random variable Y_h is distributed with the first parameter a_h and the second parameter b_h, denoted by $Y_h \sim Beta(a_h, b_h)$; the Beta distribution, denoted by $Y_h \sim Beta(\delta_h, \gamma_h)$, is also displayed. Thus, these two indices can be shown as follows:

$$I_{Sh} = \frac{\alpha_h}{\alpha_h + \beta_h} \text{(Satisfaction Index)}; \quad (1)$$

$$I_{Ih} = \frac{a_h}{a_h + b_h} \text{(Importance Index)}. \quad (2)$$

For the convenience of explanation, according to the characteristics of the Beta distribution, we set the level of satisfaction higher than 50% as high satisfaction and the level of satisfaction lower than 50% as low satisfaction. When $I_{Sh} = 1/2$, we have $\alpha_h/(\alpha_h + \beta_h) = 1/2$, and then we make the conclusion $\alpha_h = \beta_h$, showing that the level of high satisfaction is equal to the level of low satisfaction. When $I_{Sh} > 1/2$, we have $\alpha_h/(\alpha_h + \beta_h) > 1/2$, and then we make the conclusion $\alpha_h > \beta_h$; at this time, the level of high satisfaction is higher than that of low satisfaction. When $I_{Sh} < 1/2$, we have $\alpha_h/(\alpha_h + \beta_h) < 1/2$ and then we make the conclusion $\alpha_h < \beta_h$, meaning that the level of high satisfaction is lower than that of low satisfaction. When the value of Satisfaction Index I_{Sh} is lower, the level of high satisfaction is also lower. In addition, the Importance Index has the same property as the Satisfaction Index. The higher the value of Importance Index I_{Ih}, the higher the level of high importance. Similarly, as the value of Importance Index I_{Ih} is lower, the level of high importance is lower as well.

Obviously, Satisfaction Index and Importance Index are two important elements of the performance evaluation matrix. The purpose of this paper is to use the unilateral confidence intervals of these two important indicators to conduct a fuzzy test, so that the fuzzy evaluation criteria of the performance evaluation matrix can be developed. Next, according to the suggestion made by Lin et al. [19], a fuzzy performance evaluation chart is created to assist businesses with their managment and decision-making.

The remainder of this paper is organized as follows. In Section 2, we derive the $100(1-\alpha)\%$ confidence interval of Satisfaction Index and develop its fuzzy evaluation rules. In Section 3, we derive the $100(1-\alpha)\%$ confidence interval of Importance Index and make its fuzzy evaluation rules. In Section 4, we use a case study to illustrate an application of the model proposed by this study, demonstrating how to identify service items requiring improvement as well as how to prioritize them for improvement as resources are limited. In Section 5, conclusions, research limitations, and future research directions are presented and explored.

2. Fuzzy Evaluation Rules for Satisfaction Index

As mentioned earlier, this study, built on existing literature, assumes that both distributions of customer satisfaction and importance follow the Beta distribution [19]. Let $\left(X_{h,1},\ldots,X_{h,j},\ldots,X_{h,n}\right)$ be the sample data of customer satisfaction for Service Item h with a size of n, where $h = 1, 2,\ldots, k$. Then, the unbiased estimator of Satisfaction Index I_{Sh} is expressed as follows:

$$\hat{I}_{Sh} = \frac{1}{n} \times \sum_{j=1}^{n} X_{h,j}. \tag{3}$$

The expected value of the unbiased estimator $\hat{I}_{Sh}$ is equal to I_{Sh}, denoted by $E\left[\hat{I}_{Sh}\right] = I_{Sh}$. Let random variable Z_{Sh} be defined as follows:

$$Z_{Sh} = \frac{\hat{I}_{Sh} - I_{Sh}}{S_{Xh}/\sqrt{n}}, \tag{4}$$

where S_{Xh} is the sample standard deviation, given by:

$$S_{Xh} = \sqrt{\frac{1}{n-1} \times \sum_{j=1}^{n}\left(X_{h,j} - \overline{X}_h\right)^2}. \tag{5}$$

Plenty of studies have revealed that the fuzzy test method based on the confidence interval must be able to derive the two-tailed confidence interval of the indicator, so that the subsequent fuzzy testing procedure can be completed [11,12,17,18]. Accordingly, if the two-tailed confidence interval of the indicator cannot be derived, then it must be completed by the Central Limit Theorem. Nevertheless, when the sample size is not large enough, it will lead to larger sampling errors. Considering the customer satisfaction questionnaire survey, the sample size is relatively large. Thus, in this paper, on the basis of the Central Limit Theorem, the distribution of random variable Z_{Sh} approximates the standard normal distribution for large sample size n [21], expressed as follows:

$$Z_{Sh} = \frac{\hat{I}_{Sh} - I_{Sh}}{S_{Xh}/\sqrt{n}} \approx N(0,1). \tag{6}$$

Based on the above-mentioned, we have $1-\alpha = p(-z_{\alpha/2} \leq Z_{Sh} \leq z_{\alpha/2})$, where z_α is the upper α quantile of the standard normal distribution. As noted by Lin et al. [19], Satisfaction Index I_{Sh} is set as the x-axis to form the PEM. Let $\left(x_{h,1},\ldots,x_{h,j},\ldots,x_{h,n}\right)$ be the observed value of $\left(X_{h,1},\ldots,X_{h,j},\ldots,X_{h,n}\right)$. Then $\hat{I}_{Sh0}$ is the observed value of $\hat{I}_{Sh}$, written as follows:

$$\hat{I}_{Sh0} = \frac{1}{n} \times \sum_{j=1}^{n} x_{h,j}. \tag{7}$$

The average of $\hat{I}_{Sh0}$ is expressed as follows:

$$I_{S0} = \frac{1}{k}\sum_{h=1}^{k} \hat{I}_{Sh0}. \tag{8}$$

According to Yu et al. [17], when the value of Satisfaction Index for Service Item h is lower than the average value ($I_{Sh} \leq I_{S0}$), then Service Item h must be improved. The hypotheses of the statistical test for the Satisfaction Index of Service Item h are written as follows:

$$\text{null hypothesis } H_0 : \ I_{Sh} \geq I_{S0}; \tag{9}$$

$$\text{alternative hypothesis } H_1 : \ I_{Sh} < I_{S0}. \tag{10}$$

The significance level of the test is β and the critical region can be defined as $CR_{Sh} = \left\{\hat{I}_{Sh0} < C_{h0}\right\} = \left\{Z_{Sh} < \sqrt{n}(C_{h0} - I_{S0})/S_{Xh}\right\}$. Therefore, the critical value of C_{h0} is determined by

$$p\left\{Z_{Sh} < \frac{C_{h0} - I_{S0}}{S_{Xh0}/\sqrt{n}}\right\} = \beta, \tag{11}$$

where S_{Xh0} is the observed value of S_{Xh}, written as follows:

$$S_{Xh0} = \sqrt{\frac{1}{n-1} \times \sum_{j=1}^{n}\left(x_{h,j} - \overline{x}_h\right)^2}. \tag{12}$$

Thus, the critical value is denoted by $C_{h0} = I_{S0} - z_\beta S_{Xh0}/\sqrt{n}$. Obviously, we have $p\left\{\hat{I}_{Sh} > C_{h0}\,|I_{Sh} = I_{S0}\right\} = p\left\{Z_{Sh} < -z_\beta | I_{Sh} = I_{S0}\right\} = \beta$. Let the observed value of Z_{Sh} be Z_{Sh0}. Then

$$Z_{Sh0} = \frac{\hat{I}_{Sh0} - I_{S0}}{S_{Xh0}/\sqrt{n}}. \tag{13}$$

Thus, we can replace $\hat{I}_{Sh0}$ with Z_{Sh0} as the testing statistic and replace C_{h0} with z_β as the critical value. The *α-cuts* of the quasi-triangular fuzzy number $\tilde{z}_\beta$ is expressed as follows:

$$\tilde{z}_\beta[\alpha] = \begin{cases} [z_{\beta 1}(\alpha), z_{\beta 2}(\alpha)] = [-z_\beta - z_{\alpha/2}, -z_\beta + z_{\alpha/2}],\ 0.01 \leq \alpha \leq 1 \\ [z_{\beta 1}(\alpha), z_{\beta 2}(\alpha)] = [-z_\beta - z_{0.005}, -z_\beta + z_{0.005}],\ 0 \leq \alpha \leq 0.01 \end{cases}. \tag{14}$$

Obviously, when $\alpha = 1$, then $z_{\alpha/2} = 0$ and $z_{\beta 1}(1) = z_{\beta 2}(1) = -z_\beta$. The fuzzy number is $\tilde{z}_\beta = \left(-z_{\beta L}, -z_{\beta M}, -z_{\beta R}\right) = \left(-z_\beta - z_{0.005}, -z_\beta, -z_\beta + z_{0.005}\right)$. In fact, the fuzzy evaluation technique proposed by this study based on the atypical fuzzy evaluation method suggested by Buckley [22] belongs to the type-2 of fuzzy sets [23]. A quasi-triangular fuzzy membership function is mainly constructed by the two-tailed confidence interval of the parameters that need to be evaluated. The quasi- triangular fuzzy membership function of $\widetilde{z}_\beta$ is as follows:

$$\eta(x) = \begin{cases} 0, \ if \ x < -z_\beta - z_{0.005} \\ 2 \times (1 - \Phi(-z_\beta - x)), if - z_\beta - z_{0.005} \le x < -z_\beta \\ 1, \ if \ x = -z_\beta \\ 2 \times (1 - \Phi(x + z_\beta)), if - z_\beta < x \le -z_\beta + z_{0.005} \\ 0, \ if - z_\beta + z_{0.005} < x \end{cases}. \tag{15}$$

Based on Equation (15), membership function $\eta(x)$ with the vertical line of $x = \hat{I}_{Sh0}$ is depicted in Figure 1.

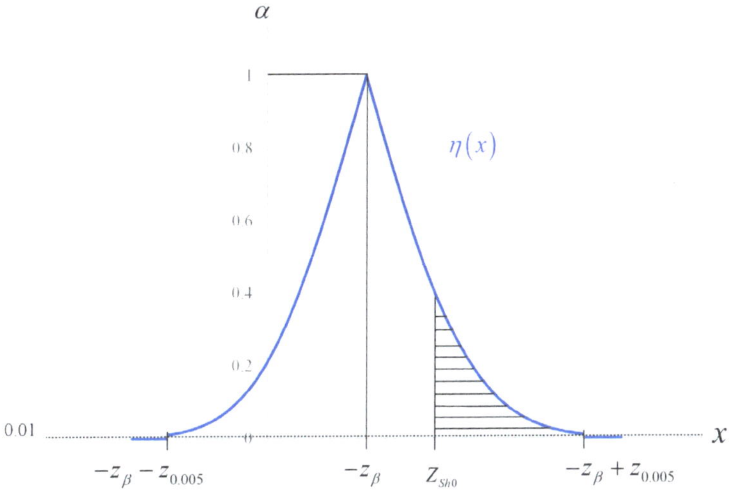

Figure 1. Membership function $\eta(x)$ with the vertical line of $x = Z_{Sh0}$.

Based on the concept of Yu et al. [24], let $d_{hR} = Z_{Sh0} - z_{\beta R} = Z_{Sh0} - (-z_\beta + z_{0.005})$ and $d_T = z_{\beta R} - z_{\beta L} = 2 \times z_{0.005}$. Then d_{hR}/d_T is expressed as follows:

$$d_{hR}/d_T = \frac{-z_\beta + z_{0.005} - Z_{Sh0}}{2 \times z_{0.005}}. \tag{16}$$

Let the decision value be d_{Sh0}, such that

$$d_{hR}/d_T = \frac{-z_\beta + z_{0.005} - d_{Sh0}}{2 \times z_{0.005}} = \phi. \tag{17}$$

Therefore, we have

$$d_{Sh0} = (1 - 2\phi)z_{0.005} - z_\beta. \tag{18}$$

According to Chen et al. [25], we let $0 < \phi < 0.5$, and the decision rules of the fuzzy two-tailed testing model are listed below:

(1) If $Z_{Sh0} < d_{Sh0}$, then $d_{hR}/d_T < \phi$. Therefore, we reject H_0 and draw the conclusion $I_{Sh} < I_{S0}$. Thus, Service Item h needs to be improved.
(2) If $Z_{Sh0} \ge d_{Sh0}$, then $d_{hR}/d_T \ge \phi$. Thus, we do not reject H_0 and draw the conclusion $I_{Sh} \ge I_{S0}$. Thus, Service Item h does not need to be improved.

3. Fuzzy Evaluation Rules for Importance Index

Let $\left(Y_{h,1}, \ldots, Y_{h,j}, \ldots, Y_{h,n}\right)$ be the sample data of customer importance for Service Item h with a size of n, where $h = 1, 2, \ldots, k$. Then, the unbiased estimator of Important Index I_{Ih} is defined below:

$$\hat{I}_{Ih} = \frac{1}{n} \times \sum_{j=1}^{n} Y_{h,j}. \tag{19}$$

The expected value of unbiased estimator $\hat{I}_{Ih}$, equal to I_{Ih}, is denoted by $E\left[\hat{I}_{Ih}\right] = I_{Ih}$. Let random variable Z_{Ih} be defined as:

$$Z_{Ih} = \frac{\hat{I}_{Ih} - I_{Ih}}{S_{Yh}/\sqrt{n}}, \tag{20}$$

where S_{Yh} is the sample standard deviation, written as:

$$S_{Yh} = \sqrt{\frac{1}{n-1} \times \sum_{j=1}^{n} \left(Y_{h,j} - \overline{Y}_{h}\right)^{2}}. \tag{21}$$

According to the Central Limit Theorem, the distribution of the random variable Z_{Ih} approximates the standard normal distribution, expressed as follows:

$$Z_{Ih} = \frac{\hat{I}_{Ih} - I_{Ih}}{S_{Yh}/\sqrt{n}} \overset{n \to \infty}{\to} N(0,1). \tag{22}$$

Based on the above-mentioned, we have $1 - \alpha = p(-z_{\alpha/2} \le Z_{Ih} \le z_{\alpha/2})$. Similarly, the Important Index I_{Ih} is set as the y-axis to form the PEM. Let $\left(y_{h,1}, \ldots, y_{h,j}, \ldots, y_{h,n}\right)$ be the observed value of $\left(Y_{h,1}, \ldots, Y_{h,j}, \ldots, Y_{h,n}\right)$. Then $\hat{I}_{Ih0}$ is the observed value of $\hat{I}_{Ih}$, expressed as follows:

$$\hat{I}_{Ih0} = \frac{1}{n} \times \sum_{j=1}^{n} y_{h,j}. \tag{23}$$

Then, the average of $\hat{I}_{Ih0}$ is defined as follows:

$$I_{I0} = \frac{1}{k} \sum_{h=1}^{k} \hat{I}_{Ih0}. \tag{24}$$

According to Yu et al. [17], when the Importance Index of Service Item h is smaller than the average value ($I_{Ih} \le I_{I0}$), the improvement priority of Service Item h is low. On the contrary, when the Importance Index of Service Item h is greater than the average value ($I_{Ih} > I_{I0}$), the improvement priority of Service Item h is high. Then, the hypotheses of the statistical test for Important Index h are written as follows:

$$\text{null hypothesis } H_0 : I_{Ih} \le I_{I0}; \tag{25}$$

$$\text{alternative hypothesis } H_1 : I_{Ih} > I_{I0}. \tag{26}$$

The significance level of the test is β', and the critical region is defined as $CR_{Ih} = \left\{\hat{I}_{Ih} > C'_{h0}\right\} = \left\{Z_{Ih} > \sqrt{n}\left(C'_{h0} - I_{Ih}\right)/S_{Yh0}\right\}$, where C'_{h0} is determined by

$$p\left\{Z_{Ih} > \frac{C'_{h0} - I_{I0}}{S_{Yh0}/\sqrt{n}}\right\} = \beta'. \tag{27}$$

Therefore, the critical value is written as $C'_{h0} = I_{I0} + z_{\beta'} S_{Yh} / \sqrt{n}$. Obviously, $p\{\hat{I}_{Ih} > C'_{h0} \, | I_{Ih} = I_{I0}\} = p\{Z_{Ih} > z_{\beta'} | I_{Ih} = I_{I0}\} = \beta'$. Let the observed value of Z_{Ih} be Z_{Ih0}, written as follows:

$$Z_{Ih0} = \frac{\hat{I}_{Ih0} - I_{I0}}{S_{Yh0}/\sqrt{n}}, \tag{28}$$

where S_{Yh0} is the observed value of S_{Yh} as follows:

$$S_{Yh0} = \sqrt{\frac{1}{n-1} \times \sum_{j=1}^{n} \left(Y_{h,j} - \overline{Y}_h\right)^2}. \tag{29}$$

Thus, we can then replace $\hat{I}_{Ih0}$ with Z_{Ih0} as the testing statistic and replace C'_{h0} with $-z_{\beta'}$ as the critical value. The α-*cuts* of the triangular fuzzy number $\tilde{z}_{\beta'}$ is

$$\tilde{z}_{\beta'}[\alpha] = \begin{cases} \left[z_{\beta'1}(\alpha), z_{\beta'2}(\alpha)\right] = \left[z_{\beta'} - z_{\alpha/2}, z_{\beta'} + z_{\alpha/2}\right], \; 0.01 \le \alpha \le 1 \\ \left[z_{\beta'1}(\alpha), z_{\beta'2}(\alpha)\right] = \left[z_{\beta'} - z_{0.005}, z_{\beta'} + z_{0.005}\right], \; 0 \le \alpha \le 0.01 \end{cases}. \tag{30}$$

As noted by Chen and Yu [18], when $\alpha = 1$, then $z_{\beta'1}(1) = z_{\beta'2}(1) = z_{\beta'}$. Therefore, the fuzzy number is $\tilde{z}_{\beta'} = \left(z_{\beta'L}, z_{\beta'M}, z_{\beta'R}\right) = \left(z_{\beta'} - z_{0.005}, z_{\beta'}, z_{\beta'} + z_{0.005}\right)$, and the membership function of $\tilde{z}_{\beta'}$ is

$$\eta'(x) = \begin{cases} 0 & , if \; x \le z_{\beta'} - z_{0.005} \\ 2 \times \left(1 - \Phi(z_{\beta'} - x)\right) & , if \; z_{\beta'} - z_{0.005} < x < z_{\beta'} \\ 1 & , if \; x = z_{\beta'} \\ 2 \times \left(1 - \Phi(x - z_{\beta'})\right) & , if \; z_{\beta'} < x < z_{\beta'} + z_{0.005} \\ 0 & , if \; z_{\beta'} + z_{0.005} \le x \end{cases}. \tag{31}$$

Following Equation (31), the diagram of membership function $\eta'(x)$ with the vertical line *of* $x = Z_{Ih0}$ is presented in Figure 2.

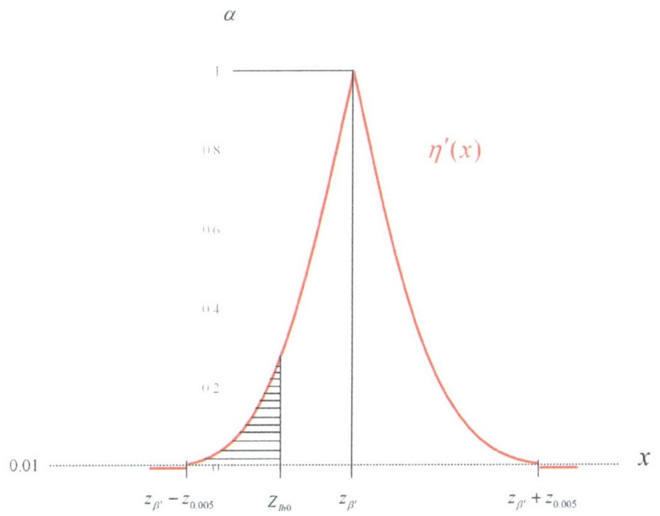

Figure 2. Membership function $\eta'(x)$ with the vertical line of $x = Z_{Ih0}$.

According to Yu et al. [24], let $d'_{hR} = z_{\beta' R} - Z_{Ih0} = (z_{\beta'} + z_{0.005}) - Z_{Ih0}$ and $d'_T = z_{\beta' R} - z_{\beta' L} = 2 \times z_{0.005}$. Then d'_{hR}/d'_T is defined as follows:

$$d'_{hR}/d'_T = \frac{Z_{Ih0} - z_{\beta'} + z_{0.005}}{2 \times z_{0.005}}. \quad (32)$$

Let d_{Ih0}, such that

$$d'_{hR}/d'_T = \frac{d_{Ih0} - z_{\beta'} + z_{0.005}}{2 \times z_{0.005}} = \phi. \quad (33)$$

Thus, we have

$$d_{I0} = (2\phi - 1)z_{0.005} + z_{\beta'}. \quad (34)$$

On the basis of the study of Chen et al. [25], we let $0 < \phi < 0.5$, and the decision-making rules of the fuzzy two-tailed testing model are displayed below:

(1) If $Z_{Ih0} > d_{Ih0}$, then $d'_{hR}/d'_T < \phi$. Therefore, we reject H_0 and draw the conclusion $I_{Ih} > I_{I0}$. Consequently, Service Item h has a high priority for improvement.
(2) If $Z_{Ih0} \leq d_{Ih0}$, then $d'_{hR}/d'_T \geq \phi$. Therefore, we do not reject H_0 and draw the conclusion $I_{Ih} \leq I_{I0}$. Consequently, Service Item h has a low priority for improvement.

4. An Application Example

In order to explain the application of the above model, this paper adopted a foreign language teaching satisfaction questionnaire made by Yu et al. [26], including five dimensions to reflect the services provided by foreign language teaching. Among them, Dimension 1 is Teaching Preparation, containing 4 teaching service items; Dimension 2 is Teaching Attitude, including 5 teaching service items; Dimension 3 is Teaching Capability, containing 2 teaching service items; Dimension 4 is Teaching Management, containing 3 teaching service items; finally, Dimension 5 is Coursework and Evaluation, containing 2 teaching service items. These five dimensions include a total of 16 question items, and each question item has two subquestions about importance and satisfaction. Therefore, the entire questionnaire has a total of 32 questions that need to be answered. A total of 16 questions ($k = 16$) in these 5 dimensions are depicted as follows:

Dimension 1: Teaching Preparation

1. The course material that the teacher has prepared is at an adequate level of complexity (X_1, Y_1).
2. The quantity of the material is appropriate (X_2, Y_2).
3. The content of the material helps improve my foreign language proficiency (X_3, Y_3).
4. The teacher has prepared thoroughly for the class (X_4, Y_4).

Dimension 2: Teaching Attitude

5. The teacher emphasizes conversation practice in a foreign language (X_5, Y_5).
6. The teacher values the opinions of students (X_6, Y_6).
7. Student-teacher interaction in class is intensive (X_7, Y_7).
8. The teacher is happy to help students solve problems (X_8, Y_8).
9. The teacher treats all students fairly (X_9, Y_9).

Dimension 3: Teaching Capability

10. The teacher speaks the foreign language clearly (X_{10}, Y_{10}).
11. The teacher expresses himself/herself logically (X_{11}, Y_{11}).

Dimension 4: Teaching Management

12. The teacher uses a variety of teaching methods (X_{12}, Y_{12}).
13. The teacher interests me in learning the foreign language (X_{13}, Y_{13}).
14. The teacher adequately controls the pace of teaching (X_{14}, Y_{14}).

Dimension 5: Coursework and Evaluation

15. The coursework or evaluation helps me improve my foreign language proficiency (X_{15}, Y_{15}).
16. Evaluation is at an adequate level of complexity (X_{16}, Y_{16}).

As mentioned above, the performance of running foreign language learning curriculums is the foundation for students who intend to increase their foreign language proficiency since foreign language proficiency is often one of the key indicators for recruitment in the corporate world [27–29]. It can not only help students improve their learning efficiency in other professional subjects but also enhance their competitiveness for more advanced studies or employment. Therefore, foreign language learning curriculums are listed in important teaching enhancement plans promoted by various universities [26].

Based on the above-mentioned 16 questions in the foreign language teaching satisfaction questionnaire, a total of 350 copies of questionnaire were given to students in the case-study school and returned on the spot; in total, 324 copies of questionnaire were collected, yielding a response rate of 92.5%.

First, we calculate observed values $\hat{I}_{Sh0}$ and S_{Xh0} for each service item according to Equations (7) and (12). Following Equation (8), we calculate the average of $\hat{I}_{Sh0}$ as follows:

$$I_{S0} = \frac{1}{16}\sum_{h=1}^{16} \hat{I}_{Sh0} = 0.645.$$

Therefore, the hypotheses of the fuzzy test for Satisfaction Index h are written as follows:

$$\text{null hypothesis } H_0 : \; I_{Sh} \geq 0.645;$$

$$\text{alternative hypothesis } H_1 : \; I_{Sh} < 0.645.$$

This study sets the significance level β as 0.05, then the membership function $\eta(x)$ with significance level $\beta = 0.05$ is expressed as follows:

$$\eta(x) = \begin{cases} 0, \; if \; x < -4.221 \\ 2 \times (1 - \Phi(-1.645 - x)), if - 4.221 \leq x < -1.645 \\ 1, \; if \; x = -1.645 \\ 2 \times (1 - \Phi(x + 1.645)), if - 1.645 < x \leq 0.931 \\ 0, \; if \; 0.931 < x \end{cases}.$$

Let $\phi = 0.4$. Then the decision value is $d_{S0} = (1 - 2\phi)z_{0.005} - z_{0.05} = 0.6 \times 2.576 - 1.645 = -0.10$. Following Equation (15), the values of Z_{Sh0} for all service items are calculated and shown in Table 1.

Similar to the Satisfaction Index, we calculate observed values $\hat{I}_{Ih0}$ and S_{Yh0} for each service item according to Equations (23) and (29). Following Equation (24), we calculate the average of $\hat{I}_{Ih0}$ as follows:

$$I_{I0} = \frac{1}{16}\sum_{h=1}^{16} \hat{I}_{Ih0} = 0.682.$$

Therefore, the hypotheses of the fuzzy test for Important Index of Service Item h are written as follows:

$$\text{null hypothesis } H_0' : \; I_{Ih} \leq 0.682;$$

$$\text{alternative hypothesis } H_1' : \; I_{Ih} > 0.682.$$

This study sets the significance level β' as 0.05 and $\phi = 0.4$, then the decision value is $d_{I0} = (2\phi - 1)z_{0.005} + z_{0.05} = -0.6 \times 2.576 + 1.645 = 0.10$. According to Equation (28), the values of Z_{Ih0} for all service items are calculated and displayed in Table 2.

Table 1. The fuzzy improvement decision table for satisfaction of service items.

Dimensions/Item	$\hat{I}_{Sh0}$	Z_{Sh0}	Remark
Dimension 1: Teaching Preparation			
1. The course material that the teacher has prepared is at an adequate level of complexity.	0.72	1.95	
2. The quantity of the material is appropriate.	0.69	0.89	
3. The content of the material helps improve my foreign language proficiency.	0.76	2.24	
4. The teacher has prepared thoroughly for the class.	0.52	−2.79	Improve
Dimension 2: Teaching Attitude			
5. The teacher emphasizes conversation practice in a foreign language.	0.67	0.45	
6. The teacher values the opinions of students.	0.49	−2.76	Improve
7. Student-teacher interaction in class is intensive.	0.68	0.54	
8. The teacher is happy to help students solve problems.	0.51	−3.27	Improve
9. The teacher treats all students fairly.	0.69	0.82	
Dimension 3: Teaching Capability			
10. The teacher speaks the foreign language clearly.	0.71	1.22	
11. The teacher expresses himself/herself logically.	0.73	1.61	
Dimension 4: Teaching Management			
12. The teacher uses a variety of teaching methods.	0.69	0.73	
13. The teacher interests me in learning a foreign language.	0.65	−0.11	Improve
14. The teacher adequately controls the pace of teaching.	0.54	−2.37	Improve
Dimension 5: Coursework and Evaluation			
15. The coursework or evaluation helps me improve my foreign language proficiency.	0.72	1.40	
16. Evaluation is at an adequate level of complexity.	0.69	0.92	

Remark: If $Z_{Sh0} < d_{Sh0} = -0.10$, then Service Item h needs to be improved.

In the fuzzy improvement decision table for satisfaction of service items, the five service items requiring improvement are as follows: "The teacher has prepared thoroughly for the class" (Item 4), "The teacher values the opinions of students" (Item 6), "The teacher is happy to help students solve problems" (Item 8), "The teacher interests me in learning a foreign language" (Item 13), and "The teacher adequately controls the pace of teaching" (Item 14). In the fuzzy decision list of improvement prioritization, improvement priority of Service Items 4, 8, and 13 is low, whereas that of Service Items 6 and 14 is high. Improvement involves various concerns, such as the cost of hiring high-quality teachers, teaching training time and methods, expenses related to formulating various reward and punishment systems, and other associated costs. Consequently, it is recommended that Service Items 6 and 14 should be listed in as the top priority for improvement within the constraints of limited resources and time to enhance the effectiveness of improvement efforts.

As highlighted by a number of studies, the fuzzy test based on the confidence interval tends to be more practical than the statistical test in real-world scenarios [11,12]. The value of Z_{Sh0} for Service Item 13 is −0.11 in the above-mentioned case, which exceeds the critical value $-z_{0.05}$. Had the statistical test been employed, then null hypothesis would not have been rejected, potentially leading us to miss opportunities for improvement. Furthermore, this model takes the prioritization of service item improvements into account, offering a basis for decision making when resources are limited. Lastly, with standardized decision-making values and critical values, compared with other existing methods based on fuzzy tests and confidence intervals, this model is more convenient for decision makers [30,31].

Table 2. Fuzzy decision list of improvement prioritization.

Dimensions/Item	$\hat{I}_{Ih0}$	Z_{Ih0}	Priority
Dimension 1: Teaching Preparation			
1. The course material that the teacher has prepared is at an adequate level of complexity.	0.75	1.68	
2. The quantity of the material is appropriate.	0.68	−0.05	
3. The content of the material helps improve my foreign language proficiency.	0.73	2.03	
4. The teacher has prepared thoroughly for the class.	0.59	−2.15	Low
Dimension 2: Teaching Attitude			
5. The teacher emphasizes conversation practice in a foreign language.	0.58	−3.00	
6. The teacher values the opinions of students.	0.78	1.23	High
7. Student-teacher interaction in class is intensive.	0.72	0.84	
8. The teacher is happy to help students solve problems.	0.56	−2.86	Low
9. The teacher treats all students fairly.	0.57	−2.84	
Dimension 3: Teaching Capability			
10. The teacher speaks the foreign language clearly.	0.82	2.80	
11. The teacher expresses himself/herself logically.	0.71	0.56	
Dimension 4: Teaching Management			
12. The teacher uses a variety of teaching methods.	0.66	−0.52	
13. The teacher interests me in learning a foreign language.	0.61	−1.83	Low
14. The teacher adequately controls the pace of teaching.	0.80	2.73	High
Dimension 5: Coursework and Evaluation			
15. The coursework or evaluation helps me improve my foreign language proficiency.	0.74	1.32	
16. Evaluation is at an adequate level of complexity.	0.61	−1.86	

5. Conclusions, Research Limitations and Future Research

The performance evaluation matrix can evaluate all service items of the service operating system simultaneously. The Importance Index and Satisfaction Index of the service items serve as two significant evaluation indicators of the performance evaluation matrix. To cater to enterprises' needs for rapid decision-making, this paper initially derived the expected value and standard deviation of the Satisfaction Index estimation formula and set the random variable Z_{Sh} equal to the standardized statistic of Satisfaction Index for Service Item h. According to the Central Limit Theorem, Z_{Sh} followed the standard normal distribution, and then the β lower quantile of the standard normal distribution was used as the critical value to establish the fuzzy test of the Satisfaction Index. Given the complexity of the fuzzy test, this paper obtained the fuzzy critical value d_{S0} following the fuzzy testing rules. Managers only need to compare the Z_{Ih} value and fuzzy critical value d_{I0} of Service

Item h to make decisions on whether to make improvements. Subsequently, this paper derived the expected value and standard deviation of the Important Index estimate and set the random variable Z_{Ih} equal to the standardized statistic of the Important Index. Similarly, following the Central Limit Theorem, Z_{Ih} also followed the standard normal distribution, and then the fuzzy test of the Important Index was established using the β' upper quantile of the standard normal distribution as the critical value. Since the fuzzy test was relatively complex, this paper received the fuzzy critical value d_{I0} following the fuzzy testing rules. Managers only need to compare the values of Z_{Ih} and d_{I0} for each service item to determine whether the service item should be prioritized for improvement. In fact, the fuzzy evaluation model proposed in this paper can maintain evaluation accuracy in cases of small samples by incorporating past accumulated data experience. In addition, the fuzzy critical value can be derived by the fuzzy testing rules, and the decision-making rules can be established by the fuzzy critical value, which can facilitate industrial applications.

The fuzzy evaluation model presented in this paper is built on the confidence intervals of indices. Moreover, its importance lies in its ability to integrate past accumulated data and experts' experiences to make the evaluation more authentic in practical settings [18,24]. However, this study has its limitations, including insufficient accumulation of past data and immature analysis techniques for experts' experiences. In future research, we can focus on the management's need for sophisticated techniques of data analysis as well as develop analysis and decision-making models based on the accumulated data. Additionally, the method proposed in this paper must be capable of deriving the two-tailed confidence interval of the indicator to complete the subsequent fuzzy testing process. If the two-tailed confidence interval of the indicator cannot be obtained, it may require a larger sample size and application of the Central Limit Theorem to reduce sampling errors caused by insufficient sample size. Therefore, future research could also explore the incorporation of other decision-making methods, such as the Multi-Criteria Decision-Making Fuzzy Methods proposed by Al-shami and Mhemdi [32].

Author Contributions: Conceptualization, K.-S.C. and C.-P.C.; methodology, K.-S.C., C.-P.C.; software, T.-H.H. (Tsung-Hua Hsieh); validation, K.-C.Y. and T.-H.H. (Tsun-Hung Huang); formal analysis, K.-S.C. and C.-P.C.; resources, T.-H.H. (Tsung-Hua Hsieh); data curation, T.-H.H. (Tsung-Hua Hsieh); writing—original draft preparation, K.-S.C., T.-H.H. (Tsung-Hua Hsieh), C.-P.C., K.-C.Y. and T.-H.H. (Tsun-Hung Huang); writing—review and editing, K.-S.C. and C.-P.C.; visualization, T.-H.H. (Tsung-Hua Hsieh); supervision, K.-S.C.; project administration, C.-P.C.; All authors have read and agreed to the published version of the manuscript.

Funding: This research received no external funding.

Data Availability Statement: The data that support the findings of this study are available from the corresponding author upon reasonable request.

Conflicts of Interest: The authors declare no conflict of interest.

References

1. Li, Y.; Wang, L.; Li, F. A data-driven prediction approach for sports team performance and its application to national basketball association. *Omega* **2021**, *98*, 102123. [CrossRef]
2. Jeng, M.Y.; Yeh, T.M.; Pai, F.Y. A Performance Evaluation Matrix for Measuring the Life Satisfaction of Older Adults Using eHealth Wearables. *Healthcare* **2022**, *10*, 605. [CrossRef] [PubMed]
3. Kim, J.; Kim, E.; Hong, A. Ott streaming distribution strategies for dance performances in the post-COVID-19 age: A modified importance-performance analysis. *Int. J. Environ. Res. Public Health* **2022**, *19*, 327. [CrossRef] [PubMed]
4. Wong, C.P.; Yang, L.; Szeto, W.Y. Comparing passengers' satisfaction with fixed-route and demand-responsive transport services: Empirical evidence from public light bus services in Hong Kong. *Travel Behav. Soc.* **2023**, *32*, 100583. [CrossRef]
5. Askari, S.; Peiravian, F.; Tilahun, N.; Yousefi Baseri, M. Determinants of users' perceived taxi service quality in the context of a developing country. *Transp. Lett.* **2021**, *13*, 125–137. [CrossRef]
6. Mustafa, H.; Omar, B.; Mukhiar, S.N.S. Measuring destination competitiveness: An importance-performance analysis (IPA) of six top island destinations in south east asia. *Asia Pac. J. Tour. Res.* **2020**, *25*, 223–243. [CrossRef]
7. Kumar, S.; Janardhanan, A.K.; Khanna, S.; William, R.M.; Saha, S. Students' Satisfaction with Remote Learning During the COVID-19 Pandemic: Insights for Policymakers. *Prabandhan Indian J. Manag.* **2023**, *16*, 43–60. [CrossRef]

8. Nam, S.; Lee, H.C. A text analytics-based importance performance analysis and its application to airline service. *Sustainability* **2019**, *11*, 6153. [CrossRef]
9. Sumrit, D.; Sowijit, K. Winning customer satisfaction toward omnichannel logistics service quality based on an integrated importance-performance analysis and three-factor theory: Insight from Thailand. *Asia Pac. Manag. Rev.* **2023**, *28*, 531–543. [CrossRef]
10. Wu, J.; Yang, T. Service attributes for sustainable rural tourism from online comments: Tourist satisfaction perspective. *J. Destin. Mark. Manag.* **2023**, *30*, 100822. [CrossRef]
11. Chen, H.Y.; Lin, K.P. Fuzzy supplier selection model based on lifetime performance index. *Expert Syst. Appl.* **2022**, *208*, 118135. [CrossRef]
12. Lo, W.; Yang, C.M.; Lai, K.K.; Li, S.Y.; Chen, C.H. Developing a novel fuzzy evaluation model by one-sided specification capability indices. *Mathematics* **2021**, *9*, 1076. [CrossRef]
13. Lin, C.J.Y. General education competencies from students' perspectives: A case study of a sports university in Taiwan. *Hum. Soc. Sci. Commun.* **2023**, *10*, 848. [CrossRef]
14. Ramírez-Hurtado, J.M.; Hernández-Díaz, A.G.; López-Sánchez, A.D.; Pérez-León, V.E. Measuring online teaching service quality in higher education in the COVID-19 environment. *Int. J. Environ. Res. Public Health* **2021**, *18*, 2403. [CrossRef] [PubMed]
15. Liu, S.H.; Tsai, C.H. Service quality of ski group package tours: A modified importance-performance analysis. *Int. J. Serv. Oper. Manag.* **2023**, *46*, 35–59. [CrossRef]
16. Oey, E.; Librianne, G.; Elvira, E.; Irawan, D.A. Comparing two ways of integrating fuzzy Kano and importance performance analysis—With a case study in a beauty clinic. *Int. J. Product. Qual. Manag.* **2023**, *38*, 285–311. [CrossRef]
17. Yu, C.M.; Li, S.Y.; Yu, C.H.; Chen, Y.P. The Performance Evaluation Model of Bus APP. *Int. J. Reliab. Qual. Saf. Eng.* **2023**, *30*, 2350023. [CrossRef]
18. Chen, K.S.; Yu, C.M. Fuzzy test model for performance evaluation matrix of service operating systems. *Comput. Ind. Eng.* **2020**, *140*, 106240. [CrossRef]
19. Lin, T.C.; Chen, H.H.; Chen, K.S.; Chen, Y.P.; Chang, S.H. Decision-Making Model of Performance Evaluation Matrix Based on Upper Confidence Limits. *Mathematics* **2023**, *11*, 3499. [CrossRef]
20. Yu, C.M.; Zhuo, Y.J.; Lee, T.S. A Novel Social Media App Performance Evaluation Model. *J. Econ. Manag.* **2020**, *16*, 69–81.
21. Pishro-Nik, H. *Introduction to Probability, Statistics, and Random Processes*; Kappa Research LLC: Sunderland, MA, USA, 2014; Available online: https://www.probabilitycourse.com (accessed on 1 March 2024).
22. Buckley, J.J. Fuzzy statistics: Hypothesis testing. *Soft Comput.* **2005**, *9*, 512–518. [CrossRef]
23. Dereli, T.; Baykasoglu, A.; Altun, K.; Durmusoglu, A.; Türksen, I.B. Industrial applications of type-2 fuzzy sets and systems: A concise review. *Comput. Ind.* **2011**, *62*, 125–137. [CrossRef]
24. Yu, C.M.; Lai, K.K.; Chen, K.S.; Chang, T.C. Process-quality evaluation for wire bonding with multiple gold wires. *IEEE Access* **2020**, *8*, 106075–106082. [CrossRef]
25. Chen, K.S.; Wang, C.H.; Tan, K.H. Developing a fuzzy green supplier selection model using Six Sigma quality indices. *Int. J. Prod. Econ.* **2019**, *212*, 1–7. [CrossRef]
26. Yu, C.M.; Chang, H.T.; Hsu, S.Y. An assessment of quality and quantity for foreign language training course to enhance students' learning effectiveness. *Int. J. Inf. Manag. Sci.* **2017**, *28*, 53–66.
27. Li, R. Research trends of blended language learning: A bibliometric synthesis of SSCI-indexed journal articles during 2000–2019. *ReCALL* **2022**, *34*, 309–326. [CrossRef]
28. Li, X.; Huang, X. Improvement and Optimization Method of College English Teaching Level Based on Convolutional Neural Network Model in an Embedded Systems Context. *Comput. Aided Des. Appl.* **2024**, *21*, 212–227. [CrossRef]
29. Graham, K.M.; Yeh, Y.F. Teachers' implementation of bilingual education in Taiwan: Challenges and arrangements. *Asia Pacific Educ. Rev.* **2023**, *24*, 461–472. [CrossRef]
30. Chiou, K.C. Building up of Fuzzy Evaluation Model of Life Performance Based on Type-II Censored Data. *Mathematics* **2023**, *11*, 3686. [CrossRef]
31. Huang, C.C.; Chang, T.C.; Chen, B.L. Fuzzy assessment model to judge quality level of machining processes involving bilateral tolerance using crisp data. *J. Chin. Inst. Eng.* **2021**, *44*, 1–10. [CrossRef]
32. Al-shami, T.M.; Mhemdi, A. Generalized Frame for Orthopair Fuzzy Sets: (m,n)-Fuzzy Sets and Their Applications to Multi-Criteria Decision-Making Methods. *Information* **2023**, *14*, 56. [CrossRef]

axioms

MDPI

Article

An Emotionally Intuitive Fuzzy TODIM Methodology for Decision Making Based on Online Reviews: Insights from Movie Rankings

Qi Wang [1], Xuzhu Zheng [1] and Si Fu [2,*]

[1] School of Industrial Design, Hubei University of Technology, Wuhan 430068, China; wq20201103@hbut.edu.cn (Q.W.); 202111440@hbut.edu.cn (X.Z.)
[2] China-Korea Institute of New Media, Zhongnan University of Economics and Law, Wuhan 430073, China
* Correspondence: z0004435@zuel.edu.cn; Tel.: +86-159-0272-0781

Abstract: With the burgeoning growth of the internet, online evaluation systems have become increasingly pivotal in shaping consumer decision making. In this context, this study introduces an intuitionistic fuzzy TODIM (an acronym in Portuguese for interactive and multicriteria decision making) methodology to rank products based on online reviews. Our approach aims to enhance user decision making efficiency and address the prevalent issue of information overload. Initially, we devised a product attribute emotion quantification framework within the confines of the intuitionistic fuzzy paradigm. This allows for the transformation of online reviews into exact functional outputs via our advanced intuitionistic fuzzy scoring mechanism and its associated precise function. Following this, we take into account the inherent correlation among product attributes, leading to the development of an attribute-associated intuitionistic fuzzy model. This model further ascertains the dominance degree of alternative products. Moreover, by integrating the risk aversion factor, we can derive a hierarchical structure for alternative products, aiding in the prioritization process. Finally, this paper validates the proposed method using movie sequencing as a case study. The results show that the proposed method, which takes into account the emotional tendencies of different attributes in a movie and the different preferences of viewers in the attribute weighting and movie selection process, is more reasonable than methods proposed in previous studies.

Keywords: online reviews; movie sorting; multi-attribute decision making; sentiment analysis; intuitionistic fuzzy sets

MSC: 62C86; 91B06

Citation: Wang, Q.; Zheng, X.; Fu, S. An Emotionally Intuitive Fuzzy TODIM Methodology for Decision Making Based on Online Reviews: Insights from Movie Rankings. *Axioms* **2023**, *12*, 972. https://doi.org/10.3390/axioms12100972

Academic Editors: Amit K. Shukla, Ta-Chung Chu and Wei-Chang Yeh

Received: 28 August 2023
Revised: 3 October 2023
Accepted: 11 October 2023
Published: 16 October 2023

1. Introduction

The unprecedented growth of the Internet and social media platforms has led to the emergence of specialized websites, such as those dedicated to books and music, as well as e-commerce sites. As a result, user reviews have burgeoned as a primary conduit for information dissemination. A significant portion of the public now gravitates towards online platforms to voice their opinions, with experiential products like movies particularly benefiting from this trend [1]. Websites like Douban and Rotten Tomatoes have evolved into primary sources for audiences for movie details and reviews. Concurrently, there has been a noticeable uptick in the quality, expertise, and social interactivity of these reviews. The emotional slant of these reviews wields considerable influence over potential consumers' decisions [2]. Factors such as online reviews, ratings, and extensive feedback play a pivotal role in influencing consumer decisions.

However, the deluge of reviews brings its own challenges, notably the phenomenon of information overload. Current recommendation systems falter in tailoring suggestions to individual genre preferences, highlighting the pressing need to deftly extract and harness the emotional nuances from multitudes of online reviews [3]. This extraction process aims to guide consumers towards more informed decisions.

While some review platforms showcase aggregate product ratings to circumvent the necessity of sifting through individual reviews, the text within reviews remains invaluable for potential consumers [4]. Therefore, efficiently pinpointing crucial information within these reviews is essential for refining consumer decision making. Notably, given that movies are experiential products, there is a surprising dearth of research on movie sorting. This paper seeks to bridge this gap and validate the proposed method by using movies as an example.

Historically, ranking methodologies grounded in online reviews have predominantly zeroed in on positive and negative sentiments, overlooking the nuances of neutral emotions [5–10]. Such oversights can lead to the omission of crucial information. Recent academic endeavors have harnessed multi-attribute decision making (MADM) methodologies for online product categorization. For instance, Fan et al. [11] derived a comparative superiority degree for various alternatives using the distribution percentages of specific features across distinct commodities, employing the PROMETHEE II (Preference Ranking Organization Method for Enrichment of Evaluations) method for comprehensive evaluation. In another study, Fan et al. [12] leveraged user ratings from online reviews to formulate two utility functions, with the subsequent application of the TOPSIS (Technique for Order Preference by Similarity to an Ideal Solution) method for holistic evaluation and ranking. Other notable works include Lee et al.'s [13] utilization of hierarchical deep neural networks (DNNs) for product ranking, and Wang et al.'s [14] user-centric commodity recommendation model. As a result of our research, we found that many extant studies exhibit a propensity to view products monolithically, often sidelining detailed attributes and features. Some product ranking methods based on online reviews only take into account the positive and negative affective tendencies of online reviews, ignoring the fact that the affective tendencies in the reviews can be neutral, i.e., ambiguous information, which can lead to a loss of information in the decision making process. Established models like TOPSIS and PROMETHEE II, premised on the notion of the decision maker's absolute rationality, disregard the psychological intricacies underpinning the decision making process [11,12]. The TODIM method is appropriate for illustrating the psychological behavior of consumers during the product prioritization process [15,16]. Its central concept involves determining gain and loss values by comparing the characteristic values of each alternative product, and then calculating the dominance degree between every pair of alternatives and the overall prospect value of each product [17,18]. The alternative products are ranked based on their overall prospect value.

To address these lacunae, this paper embarks on an exploration of product ranking, leveraging online review data within an intuitionistic fuzzy framework. We introduce an intuitionistic fuzzy TODIM methodology, predicated on the multifaceted aspects of product reviews, particularly emphasizing the quantification of emotional tendencies across varying product attributes. This method encompasses the following:

(1) The creation of a quantitative model for product attribute sentiment within an intuitionistic fuzzy paradigm. Recognizing the diverse preferences among consumers, we harness the Cemotion library, a sentiment analysis tool rooted in Bert, to discern nuanced emotional cues from reviews. Subsequently, we introduce the emotional intuition fuzzy value (E-IFV), which intuitively demonstrates the level of support for an attribute by integrating multidimensional eigenvalues of a product attribute with its emotional tendency.
(2) An enhancement of the current intuitionistic fuzzy score function is presented. We analyze existing research to discern gaps in the current model. An augmented intuitionistic fuzzy score function, coupled with an exact function, is proposed, aiming to streamline decision making. This enhancement amalgamates concepts from hesitation allocation and voting models.

(3) The integration of the TODIM approach to classify alternative products. The correlation among attributes is acknowledged, with the DEMATEL (Decision Making Trial and Evaluation Laboratory) method determining attribute weights for distinct product genres. These weights, in tandem with the loss aversion coefficients within the TODIM model, ascertain the relative prominence of alternative products. The final result is a personalized ranking, derived from consumer preferences, and a loss aversion risk factor.

By synthesizing information from online comments, likes, and comment volumes into intuitive fuzzy values, and aligning products with consumer attribute preferences, we aim to significantly augment the consumers' decision making efficiency. In essence, our refined intuitionistic fuzzy TODIM product ranking method, rooted in multidimensional product review features, integrates diverse data sources, providing consumers with a tailored decision making guide.

2. Materials and Methods

This article introduces a novel methodology for product ranking by integrating insights from online reviews with inherent product attributes. In the intuitionistic fuzzy environment, the emotional tendency of product attributes is transformed into an emotional intuitionistic fuzzy value (E-IFV); in order to effectively compare the magnitude of intuitionistic fuzzy numbers, an intuitionistic fuzzy score function and an intuitionistic fuzzy exact function are proposed. Finally, a product ranking model based on online reviews is established by integrating the TODIM method. As this paper is based on the study of product ranking in an intuitionistic fuzzy environment, the intuitionistic fuzzy score function and the exact function are proposed, and the advantages of the intuitionistic fuzzy value (IFV) and intuitionistic fuzzy set (IFS) in terms of representing the affective tendencies of product features are considered. This section presents the literature related to intuitionistic fuzzy sets.

Zadeh [19], in 1965, introduced the fuzzy set (FS) theory. However, as fuzzy multi-attribute decision making paradigms evolved, it became evident that fuzzy sets were insufficient in capturing decision makers' uncertainty comprehensively [20]. Addressing this, Atanassov [21] unveiled the intuitionistic fuzzy sets theory. This enhanced approach emphasized both membership and non-membership degrees, providing a richer representation of a decision maker's hesitations. Later on, Liu et al. [22,23] built upon this theory by incorporating sentiment analysis, allowing online product reviews to be represented by intuitionistic fuzzy numbers. Furthermore, Roszkowska et al. [24–26] introduced a composite measure. This measure was tailored to the evaluation of complex social phenomena using questionnaires, refining the fuzzy set based on "objective" data. Specifically, research [24] advocated for the use of interval intuitionistic fuzzy sets (I-VIFS) to articulate the data from questionnaires. They crafted the I-VIFS composite measure and utilized the outcomes to stipulate the optimistic coefficients, thereby defining the bounds of the interval for the I-VIFS parameters. Meanwhile, other reserach [25] employed the intuitionistic fuzzy synthesis measure (IFSM) based on pattern object distance. They suggested translating ordered data using intuitionistic fuzzy sets and juxtaposed the results with traditional methods. Çalı and Balaman [27] used IFS to represent online ratings of hotel customers and used IF-ELECTRE to rank alternative hotels integrated with VIKOR. In this paper, we draw inspiration from the literature [24] and aim to augment the precision of product attributes tied to intuitionistic fuzzy values by formulating multidimensional eigenvalues for these attributes.

Upon this proposition, our subsequent focus revolved around refining the intuitionistic fuzzy multi-attribute decision theory to bolster decision making efficacy. Of paramount importance here is the intuitionistic fuzzy score function, pivotal for comparing and ranking intuitionistic fuzzy numbers. Despite significant contributions from scholars like Chen and HONG [28,29], certain aspects of these functions, such as the hesitancy degree's impact on scheme ranking, remain under-explored.

While significant strides have been made in areas like text sentiment analysis and multi-attribute decision making using online reviews, specific challenges persist:

(1) The existing research often overlooks neutral and ambiguous sentiments in reviews.
(2) Most research presumes attribute independence, neglecting potential inter-attribute correlations.
(3) Models like TOPSIS tend to assume complete decision maker rationality, sidelining psychological influences.
(4) Research surrounding intuitive fuzzy score functions requires bolstering to achieve improved accuracy in intuitive fuzzy number discrimination.

Addressing these challenges, this paper presents a product-ranking model anchored in consumer preferences. By amalgamating sentiment analysis with intuitionistic fuzzy sets and integrating the TODIM multi-attribute decision making method, we aim to elevate decision making efficiency for consumers, as exemplified using data from the Douban (https://movie.douban.com (accessed on 13 May 2023)) platform.

3. Problem Description

Prior to finalizing a product selection, consumers frequently consult various online indicators including reviews, ratings, aggregate review counts, and other pertinent metrics. A substantial body of research corroborates the utility of these scoring data in facilitating informed decisions [30–33].

The primary challenge addressed in this study is to formulate a product ranking system that serves as a robust decision making tool for consumers. This ranking integrates data from online reviews and the associated engagement metrics—such as "likes" and comment counts—and takes into consideration consumers' preferences related to specific product attributes. The overarching goal is to improve the decision making efficiency for potential consumers. The subsequent sections detail the representations and precise definitions of the sets and variables relevant to this problem.

Let $A = \{A_1, A_2, \cdots, A_m\}$ be the set of products that the consumer is interested in choosing, where A_i denotes the ith product, $i = 1, 2, \cdots, m$.

Let $F = \{f_1, f_2, \cdots, f_n\}$ be the set of n attributes of the alternative product, where f_j denotes the jth attribute, $j = 1, 2, \cdots, n$.

Let $w = [w_1, w_2, \cdots, w_n]$ be a vector of product attribute weights, where w_j is the weight corresponding to attribute f_j. $w_j \geq 0$, $\sum\limits_{j=1}^{n} w_j = 1$. The weights represent the differences in consumer preferences for product attributes.

Let $Q = \{q_1, q_2, \cdots, q_n\}$ be the number of online reviews for the alternative product, where q_i is the number of online reviews for product A_i, $i = 1, 2, \cdots, n$.

The following collections will be described below:

Suppose z^j_{p,q_i} denotes the number of "likes" for the qth comment under the jth attribute of product A_i. If the sentiment tendency of the qth comment is positive, then $z^j_{p,q_i} \in z^{f_j}_{pos}$, otherwise $z^j_{p,q_i} = 0$;

Suppose z^j_{n,q_i} denotes the number of "likes" for the qth comment under the jth attribute of product A_i. If the sentiment tendency of the qth comment is negative, then $z^j_{n,q_i} \in z^{f_j}_{neg}$, otherwise $z^j_{n,q_i} = 0$;

Suppose t^j_{p,q_i} denotes the number of words of the qth comment under the jth attribute of product A_i. If the sentiment tendency of the qth comment is positive, then $t^j_{p,q_i} \in t^{f_j}_{pos}$, otherwise $t^j_{p,q_i} = 0$;

Suppose t^j_{n,q_i} denotes the number of words of the qth comment under the jth attribute of product A_i. If the sentiment tendency of the qth comment is negative, then $t^j_{n,q_i} \in t^{f_j}_{neg}$, otherwise $t^j_{n,q_i} = 0$;

Let $z_{pos}^{f_j} = \left\{z_{p,1}^j, z_{p,2}^j, \cdots, z_{p,q_i}^j\right\}$ be the set consisting of the number of "likes" by consumers for the comments of product A_i on attribute f_j with a positive emotional tendency, where $z_{p,k}^j$ denotes the data of likes for the kth comment on product A_i on attribute f_j, $k = 1,2,\cdots,q_i, i = 1,2,\cdots,m$.

Let $z_{neg}^{f_j} = \left\{z_{n,1}^j, z_{n,2}^j, \cdots, z_{n,q_i}^j\right\}$ be the set consisting of the number of "likes" by consumers for the comments of product A_i on attribute f_j with a negative emotional tendency, where $z_{n,k}^j$ denotes the data of likes for the kth comment on product A_i on attribute f_j, $k = 1,2,\cdots,q_i, i = 1,2,\cdots,m$.

Let $t_{pos}^{f_j} = \left\{t_{p,1}^j, t_{p,2}^j, \cdots, t_{p,q_i}^j\right\}$ be the set consisting of the word counts of the consumer comments on the product A_i on attribute f_j with a positive affective tendency, where $t_{p,k}^j$ denotes the word count of the kth comment on the product A_i on attribute f_j,$k = 1,2,\cdots,q_i, i = 1,2,\cdots,m$.

Let $t_{neg}^{f_j} = \left\{t_{n,1}^j, t_{n,2}^j, \cdots, t_{n,q_i}^j\right\}$ be the set consisting of the word counts of the consumer's comments on the product A_i on attribute f_j with a negative affective tendency, where $t_{n,k}^j$ denotes the word count of the kth comment on the product A_i on attribute f_j, $k = 1,2,\cdots,q_i, i = 1,2,\cdots,m$.

The process of problem solving in this research is divided into two primary segments:

- Firstly, this investigation introduces a model tailored to the quantification of emotions within the realm of intuitionistic fuzzy contexts. The preliminary step involves the transformation of online product reviews into intuitionistic fuzzy numbers. Subsequently, using these comments, we propose an improved intuitionistic method for fuzzy and exact functions. This leads to the derivation of intuitionistic fuzzy exact functions anchored to specific product attributes.
- Secondly, the ranking of alternative products is executed via the enhanced intuitionistic fuzzy TODIM methodology, emphasizing attribute associations. Acknowledging the interdependencies among attributes, the DEMATEL approach is deployed to ascertain attribute weights. The dominance hierarchy amongst products is determined utilizing the intuitionistic fuzzy TODIM methodology. This hierarchy, when integrated with the risk tolerance parameters of the consumer, culminates in a bespoke product ranking schema.

The flowchart for solving the problem is shown in Figure 1.

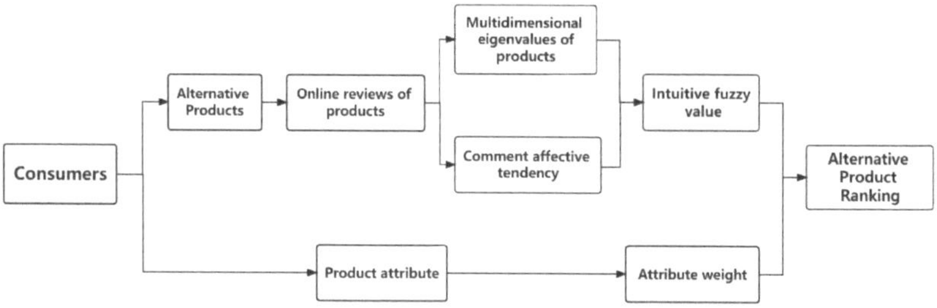

Figure 1. The process of problem solving.

4. A Quantitative Model of Product Attribute Sentiment in an Intuitively Ambiguous Environment

In order to solve the above problems, this section proposes a quantitative model of emotion based on product attributes in an intuitionistic fuzzy environment. The methodology comprises three parts:

- Identification of emotional tendencies in online product reviews;
- Multidimensional eigenvalue computation based on product attributes;
- Calculation of emotional intuition fuzzy values based on product attributes.

4.1. Identification of Emotional Tendencies in Online Product Reviews

This study uses Cemotion, a Chinese sentiment tendency analysis library based on Bert, to identify positive, neutral, and negative sentiment tendencies regarding alternative product attributes in online product reviews. Cemotion's model is trained by a recurrent neural network, which returns a confidence level between 0 and 1 for the sentiment tendency of Chinese text and can accurately identify the sentiment tendency of online product reviews by linking them to the context of the online reviews.

Using the movie *Titanic* as an example, Table 1 presents some of the identified movie review data. In Figure 2, (a) illustrates the distribution of various attribute comments among online reviews of the movie *Titanic*, while (b) shows the proportion of the emotional tendency of the "Frame" attribute in those reviews.

Table 1. Partial movie review data obtained from emotional tendency recognition.

Review Text	Emotion Score	Analysis Result	Number of Favorable Reviews
The old couple who had no fear of death, the band who didn't let the outside world interfere, the man who pretended to be a father for a living, the woman who whistled for her lover. All for a kind of spiritual attachment.	0.9961	positive	15,204
It will always be the film I have seen the most, the most moving and the best in the cinema	0.9942	positive	12,114
In the film organized by the school, people seemed to be curious about the love process between rose and jack. However, the scene that touched me was that the ship was about to sink, and the gentlemen of the sea band arranged their bow ties and played the last song solemnly. At that moment, it seemed to hear a soulless song ringing.	0.9996	positive	20,726

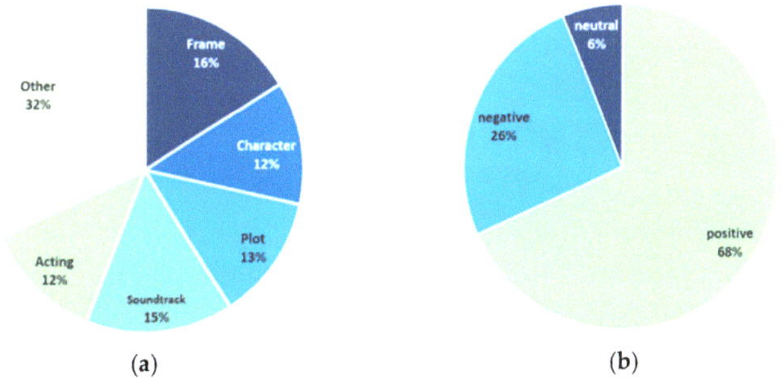

Figure 2. Relevant analysis of online reviews of the movie *Titanic*. (**a**) Distribution of attributes in reviews of the movie *Titanic*. (**b**) Percentage of emotional tendencies of "Frame" attributes of reviews of the movie *Titanic*.

4.2. Multidimensional Eigenvalue Computation Based on Product Attributes

Consumers usually refer to existing online reviews, ratings, the number of likes, the number of comments, and other information to make a comparative choice of products before making a decision [34,35]. Therefore, in this paper, multidimensional eigenvalues of products are introduced to improve the reasonableness of the ranking [36,37]. Since the credibility of a comment can be determined by the number of likes it receives [31], the length of the text [36,38,39], and the emotion it conveys, this study used the number of likes and the number of words in the text as indicators to evaluate the satisfaction of consumers with the product characteristics. The indicators will be described in more detail in the following sections.

Suppose there are m alternative products A_i and n decision attributes $F = \{f_1, f_2, \cdots, f_n\}$.

Definition 1. *"Positive Likes" is the ratio of the number of likes in the comments with positive emotional tendencies to the total number of likes in the comments of the corresponding attribute f_j of the alternative product A_i.*

- Positive liking rate:

$$Z_{pos}^{f_j} = \frac{1}{z_{f_j}} \sum_{k=1}^{q_i} z_{p,k}^{j}, \tag{1}$$

where $Z_{pos}^{f_j}$ is the positive liking rate of the corresponding attribute f_j of product A_i, and z_{f_j} is the total number of likes of the alternative product A_i on attribute f_j.

Definition 2. *"Negative Likes" is the ratio of the number of likes in the comments with negative emotional tendencies to the total number of likes in the comments of the corresponding attribute f_j of the alternative product A_i.*

- Negative liking rate:

$$Z_{neg}^{f_j} = \frac{1}{z_{f_j}} \sum_{k=1}^{q_i} z_{n,k}^{j}, \tag{2}$$

where $Z_{neg}^{f_j}$ is the negative liking rate of the corresponding attribute f_j of product A_i, and z_{f_j} is the total number of likes of the alternative product A_i on attribute f_j.

Definition 3. *"The Positive Text Rate" is the ratio of the number of words of text in the comments with positive emotional tendencies to the total number of words of text in the comments corresponding to attribute f_j of alternative product A_i.*

- Positive text rate:

$$T_{pos}^{f_j} = \frac{1}{t_{f_j}} \sum_{k=1}^{q_i} t_{p,k}^{j}, \tag{3}$$

where $T_{pos}^{f_j}$ is the positive text rate of the corresponding attribute f_j of product A_i, and t_{f_j} is the total number of words for the alternative product A_i on attribute f_j.

Definition 4. *"The Negative Text Rate" is the ratio of the number of words of text in the comments with negative emotional tendencies to the total number of words of text in the comments corresponding to attribute f_j of alternative product A_i.*

- Negative text rate:

$$T_{neg}^{f_j} = \frac{1}{t_{f_j}} \sum_{k=1}^{q_i} t_{n,k}^{j}, \tag{4}$$

where $T_{neg}^{f_j}$ is the negative text rate of the corresponding attribute f_j of product A_i, and t_{f_j} is the total number of words for the alternative product A_i on attribute f_j.

From Equations (1)–(4), we can obtain the text rate and like rate of the product A_i corresponding to the attribute f_j. The multidimensional eigenvalue for the product attribute f_j is obtained through further calculation:

Definition 5. *"The Positive Multidimensional Eigenvalue" $D_{pos}^{f_j}$ is the average of "The Positive Liking Rate" $Z_{pos}^{f_j}$ and "The Positive Text Rate" $T_{pos}^{f_j}$ of the attribute f_j corresponding to product A_i.*

- Positive multidimensional eigenvalues:

$$D_{pos}^{f_j} = \frac{Z_{pos}^{f_j}+T_{pos}^{f_j}}{2}, \tag{5}$$

Definition 6. *"The Negative Multidimensional Eigenvalue" $D_{neg}^{f_j}$ is the average of "The Negative Liking Rate" $Z_{neg}^{f_j}$ and "The Negative Text Rate" $T_{neg}^{f_j}$ of the attribute f_j corresponding to product A_i.*

- Negative multidimensional eigenvalues:

$$D_{neg}^{f_j} = \frac{Z_{neg}^{f_j}+T_{neg}^{f_j}}{2}, \tag{6}$$

4.3. Calculation of Sentiment Means for Product Attributes

The identification of affective tendencies for the products' online product reviews through Section 4.1 yields positive, neutral, and negative affective tendencies regarding the attributes of the alternative products. The identified affective tendencies are further explained below.

Let $\alpha_{\text{ik}}^{\text{j}}$, $\beta_{\text{ik}}^{\text{j}}$, and $\nu_{\text{ik}}^{\text{j}}$ be the positive, negative, and neutral sentiment strengths of the kth comment by the consumer on attribute f_j of the alternative product A_i, respectively. Since this study uses Cemotion, a Bert-based Chinese affective tendency analysis library, for identification, Cemotion returns an affective tendency confidence level between 0 and 1 for the Chinese text, so $\alpha_{\text{ik}}^{\text{j}}, \beta_{\text{ik}}^{\text{j}}, \nu_{\text{ik}}^{\text{j}} \in [0,1]$, and $\alpha_{\text{ik}}^{\text{j}} + \beta_{\text{ik}}^{\text{j}} + \nu_{\text{ik}}^{\text{j}} \in [0,1]$, $k = 1,2,\cdots,q_i$, $i = 1,2,\cdots,m$.

In the following, the mean values of positive, negative, and neutral emotions corresponding to attribute f_j of alternative product A_i will be calculated as follows:

$$p_{ij}^{pos} = \frac{1}{q_{f_j}}\sum_{k=1}^{q_i} \alpha_{ik}^{j}, i = 1,2,\cdots,n, j = 1,2\cdots,m, \tag{7}$$

$$p_{ij}^{neg} = \frac{1}{q_{f_j}}\sum_{k=1}^{q_i} \beta_{ik}^{j}, i = 1,2,\cdots,n, j = 1,2\cdots,m, \tag{8}$$

$$p_{ij}^{neu} = \frac{1}{q_{f_j}}\sum_{k=1}^{q_i} \nu_{ik}^{j}, i = 1,2,\cdots,n, j = 1,2\cdots,m. \tag{9}$$

where q_{f_i} is the total number of comments on attribute f_j for alternative product A_i;
p_{ij}^{pos} is the mean value of positive sentiment about attribute f_j for alternative product A_i;
p_{ij}^{neg} is the mean value of negative sentiment about attribute f_j for alternative product A_i;
and p_{ij}^{neu} is the mean value of neutral sentiment about attribute f_j for alternative product A_i.

4.4. Calculation of Intuitional Fuzzy Values

In this section, leveraging the unique characteristics of product reviews and integrating the foundational principles of IFV, we propose the emotional intuitionistic fuzzy value (E-IFV) model for product attributes. The E-IFV serves as an intuitive representation of the emotional tendencies associated with attributes in product reviews.

The concept of the emotional intuitionistic fuzzy value introduced in this paper is rooted in the theory of intuitionistic fuzzy value, which will be elaborated upon in subsequent sections.

Definition 7. *Suppose X is an argument. If there are two mappings $\mu_A : X \to [0,1]$ and $\nu_A : X \to [0,1]$ above X so that*

$$x \in X| \to \mu_A(x) \in [0,1], \tag{10}$$

and

$$x \in X| \to \nu_A(x) \in [0,1], \tag{11}$$

simultaneously satisfy condition

$$0 \le \mu_A(x) + \nu_A(x) \le 1, \tag{12}$$

μ_A and ν_A are said to determine an intuitionistic fuzzy set on X, which can be denoted as

$$A = \{\langle x, \mu_A(x), \nu_A(x)\rangle | x \in X\}, \tag{13}$$

where $\mu_A(x)$ and $\nu_A(x)$ are referred to as the degree of affiliation and non-affiliation of x. The degree of hesitancy is defined as the following equation:

$$\pi_A = 1 - \mu_A(x) - \nu_A(x). \tag{14}$$

For any $x \in X$, there is $0 \le \pi_A \le 1$.

To facilitate the understanding and application of intuitionistic fuzzy sets, Xu [40] defines $\alpha = (\mu, \nu)$ as an intuitionistic fuzzy number, where $\mu \ge 0$, $\nu \ge 0$, and $\mu + \nu \le 1$, and the degree of hesitation of intuitionistic fuzzy number α is $\pi_\alpha = 1 - \mu - \nu$.

From the aforementioned definition, it is evident that the intuitionistic fuzzy set (IFS) can encapsulate affirmative, negative, and hesitant attitudes simultaneously, minimizing the loss of emotional information. This makes it a robust tool for representing ambiguous and uncertain data. Given the intricate and subjective nature of emotions in online product reviews, factors such as online reviews, ratings, and particularly the emotional sentiments within these reviews, significantly influence consumers' inclination to select alternative products. To enhance the applicability of the intuitionistic fuzzy value, this paper introduces the emotional intuitionistic fuzzy value (E-IFV). This is achieved by integrating the emotional mean with the multidimensional eigenvalue of product attributes, building upon the foundational intuitionistic fuzzy value. The subsequent sections detail the calculation methodology for E-IFV.

$$\mu_{ij} = p_{pos}^{ij} \cdot D_{pos}^{ij}, \tag{15}$$

$$\nu_{ij} = p_{neg}^{ij} \cdot D_{neg}^{ij}, \tag{16}$$

$$\pi_{ij} = 1 - \mu_{ij} - \nu_{ij}, \tag{17}$$

$x_{ij} = (\mu_{ij}, \nu_{ij})$ denotes the E-IFV of alternative product A_i on the attribute word f_j.

Where μ_{ij}, ν_{ij}, and π_{ij} are the degree of affiliation, non-affiliation, and hesitation of the alternative product A_i on attribute word f_j, i.e., the consumers' support, opposition, and neutrality to attribute f_j to alternative product A_i.

5. A Study on the Improved Intuitionistic Fuzzy TODIM Model Based on Attribute Association for Product Ranking

This paper proposes s an improved intuitionistic fuzzy TODIM model for the correlation of attributes to rank alternative products, taking into account that consumer attributes vary person-to-person and that correlations exist between them. The model comprises two primary segments:

- An improved intuitionistic fuzzy score function is proposed, based on which an exact function is determined to improve the accuracy of comparing intuitionistic fuzzy numbers.

- Since the attributes are correlated, instead of being independent of each other, the weights for different product types' attributes are determined using the DEMATEL method. This study employs the TODIM method to determine the superiority of alternative products. The loss aversion coefficient θ, which reflects the consumer's risk appetite, is combined with the ranking to provide personalized decision making suggestions for the consumer.

5.1. The Available Intuitionistic Fuzzy Score Function

Establishing relative dominance (superiority and inferiority relationships) between intuitionistic fuzzy numbers is pivotal to intuitionistic fuzzy multi-attribute decision making. Existing research offers multiple methodologies for computing the score function of intuitionistic fuzzy values [17,18,28,29,41–46], such as

$$S(\alpha) = \frac{\mu_\alpha}{2} + \frac{3v_\alpha}{2} - 1, \tag{18}$$

$$S_1(\alpha) = \mu_\alpha - v_a - \frac{1 - \mu_\alpha - v_a}{2} = \frac{3\mu_\alpha - v_a - 1}{2}, \tag{19}$$

$$S_2(\alpha) = \mu_\alpha(1 + \pi_\alpha) - {\pi_\alpha}^2, \tag{20}$$

etc, where Equations (18)–(20) belong to the literature [44–46] respectively.

These studies provide new ideas for the improvement of intuitionistic fuzzy functions, but at the same time, there are shortcomings.

Counterexample 1. *Taking Equation (18) [44] as an example, suppose $\alpha = (\mu_\alpha, v_\alpha)$ is an intuitionistic fuzzy number, then let*

$$S(\alpha) = \frac{\mu_\alpha}{2} + \frac{3v_\alpha}{2} - 1, \tag{21}$$

be the score value of α and $S(\alpha)$ be the score function of α. For any two intuitionistic fuzzy numbers α_1 and α_2, there is then

If $S(\alpha_1) < S(\alpha_2)$, $\alpha_1 \prec \alpha_2$;
If $S(\alpha_1) > S(\alpha_2)$, $\alpha_1 \succ \alpha_2$;
If $S(\alpha_1) = S(\alpha_2)$, $\alpha_1 \sim \alpha_2$.

Suppose two intuitionistic fuzzy numbers $\alpha_1 = (0, 0.2)$ and $\alpha_2 = (0, 0.3)$ exist, which can be obtained using Equation (18), $S(\alpha_1) = -0.7$, $S(\alpha_2) = -0.55$. Obviously, $S(\alpha_1) < S(\alpha_2)$, i.e., $\alpha_1 \prec \alpha_2$. However, in the actual decision making process, people tend to choose a small degree of opposition to α_1. At this time, Equation (18) cannot judge the size of the two intuitionistic fuzzy numbers. (For a comparison of the arithmetic examples of different intuitionistic fuzzy score function ranking methods, see Table 2 in Summary 5.2).

From this, it can be found that the existing intuitionistic fuzzy score function has the following problems:

(1) The same result is obtained when calculating two different intuitionistic fuzzy numbers, and it is impossible to judge the size of two intuitionistic fuzzy numbers.
(2) Owing to the inherent constraints of the score function, the derived results occasionally contradict real-world decision making scenarios.

To enhance the comparative efficacy of intuitionistic fuzzy numbers and, in turn, refine the product ranking model, this study introduces a refined intuitionistic fuzzy score function and an exact function. Subsequently, we provide proof for the formula of this newly formulated intuitionistic fuzzy score function.

Table 2. Comparison of different intuitionistic fuzzy score functions.

Example	Sorting Methods	Sorting Results
$\alpha_1 = (0,0)$, $\alpha_2 = (0.5,0.5)$	[29]	$\alpha_1 \sim \alpha_2$
	[28]	
	[45]	
	[44]	$\alpha_1 \prec \alpha_2$
	[46]	
	Methodology of the paper	$\alpha_1 \prec \alpha_2$
$\alpha_3 = (0.5,0.2)$, $\alpha_4 = (0.3,0)$	[28]	$\alpha_3 \sim \alpha_4$
	[29]	$\alpha_3 \succ \alpha_4$
	[45]	
	[44]	
	[46]	
	Methodology of the paper	$\alpha_3 \prec \alpha_4$
$\alpha_5 = (0.9,0.1)$, $\alpha_6 = (0.8,0)$	[28]	$\alpha_5 \sim \alpha_6$
	[29]	$\alpha_5 \succ \alpha_6$
	[45]	
	[44]	
	[46]	$\alpha_5 \prec \alpha_6$
	Methodology of the paper	$\alpha_5 \prec \alpha_6$
$\alpha_7 = (0,0.2)$, $\alpha_8 = (0,0.3)$	[29]	$\alpha_7 \succ \alpha_8$
	[28]	
	[45]	
	[44]	$\alpha_7 \prec \alpha_8$
	[46]	
	Methodology of the paper	$\alpha_7 \succ \alpha_8$
$\alpha_9 = (0.5,0.2)$, $\alpha_{10} = (0.5,0.3)$	[29]	$\alpha_9 \succ \alpha_{10}$
	[28]	
	[45]	
	[44]	$\alpha_9 \prec \alpha_{10}$
	[46]	$\alpha_9 \sim \alpha_{10}$
	Methodology of the paper	$\alpha_9 \succ \alpha_{10}$

5.2. Improved Intuitionistic Fuzzy Score Function

In this paper, we introduce the concept of the improved intuitionistic fuzzy score function using a voting model as an illustrative example.

In real life, when faced with indecision, people often choose to wait and see what others do before making a final decision. Thus, for the intuitionistic fuzzy number $\alpha = (\mu_\alpha, \nu_\alpha)$, suppose that during the first vote, μ represents the proportion of those who voted in favor, ν represents the proportion of those who voted against, and π represents the proportion of those who voted neutrally; and during the second vote, since the hesitant part of the first vote is affected by the first vote, that during the second vote π_μ represents the part of the vote that voted in favor, π_ν is the part of the vote that voted against, and π^2 is the part of the vote that continued to be neutral. The cycle continues in rounds, calculating the sum of the proportional parts in favor. The intuitionistic fuzzy score function considered in this paper is the sum of the final proportions in favor:

$$S_E = \mu + \pi\mu + \pi^2\mu + \cdots + \pi^{n-1}\mu = \frac{\mu(1-\pi^n)}{1-\pi} = \frac{\mu(1-\pi^n)}{\mu+\nu}, \tag{22}$$

The following equation is obtained by taking the limit of the formula:

$$\lim_{n\to\infty} S_E = \lim_{n\to\infty} \frac{\mu(1-\pi^n)}{\mu+\nu} = \frac{\mu}{\mu+\nu}, \tag{23}$$

Therefore, this paper proposes an improved intuitionistic fuzzy score function as shown in the following equation:

Theorem 1. *Suppose $\alpha = (\mu_\alpha, \nu_\alpha)$ is an intuitionistic fuzzy number, then*

$$S_E = \frac{\mu}{\mu+\nu}, \tag{24}$$

is the score function of the intuitionistic fuzzy number $\alpha = (\mu_\alpha, \nu_\alpha)$. In particular, when $\mu = \nu = 0$, we define $S(\alpha) = 0$.

The equation above reveals that the intuitionistic fuzzy score function proposed in this paper has limitations, since the degree of affiliation of the intuitionistic fuzzy number cannot be 0. To address this issue, the paper introduces the intuitionistic fuzzy exact function. This paper presents a study on intuitionistic fuzzy exact functions as follows: for an intuitionistic fuzzy number, a larger degree of affiliation is considered better while a smaller degree of non-affiliation is preferred. If the hesitation degree is taken into account, the smaller the hesitation degree, the better. Building on the above concepts, this paper puts forward the subsequent equation:

$$h_E = \frac{\mu-\nu}{\pi+1} + 1, \tag{25}$$

where, to avoid the case of $\pi = 0$, the denominator of the formula is taken as $\pi + 1$.

Further simplifying the formula, the new intuitionistic fuzzy exact function given in this paper is defined as follows:

Theorem 2. *Suppose $\alpha = (\mu, \nu)$ is an intuitionistic fuzzy number, then*

$$h_E = \frac{2-2\nu}{\pi+1}, \tag{26}$$

is said to be an exact function of the intuitionistic fuzzy number $\alpha = (\mu, \nu)$.

After proposing the intuitionistic fuzzy score function and intuitionistic fuzzy exact function, this paper proposes the following new ranking method for intuitionistic fuzzy numbers:

Definition 8. *Suppose that for any two intuitionistic fuzzy numbers $\alpha_1 = (\mu_{\alpha_1}, \nu_{\alpha_1})$ and $\alpha_2 = (\mu_{\alpha_2}, \nu_{\alpha_2})$, $S_E(\alpha_1)$ and $h_E(\alpha_1)$ are the values of the score function and the exact function for α_1, and $S_E(\alpha_2)$ and $h_E(\alpha_2)$ are the values of the score function and the exact function for α_2. Then*

If $S_E(\alpha_1) < S_E(\alpha_2)$, then $\alpha_1 \prec \alpha_2$;
If $S_E(\alpha_1) > S_E(\alpha_2)$, then $\alpha_1 \succ \alpha_2$;
If $S_E(\alpha_1) = S_E(\alpha_2)$, then
when $h_E(\alpha_1) < h_E(\alpha_2)$, $\alpha_1 \prec \alpha_2$,

when $h_E(\alpha_1) > h_E(\alpha_2)$, $\alpha_1 \succ \alpha_2$,
when $h_E(\alpha_1) = h_E(\alpha_2)$, $\alpha_1 \sim \alpha_2$.

The properties of the intuitionistic fuzzy score function proposed in this paper are described and formulas are proved in the following:

Property 1. *The score function $S_E(\alpha)$ of an intuitionistic fuzzy number $\alpha = (\mu, \nu)$ is monotonically increasing with respect to the degree of affiliation μ and monotonically decreasing with respect to the degree of non-affiliation ν.*

Proof of Property 1. Since

$$\frac{\partial S_E(\mu,\nu)}{\partial \mu} = \frac{\nu}{(\mu+\nu)^2}, \tag{27}$$

and $0 \le \mu + \nu \le 1$, so $\frac{\nu}{(\mu+\nu)^2} \ge 0$, the score function of the intuitionistic fuzzy number $\alpha = (\mu, \nu)$ is monotonically increasing with respect to the degree of affiliation μ;
similarly

$$\frac{\partial S_E(\mu,\nu)}{\partial \nu} = \frac{-\mu}{(\mu+\nu)^2}, \tag{28}$$

and $0 \le \mu \le 1$ and $0 \le \mu + \nu \le 1$, so that $\frac{-\mu}{(\mu+\nu)^2} \le 0$, so the score function of the intuitionistic fuzzy number $\alpha = (\mu, \nu)$ is monotonically decreasing with respect to the unaffiliated degree ν, which is proved. □

Property 2. *Intuitionistic fuzzy score function $S_E(\alpha) \in [0,1]$.*

Proof of Property 2. Since Formula (24), and $0 \le \mu \le \mu + \nu \le 1$, $0 \le \frac{\mu}{\mu+\nu} \le 1$.

In particular,

(1) when $\mu = 0$, then $S_E(\alpha) = 0$;
(2) when $\nu = 0$, then $S_E(\alpha) = 1$;
(3) when $\mu = \nu = 0$, then $S_E(\alpha) = 0$.

Proof is completed. □

Property 3. *Suppose two intuitionistic fuzzy numbers $\alpha_1 = (\mu_1, \nu_1)$ and $\alpha_2 = (\mu_2, \nu_2)$. If $\mu_1 > \mu_2$ and $\nu_1 < \nu_2$, then $S_E(\alpha_1) > S_E(\alpha_2)$.*

Proof of Property 3. Because

$S_E(\alpha_1) - S_E(\alpha_2) = \frac{\mu_1}{\mu_1+\nu_1} - \frac{\mu_2}{\mu_2+\nu_2} = \frac{\mu_1(\mu_2+\nu_2)-\mu_2(\mu_1+\nu_1)}{(\mu_1+\nu_1)(\mu_2+\nu_2)} = \frac{\mu_1\nu_2-\mu_2\nu_1}{(\mu_1+\nu_1)(\mu_2+\nu_2)}$

and $\mu_1 > \mu_2$ and $\nu_1 < \nu_2$, then $\mu_1\nu_2 - \mu_2\nu_1 > 0$, $(\mu_1+\nu_1)(\mu_2+\nu_2) > 0$; then, $\frac{\mu_1\nu_2-\mu_2\nu_1}{(\mu_1+\nu_1)(\mu_2+\nu_2)} > 0$, i.e., $S_E(\alpha_1) - S_E(\alpha_2) > 0$.

It is clear that $S_E(\alpha_1) > S_E(\alpha_2)$. Proof is completed. □

The proposed intuitionistic fuzzy function and intuitionistic fuzzy exact function are compared and analyzed with the existing methods in the following, and the results of the comparative analysis are shown in Table 2.

5.3. DEMATEL Determines Attribute Weights

In problems involving multi-attribute decision making, calculating indicator weights through traditional methods is often based on subjective or objective criteria to reflect the attributes' characteristics, but this ignores the correlation between them. This paper utilizes the DEMATEL method to analyze the mutual influence relationship between attributes and determine their respective weights. The traditional multi-attribute decision making approach, which does not consider attribute weights in relation to each other, is addressed. Because attribute weights vary between different types of products, experts are invited to score the attributes of each product type. The average of these scores is then calculated to obtain the relative weights of the attributes for each particular product type.

Assuming that the alternative product to be evaluated $A_i(i = 1, 2, \cdots, m)$ belongs to a certain category of products, the evaluation of the product attribute is $f_j(j = 1, 2, \cdots, n)$ and that the category of the product is t, and r experts are invited to use the linguistic scale evaluation for evaluation and scoring on a five-level scale, then, the DEMATEL method is used to calculate the attribute weight matrix $W^r_{t,n}$ given by the r experts under the different categories of the product, and the specific representation is as follows:

$$W^r_{t,n} = \begin{vmatrix} w_{11,r} & w_{12,r} & \cdots & w_{1n,r} \\ w_{21,r} & \cdots & \cdots & w_{2n,r} \\ \cdots & \cdots & \cdots & \cdots \\ w_{t1,r} & \cdots & \cdots & w_{tn,r} \end{vmatrix}, \tag{29}$$

where t denotes the number of product types, n is the number of attributes of the product, and r is the number of experts.

Finally, the weights obtained by r experts are averaged to obtain the attribute weight matrix for different types of products:

$$\overline{W_{t,n}} = \frac{1}{r}\sum_{s=1}^{r} W^s_{t,n}, \tag{30}$$

where r denotes the number of experts.

The alternative product average weight vector is calculated as follows:

$$\overline{W} = \frac{1}{t}\sum_{z=1}^{t} \overline{W_{z,n}}, \tag{31}$$

where $\overline{W}$ denotes the vector of average weights of alternative products and t is the number of categories of products.

5.4. A Product Ranking Method Based on the Intuitionistic Fuzzy TODIM Model

Multi-attribute decision making refers to a process where a decision maker identifies the best solution among various alternatives based on selected attributes; the method ranks and selects solutions by calculating their perceived superiority relative to each other. When selecting a product, consumers often consider various attributes to make alternative choices. This paper presents an emotion quantification model for online reviews of products, constructed in an intuitionistic fuzzy environment. Product attributes are transformed into intuitionistic fuzzy values. The differences between attributes of different types of products and the correlation relationship between them are considered. The DEMATEL method is used to obtain the attribute weights of the correlations. Finally, identifying the degree of superiority of alternative products based on consumer preferences is achieved using the intuitionistic fuzzy TODIM model. A detailed description of the decision making steps is provided below.

- Decision step:

Assume that $A = \{A_i | i \in M\}$ is a limited number of alternative product scenarios and that $f = \{f_j | j \in N\}$ is a finite set of attributes, where $N = \{1, 2, 3, \cdots, n\}$ and the weight of each attribute is $w = (w_1, w_2, \cdots, w_n)^T$, where $\sum_{i=1}^{n} w_i = 1$. The alternative product A_i has an evaluation value of I_{ij} under the product attribute f_j, where I_{ij} expresses the intuitionistic fuzzy set. This results in a decision matrix of m decision scenarios under n decision attributes.

$$D = \begin{bmatrix} I_{11} & I_{12} & \cdots & I_{1n} \\ I_{21} & I_{22} & \cdots & I_{2n} \\ \vdots & \vdots & \vdots & \vdots \\ I_{m1} & I_{m2} & \cdots & I_{mn} \end{bmatrix}, \tag{32}$$

the specific decision making steps are as follows:

normalize the original decision matrix $D = [I_{ij}]_{m\times n}$ to obtain $X = [x_{ij}]_{n\times n}$, where $M = \{1,2,3,\cdots,m\}$, $N = \{1,2,3,\cdots,n\}$, and $i \in M$, $j \in N$.

Determine the attribute with the largest value as the reference attribute f_j and calculate the ratio of each attribute relative to the reference attribute w_{jr}. The formula is

$$w_{jr} = \frac{f_j}{f_r}, \tag{33}$$

where $w_{jr} = \max\{w_j | j \in N\}$.

The degree of dominance of Scenario A_i over Scenario A_k when the attribute is f_j is calculated. The formula is

$$\Phi_j(A_i, A_k) = \begin{cases} \sqrt{\dfrac{(x_{ij}-x_{kj})w_{jr}}{\sum\limits_{j=1}^{n} w_{jr}}} & x_{ij} - x_{kj} > 0 \\ 0 & x_{ij} - x_{kj} = 0 \\ -\frac{1}{\theta}\sqrt{\dfrac{(x_{kj}-x_{ij})\left(\sum\limits_{j=1}^{n} w_{jr}\right)}{w_{jr}}} & x_{ij} - x_{kj} < 0 \end{cases}. \tag{34}$$

Loss aversion coefficient θ can reflect the psychological behavior of decision makers; the smaller the value of θ, the higher the risk tolerance of decision makers, and the larger the value of θ, the lower the risk tolerance of decision makers. In the subsequent experiments, this paper will analyze the value of θ to confirm whether the size of θ will have an impact on the final ranking results.

Calculate the degree of dominance of Scenario A_i over Scenario A_k for all attributes; the formula is

$$\delta(A_i, A_k) = \sum_{j=1}^{n} \Phi(A_1, A_k) \; i,k \in M, \tag{35}$$

Calculate the combined degree of dominance of all alternatives A_i over the other alternatives. The formula is

$$\xi(A_i) = \frac{\sum\limits_{k=1}^{m} \delta(A_i, A_k) - \min\left\{\sum\limits_{k=1}^{m} \delta(A_i, A_k)\right\}}{\max\limits_{i\in M}\left\{\sum\limits_{k=1}^{m} \delta(A_i, A_k)\right\} - \min\limits_{i\in M}\left\{\sum\limits_{k=1}^{m} \delta(A_i, A_k)\right\}}, i \in M. \tag{36}$$

According to the above equation, the $\xi(A_i)$ of each scenario can be obtained, and the scenarios are ranked according to the size relationship of the $\xi(A_i)$. A larger value of $\xi(A_i)$ indicates a better scenario A_i.

6. Tests and Results

This section demonstrates the application of the proposed method through a case study where movies are ranked based on online reviews.

6.1. Problem Description and Data Source

With the development of the Internet, more and more viewers regularly use online reviews on relevant platforms as a decision making reference before making decisions.

A viewer chooses $(A_1, A_2, A_3, A_4, A_5, A_6)$ as an alternative movie by referring to online reviews. The software used to process the data in this paper and the dates on which the data were analyzed are shown below in Table 3.

Table 3. Software for processing data and date of analysis.

Steps	Software	Date of Data Analysis
Crawling Movie Online Reviews	Python 3.9	5.13–5.15
Sentiment Analysis of Online Reviews	Python 3.9; The Bert-model-based Cemotion library	6.10–6.14
Calculation of the Degree of Dominance	Matlab R2022a	6.20–6.25

The movies selected are *Titanic* (A_1), *Farewell My Concubine* (A_2), *The Shawshank Redemption* (A_3), *This Killer Is Not Too Cold* (A_4), *Green Book* (A_5), and *Le fabuleux destin d'Amélie Poulain* (A_6). The classification of the movies reveals that the six selected films belong to three different genres, namely, romance, drama, and comedy. The evaluation panel, comprising five members, selected five attributes to assess the alternative movies by: these attributes are "Frame" (f_1), "Character" (f_2), "Plot" (f_3), "Soundtrack" (f_4), and "Acting" (f_5). Next, the proposed method is used in this study to rank the six movies mentioned above.

Step 1. We performed sentiment analysis on reviews of alternative movies to determine their tendencies of sentiments towards movie attributes. The affective tendencies of some of the movie attributes are shown in Table 1.

Step 2. The sentiment orientation of the reviews is acquired in Step 1, and the average sentiment value and multidimensional feature values of the movie attributes are calculated according to the content in Section 4.2. The results of the calculations are shown in Tables 4 and 5.

Table 4. Mean values of sentiment corresponding to each attribute of the alternative movie.

Alternative Movie	Attribute	Frame	Character	Plot	Soundtrack	Acting
Titanic	Mean positive affect	0.8934	0.8814	0.8751	0.8975	0.8551
	Mean negative affect	0.4467	0.4331	0.4261	0.4131	0.4313
Farewell my concubine	Mean positive affect	0.8813	0.8931	0.8852	0.8834	0.8821
	Mean negative affect	0.4237	0.4424	0.4324	0.4541	0.4426
The Shawshank Redemption	Mean positive affect	0.8701	0.8835	0.8857	0.8921	0.8936
	Mean negative affect	0.4351	0.4234	0.4327	0.4353	0.4313
This killer's not too cold	Mean positive affect	0.8831	0.8953	0.8924	0.8813	0.8854
	Mean negative affect	0.4353	0.4424	0.4345	0.4334	0.4313
Green book	Mean positive affect	0.8834	0.8924	0.8835	0.8834	0.8954
	Mean negative affect	0.4335	0.4324	0.4352	0.4315	0.4314
Le fabuleux destin d'Amélie Poulain	Mean positive affect	0.8741	0.8831	0.8937	0.8814	0.8764
	Mean negative affect	0.4324	0.4354	0.4334	0.4353	0.4375

Table 5. Multidimensional feature values corresponding to each attribute of the alternative movie.

Alternative Movie	Attribute	Frame	Character	Plot	Soundtrack	Acting
Titanic	Positive multidimensional eigenvalues	0.8272	0.8317	0.8264	0.8152	0.8366
	Negative multidimensional eigenvalues	0.1721	0.1673	0.1736	0.1843	0.1638
Farewell my concubine	Positive multidimensional eigenvalues	0.8253	0.8214	0.8124	0.8319	0.8132
	Negative multidimensional eigenvalues	0.1704	0.1786	0.1878	0.1631	0.1868
The Shawshank Redemption	Positive multidimensional eigenvalues	0.8232	0.8174	0.8236	0.8281	0.8233
	Negative multidimensional eigenvalues	0.1768	0.1821	0.1762	0.1718	0.1767
This killer's not too cold	Positive multidimensional eigenvalues	0.8176	0.8231	0.8289	0.8219	0.8224
	Negative multidimensional eigenvalues	0.1804	0.1763	0.1711	0.1781	0.1776
Green book	Positive multidimensional eigenvalues	0.8193	0.8165	0.8151	0.8169	0.8319
	Negative multidimensional eigenvalues	0.1803	0.1835	0.1848	0.1836	0.1681
Le fabuleux destin d'Amélie Poulain	Positive multidimensional eigenvalues	0.8185	0.8187	0.8293	0.8109	0.8303
	Negative multidimensional eigenvalues	0.1814	0.1818	0.1706	0.1899	0.1697

Step 3. The intuitionistic fuzzy values for the movie attributes are calculated using Equations (15) and (16), and are shown in Table 6.

Table 6. Intuitive fuzzy values for movie attributes.

Alternative Movie	Intuitive Fuzzy Value	Frame	Character	Plot	Soundtrack	Acting
Titanic	Membership degree u	0.7367	0.7315	0.7189	0.7259	0.7113
	Degree of non-membership v	0.0774	0.0725	0.0729	0.0756	0.0702
Farewell my concubine	Membership degree u	0.7299	0.7314	0.7146	0.7358	0.7139
	Degree of non-membership v	0.0716	0.0785	0.0807	0.0737	0.0832
The Shawshank Redemption	Membership degree u	0.7244	0.7197	0.7249	0.7373	0.7327
	Degree of non-membership v	0.0762	0.0765	0.0757	0.0739	0.0759
This killer's not too cold	Membership degree u	0.7198	0.7331	0.7377	0.7232	0.7241
	Degree of non-membership v	0.0782	0.0776	0.0735	0.0765	0.0761
Green book	Membership degree u	0.7212	0.7269	0.7173	0.7189	0.7402
	Degree of non-membership v	0.0775	0.0788	0.0794	0.0787	0.0723
Le fabuleux destin d'Amélie Poulain	Membership degree u	0.7203	0.7199	0.7381	0.7136	0.7306
	Degree of non-membership v	0.0781	0.0781	0.0733	0.0813	0.0729

Step 4. The improved intuitionistic fuzzy exact function Formula (26) is used to obtain the exact function values for each attribute of the movie, as shown in Table 7.

Table 7. Exact function values for movie attributes.

Movie \ Attribute	Frame	Character	Plot	Soundtrack	Acting
Titanic	1.5561	1.5511	1.5346	1.5426	1.5259
Farewell my concubine	1.5493	1.5481	1.5262	1.5562	1.5244
The Shawshank Redemption	1.5405	1.5343	1.5412	1.5576	1.5513
This killer's not too cold	1.5337	1.5514	1.5587	1.5388	1.5401
Green book	1.5359	1.5427	1.5301	1.5324	1.5624
Le fabuleux destin d'Amélie Poulain	1.5346	1.5341	1.5593	1.5247	1.5497

Step 5. Five experts were invited to assess various types of movies based on the methodology used to establish attribute weights in Section 5.3. Expert 1 rated the attributes in the romance movie category, as illustrated in Table 8.

Table 8. Results of expert scoring of romance movies.

	Acting	Character	Plot	Soundtrack	Frame
Expert 1	f_1	f_2	f_3	f_4	f_5
C1	0	4	3	0	3
C2	4	0	3	2	2
C3	2	2	0	2	2
C4	2	0	3	0	2
C5	1	2	3	0	0

The weights ω of the attributes of the movies in the romance category evaluated by Expert 1 were obtained using the DEMATEL method, as shown in Table 9.

Table 9. Weights of attributes scored by Expert 1 for romance movies.

	f_1	f_2	f_3	f_4	f_5
ω_j	7.6197	7.8342	7.7847	4.6162	6.2126
$\overline{\omega}_j$	0.2237	0.2299	0.2285	0.1355	0.1824

Here, ω_j is the weight of each attribute obtained via DEMATEL after Expert 1 scored the movies in the romance category, and $\overline{\omega}_j$ is the weight of each attribute of the movies in the romance category after normalization.

Since the genres of the alternative films are romance, drama, and comedy, the attribute weights of the different genres of films obtained from the scoring of the three genres by the five experts according to Equation (30) are shown in Table 10.

Table 10. Attribute weights for different types of movies in the romance, drama, and comedy genres.

Attribute / Movie Genre	Frame	Character	Plot	Soundtrack	Acting
Romance Movie	0.2026	0.2291	0.1932	0.2136	0.1615
Drama Movie	0.1956	0.2211	0.2134	0.1708	0.1991
Comedy Movie	0.2017	0.1934	0.2004	0.1936	0.2109

Step 6. Table 11 presents the average attribute weights for the three alternative movie types, as obtained using Equation (31).

Table 11. Average attribute weights for alternative movies.

Attribute / Weight	Frame	Character	Plot	Soundtrack	Acting
Average weight of alternative movies	0.2053	0.2397	0.2159	0.1506	0.1885

Step 7. Construct the decision matrix for Scheme A_i under attribute f_j based on the exact function values in Table 7, as shown in Table 12.

Table 12. Judgment matrix of Scheme A_i under attribute f_j.

	f_1	f_2	f_3	f_4	f_5
A_1	1.5562	1.5511	1.5347	1.5426	1.5259
A_2	1.5494	1.5481	1.5263	1.5562	1.5245
A_3	1.5405	1.5343	1.5412	1.5577	1.5513
A_4	1.5337	1.5511	1.5587	1.5388	1.5405
A_5	1.5359	1.5427	1.5301	1.5324	1.5624

Step 8. The total degree of dominance of scenario A_i over scenario A_I under all attributes is calculated using Equations (34) and (35). In this step, it is assumed that the loss recession factor $\theta = 1$. Finally, based on Equation (36), we get the total dominance degree of scheme A_i over scheme A_I under all attributes $\Phi(A_i, A_I)$. The results are as follows:

$$\widetilde{\Phi}(A_i, A_I) = \begin{bmatrix} 0 & -0.8868 & -3.7232 & -2.7445 & -1.0983 & -2.4641 \\ -3.3320 & 0 & -3.2042 & -3.7951 & -2.0809 & -2.8380 \\ -3.0381 & -2.4698 & 0 & -2.8097 & -1.8626 & -0.8174 \\ -2.3941 & -2.7771 & -3.5970 & 0 & -1.5087 & -1.1336 \\ -5.3305 & -4.4661 & -3.9319 & -4.0953 & 0 & -1.1236 \\ -5.3599 & -5.3583 & -4.1064 & -3.3616 & -4.1401 & 0 \end{bmatrix}. \quad (37)$$

Step 9. Normalize the total dominance degree of Scenario A:
$\widetilde{\xi}(A_1) = 1, \widetilde{\xi}(A_2) = 0.62, \widetilde{\xi}(A_3) = 0.99, \widetilde{\xi}(A_4) = 0.95, \widetilde{\xi}(A_5) = 0.29, \widetilde{\xi}(A_6) = 0.$

Based on the above results, the six alternative movies can be ranked in order: $A_6 \prec A_5 \prec A_2 \prec A_4 \prec A_3 \prec A_1$.

From the above results, it is clear that movie A_i has the largest $\widetilde{\xi}(A_1)$ and is the best alternative movie for the audience.

6.2. Impact Analysis of Different Weights

To confirm the feasibility of the proposed method, we introduce identical attribute weights and objective weights to rank six alternative movies using the method described in this paper. We then compare these results with the rankings obtained from the paper. Equivalent attribute weights can be seen as disregarding certain movie attributes in online movie reviews, while objective weights exclude the influence of movie genres and viewer personality preferences. The ranking results are shown in Table 13.

Table 13. Alternative movie ranking results with different attribute weights.

Alternate Movie Sorting Results	**Attribute Weight**				
	w_1	w_2	w_3	w_4	w_5
$A_1 \succ A_3 \succ A_4 \succ A_2 \succ A_5 \succ A_6$	0.2053	0.2397	0.2159	0.1506	0.1885
$A_1 \succ A_4 \succ A_3 \succ A_2 \succ A_5 \succ A_6$	0.1393	0.1173	0.2557	0.1936	0.2941
$A_1 \succ A_4 \succ A_3 \succ A_2 \succ A_5 \succ A_6$	0.2000	0.2000	0.2000	0.2000	0.2000

The rankings resulting from the weightings proposed in this paper produce differences compared with the other two weightings. The results achieved through ranking by the equal and objective weights are identical. However, the ranking based on the weights proposed in this paper, which consider movie genres and attributes, produce similar but distinct ranking outcomes when compared with the other two weights. A comparison of the ranking results under different weights reveals the presence of personalized attribute preferences. The final ranking results of the scheme are subject to variation, based on the preferred attributes.

6.3. Impact Analysis of Different Parameters

The loss aversion parameter θ in the intuitionistic fuzzy TODIM method can reflect the decision maker's psychological behavior. To examine whether the magnitude of θ affects the final ranking results, this section presents a sensitivity analysis of the θ parameter under the intuitionistic fuzzy TODIM method. The sensitivity analysis mainly involves selecting various values of θ, calculating whether the final program rankings are consistent under these values, and analyzing the impact of θ on the rankings. Table 14 displays the results of ranking the schemes based on different parameters θ.

Table 14. Alternative movie ranking results with different parameters θ.

Specifies the Value of Parameter θ	Scheme Sorting Result
$\theta = 0.2$	$A_3 \succ A_1 \succ A_4 \succ A_2 \succ A_5 \succ A_6$
$\theta = 0.6$	$A_3 \succ A_1 \succ A_4 \succ A_2 \succ A_5 \succ A_6$
$\theta = 1$	$A_1 \succ A_3 \succ A_4 \succ A_2 \succ A_5 \succ A_6$
$\theta = 1.5$	$A_1 \succ A_3 \succ A_4 \succ A_2 \succ A_5 \succ A_6$
$\theta = 2$	$A_1 \succ A_3 \succ A_4 \succ A_2 \succ A_5 \succ A_6$
$\theta = 2.5$	$A_1 \succ A_3 \succ A_4 \succ A_2 \succ A_5 \succ A_6$
$\theta = 3$	$A_1 \succ A_3 \succ A_4 \succ A_2 \succ A_5 \succ A_6$

The sensitivity analysis results in Table 12 indicate a difference in the final ranking of scenarios when the parameter θ value is changed from 0.2 to 3. As θ increases, the scheme is ranked $A_3 \succ A_1 \succ A_4 \succ A_2 \succ A_5 \succ A_6$ when θ is between 0.2 and 0.6 and $A_1 \succ A_3 \succ A_4 \succ A_2 \succ A_5 \succ A_6$ when θ is between 1 and 3. It can be shown that the final ranking result of the scheme is sensitive to the values of the parameters.

7. Discussion

In this study, movies are systematically ranked while considering diverse weightage criteria, which is followed by a comparative analysis of the experimental outcomes. Notably, when movies are ranked using uniform and objective weights, the results display consistency. However, alterations in movie attribute weights ω lead to discernible variations in rankings. This accentuates the pivotal role that attribute weight preferences play in influencing ranking outcomes.

Our experimental framework provides a salient methodology that empowers viewers to obtain recommendations that are meticulously tailored to their individualistic preferences concerning diverse movie attributes. Through parameter sensitivity analysis, it has been discerned that the parameter value θ oscillates between 0.2 and 3. The chosen parameter value θ plays a pivotal role, inducing specific shifts in the final ranking algorithm.

The efficacy of the decision making approach developed in this paper is not merely theoretical; it finds practical resonance in real-world decision making paradigms. The underpinning rationale is that the parameter θ encapsulates the spectrum of risk preferences among decision makers, and such heterogeneity directly modulates the final ranking. This paradigm can be analogously understood in the context of audience movie preferences: those with a more eclectic taste, displaying an openness to diverse movie genres and attributes, exhibit higher risk tolerance. They are, in essence, more resilient to potential misalignments between movie preferences and actual viewings. Consequently, for such an audience demographic, a lower parameter value θ is apt, as it resonates with their higher risk tolerance threshold, suggesting they are more amenable to movie selections that might not align perfectly with their expectations. Conversely, an audience cohort with a pronounced predilection for movie personalization demonstrates reduced resilience to discrepancies in attribute preferences. Their diminished risk tolerance suggests the advisability of a higher parameter value θ when leveraging the model proposed for movie selections.

Furthermore, our empirical observations emphasize the inherent sensitivity of the ranking model to the parameter value. Variability in parameter preferences and weightage criteria culminates in diverse ranking outcomes. This underscores the burgeoning demand for bespoke movie recommendations, aligning seamlessly with nuanced viewer predilections.

8. Conclusions

This study presents a structured approach to extracting and processing product attribute reviews from digital platforms and subsequently converting them into evaluative metrics via intuitionistic fuzzy set principles. In this context, we established a quantitative model that adeptly encapsulates the emotional nuances of product attributes. Moreover, we have worked towards refining the prevalent scoring function for intuitionistic fuzzy numbers, enhancing the model's accuracy and efficacy in decision making.

Our study further explores the relationship dynamics between product attributes, harnessing the DEMATEL methodology. This technique assists in determining attribute weights specific to individual product genres. Employing the multi-attribute decision making paradigm, based on TODIM, provides a sophisticated solution to modern product sequencing challenges and expands the use cases of the TODIM decision making model within an intuitionistic fuzzy framework. This study's innovations can be articulated through three primary conclusions:

(1) The introduction of a comprehensive product ranking methodology, attuned to the consumer's genre and attribute preferences. The resulting rankings, molded by these unique inclinations, provide consumers with a more tailored decision making tool for product selection.

(2) The systematic quantification of emotional weightage associated with product attributes in online dialogues using intuitionistic fuzzy techniques. We propose a refined quantitative model to decipher emotional tones within a fuzzy setting. The improved intuitionistic fuzzy scoring algorithm captures consumers' emotional evaluations effectively. By leveraging multidimensional eigenvalues, the ranking's foundation is further solidified. Our methodology, which prioritizes consumer feedback over traditional expert opinions, offers a more authentic and objective assessment. It presents a robust foundation for the product ranking algorithm, effectively handling vast online reviews and countering information overload.

(3) A pioneering product ranking algorithm designed for recommendation systems. Although many existing recommendation systems offer basic sorting options, they often lack depth and refinement. Our approach, recognizing the multitude of attributes that consumers consider, can be smoothly integrated into digital platforms, providing users with a nuanced multi-attribute ranking tool, thereby enriching their browsing experience.

While this research heralds a notable advancement in product rankings based on digital reviews, some aspects could benefit from further development:

(1) Scope of platforms: Although our methodology is tested on the Chinese platform Douban, there is a need to expand its applicability to global platforms like Rotten Tomatoes and IMDb, requiring adaptation to diverse linguistic contexts.

(2) Decision maker's risk aversion: Our framework incorporates a parameter reflecting the decision maker's risk aversion. Determining its optimal value is crucial for accurate ranking and clear decision making guidance.

(3) Focus on product attributes: While our method provides a comprehensive ranking based on product attributes, it may overlook potential correlations with other product features.

Building on the above, our proposed "Emotionally Intuitive Fuzzy TODIM Methodology for Decision-Making Based on Online Reviews" has distinct potential. However, its current testing is limited to movie rankings. A transition to other product arenas may necessitate deeper dives into consumer focus areas and nuanced object classifications. Future endeavors will delve into these areas, ensuring a broader applicability of our methodology.

Author Contributions: Research Methodology, Q.W., X.Z. and S.F.; Software, X.Z.; Validation, S.F. and X.Z.; Formal analysis, Q.W.; Survey, X.Z.; Data collation, X.Z.; writing—original draft preparation, X.Z.; writing—review and editing, X.Z. and S.F.; supervision, Q.W. and S.F.; project administration, Q.W.; funding acquisition, Q.W. and S.F. All authors have read and agreed to the published version of the manuscript.

Funding: This study was funded by Humanities and Social Sciences Research Project of Ministry of Education of the People's Republic of China "Research on the Reliability Evaluation and Optimization Mechanism of Online Film Reviews Based on Artificial Intelligence" (No. 21YJC760081).

Institutional Review Board Statement: Not applicable.

Informed Consent Statement: Not applicable.

Data Availability Statement: All data included in this study are available upon request by contacting the corresponding author.

Conflicts of Interest: The authors declare no conflict of interest.

References

1. Li, X.; Hitt, L.M. Self-Selection and Information Role of Online Product Reviews. *Inf. Syst. Res.* **2008**, *19*, 456–474. [CrossRef]
2. Lee, J.S.; Kim, J. The Online Word-of-Mouth Effect of Online Movie Reviews. *Korean J. Broadcast. Telecommun. Stud.* **2009**, *23*, 449–484. (In Korean)
3. Liu, P.; Teng, F. Probabilistic linguistic TODIM method for selecting products through online product reviews. *Inf. Sci.* **2019**, *485*, 441–455. [CrossRef]
4. Goscinski, A.; Brock, M. Toward dynamic and attribute based publication, discovery and selection for cloud computing. *Future Gener. Comput. Syst. Int. J. Escience* **2010**, *26*, 947–970. [CrossRef]
5. Luo, L.; Duan, S.; Shang, S.; Pan, Y. What makes a helpful online review? *Empirical evidence on the effects of review and reviewer characteristics. Online Inf. Rev.* **2021**, *45*, 614–632.
6. Liang, X.; Guo, J.; Sun, Y.; Liu, X.A. Method of Product Selection Based on Online Reviews. *Mob. Inf. Syst.* **2021**, *2021*, 1–6. [CrossRef]
7. Li, J.; Lan, Q.; Liu, L.; Yang, F. Integrated Online Consumer Preference Mining for Product Improvement with Online Reviews. *J. Syst. Sci. Inf.* **2019**, *7*, 17–36. [CrossRef]
8. Peng, Y.; Kou, G.; Li, J.A. Fuzzy PROMETHEE Approach for Mining Customer Reviews in Chinese. *Arab. J. Sci. Eng.* **2014**, *39*, 5245–5252. [CrossRef]
9. Chen, K.; Kou, G.; Shang, J.; Chen, Y. Visualizing market structure through online product reviews: Integrate topic modeling, TOPSIS, and multi-dimensional scaling approaches. *Electron. Commer. Res. Appl.* **2015**, *14*, 58–74. [CrossRef]
10. Najmi, E.; Hashmi, K.; Malik, Z.; Rezgui, A.; Khan, H.U. CAPRA: A comprehensive approach to product ranking using customer reviews. *Computing* **2015**, *97*, 843–867. [CrossRef]
11. Fan, Z.P.; Xi, Y.; Li, Y.L. Supporting the purchase decisions of consumers: A comprehensive method for selecting desirable online products. *Kybernetes* **2018**, *47*, 689–715. [CrossRef]
12. Fan, Z.P.; Xi, Y.; Liu, Y. Supporting consumer's purchase decision: A method for ranking products based on online multi-attribute product ratings. *Soft Comput.* **2018**, *22*, 5247–5261. [CrossRef]
13. Lee, H.C.; Rim, H.C.; Lee, D.G. Learning to rank products based on online product reviews using a hierarchical deep neural network. *Electron. Commer. Res. Appl.* **2019**, *36*, 100874. [CrossRef]
14. Wang, Q.; Wang, S.; Fu, S. A Sustainable Iterative Product Design Method Based on Considering User Needs from Online Reviews. *Sustainability* **2023**, *15*, 5950. [CrossRef]
15. Liu, P.D.; Zhang, P. Normal wiggly hesitant fuzzy TODIM approach for multiple attribute decision making. *J. Intell. Fuzzy Syst.* **2020**, *39*, 627–644. [CrossRef]
16. Lu, J.P.; Wei, C. TODIM method for performance appraisal on social-integration-based rural reconstruction with interval-valued intuitionistic fuzzy information. *J. Intell. Fuzzy Syst.* **2019**, *37*, 1731–1740. [CrossRef]
17. Zhang, Z.; Lin, J.; Zhang, H.; Wu, S.; Jiang, D. Hybrid TODIM Method for Law Enforcement Possibility Evaluation of Judgment Debtor. *Mathematics* **2020**, *8*, 1806. [CrossRef]
18. Zhang, Z.; Lin, J.; Zhang, H.; Wu, S.; Jiang, D. Interval type-2 fuzzy TOPSIS approach with utility theory for subway station operational risk evaluation. *J. Ambient. Intell. Humaniz. Comput.* **2021**, *8*, 1806. [CrossRef]

19. Zadeh, L.A. Fuzzy sets. *Inf. Control.* **1965**, *8*, 338–353. [CrossRef]
20. Angelova, N.A.; Atanassov, K.T. Intuitionistic Fuzzy Implications and the Axioms of Intuitionistic Logic. In Proceedings of the World Congress of the International Fuzzy Systems Association, Asturias, Spain, 30 June–3 July 2015. Conference of the European Society for Fuzzy Logic and Technology.
21. Atanassov, K.T. Intuitionistic fuzzy sets. *Fuzzy Sets Syst.* **1986**, *20*, 87–96. [CrossRef]
22. Liu, Y.; Bi, J.W.; Fan, Z.P. Ranking products through online reviews: A method based on sentiment analysis technique and intuitionistic fuzzy set theory. *Inf. Fusion* **2017**, *36*, 149–161. [CrossRef]
23. Liu, Y.; Bi, J.W.; Fan, Z.P. A Method for Ranking Products through Online Reviews Based on Sentiment Classification and Interval-Valued Intuitionistic Fuzzy TOPSIS. *Int. J. Inf. Technol. Decis. Mak.* **2017**, *16*, 1497–1522. [CrossRef]
24. Roszkowska, E.; Jefmański, B. BInterval-Valued Intuitionistic Fuzzy Synthetic Measure (I-VIFSM) Based on Hellwig's Approach in the Analysis of Survey Data. *Mathematics* **2021**, *9*, 201. [CrossRef]
25. Jefmański, B.; Roszkowska, E.; Kusterka-Jefmańska, M. Intuitionistic Fuzzy Synthetic Measure on the Basis of Survey Responses and Aggregated Ordinal Data. *Entropy* **2021**, *23*, 1636. [CrossRef]
26. Roszkowska, E.; Filipowicz-Chomko, M.; Kusterka-Jefmańska, M.; Jefmański, B. The Impact of the Intuitionistic Fuzzy Entropy-Based Weights on the Results of Subjective Quality of Life Measurement Using Intuitionistic Fuzzy Synthetic Measure. *Entropy* **2023**, *25*, 961. [CrossRef]
27. Cali, S.; Balaman, S.Y. Improved decisions for marketing, supply and purchasing: Mining big data through an integration of sentiment analysis and intuitionistic fuzzy multi criteria assessment. *Comput. Ind. Eng.* **2019**, *129*, 315–332. [CrossRef]
28. Chen, S.M.; Tan, J.M. Handling multicriteria fuzzy decision-making problems based on vague set theory. *Fuzzy Sets Syst.* **1994**, *67*, 163–172. [CrossRef]
29. Hong, D.H.; Choi, C.H. Multicriteria fuzzy decision-making problems based on vague set theory. *Fuzzy Sets Syst.* **2000**, *114*, 103–113. [CrossRef]
30. Dong, H.; Hou, Y.; Hao, M.; Wang, J.; Li, S. Method for Ranking the Helpfulness of Online Reviews Based on SO-ILES TODIM. *IEEE Access* **2021**, *9*, 1723–1736. [CrossRef]
31. Ghose, A.; Ipeirotis, P.G. Estimating the Helpfulness and Economic Impact of Product Reviews: Mining Text and Reviewer Characteristics. *Ieee Trans. Knowl. Data Eng.* **2011**, *23*, 1498–1512. [CrossRef]
32. Hennig-Thurau, T.; Gwinner, K.P.; Walsh, G.; Gremler, D.D. Electronic word-of-mouth via consumer-opinion platforms: What motivates consumers to articulate themselves on the Internet? *J. Interact. Mark.* **2010**, *18*, 38–52. [CrossRef]
33. Bickart, B.; Schindler, R. Internet forums as influential sources of consumer information. *J. Interact. Mark.* **2010**, *15*, 31–40. [CrossRef]
34. Abulaish, M.; Bhardwaj, A. OMCR: An Opinion-Based Multi-Criteria Ranking Approach. *J. Intell. Fuzzy Syst. Appl. Eng. Technol.* **2019**, *36*, 397–411. [CrossRef]
35. Song, Y.; Li, G.; Ergu, D. Recommending Products by Fusing Online Product Scores and Objective Information Based on Prospect Theory. *IEEE Access* **2020**, *8*, 58995–59006. [CrossRef]
36. Scholz, M.; Dorner, V. The Recipe for the Perfect Review? An Investigation into the Determinants of Review Helpfulness. *Bus. Inf. Syst. Eng.* **2013**, *5*, 141–151. [CrossRef]
37. Cao, Q.; Duan, W.J.; Gan, Q.W. Exploring determinants of voting for the "helpfulness" of online user reviews: A text mining approach. *Decis. Support Syst.* **2011**, *50*, 511–521. [CrossRef]
38. Mudambi, S.M.; Schuff, D. What makes a helpful online review? A study of customer reviews on am-azon.com. *Mis Q.* **2010**, *34*, 185–200. [CrossRef]
39. Pan, Y.; Zhang, J.Q. Born Unequal: A Study of the Helpfulness of User-Generated Product Reviews. *J. Retail.* **2011**, *87*, 598–612. [CrossRef]
40. Xu, Z. Intuitionistic Fuzzy Aggregation Operators. *IEEE Trans. Fuzzy Syst.* **2013**, *15*, 1179–1187.
41. Ye, J. Improved method of multicriteria fuzzy decision-making based on vague sets. *Comput. Aided Des.* **2007**, *39*, 164–169. [CrossRef]
42. Wang, J.Q.; Li, K.J.; Zhang, H.Y. Interval-valued intuitionistic fuzzy multi-criteria decision-making approach based on prospect score function. *Knowl. Based Syst.* **2012**, *27*, 119–125. [CrossRef]
43. Zhang, F.; Huang, W.; Li, Q.; Wang, S.; Tan, G. Parameterized utility functions on interval-valued intuitionistic fuzzy numbers with two kinds of entropy and their application in multi-criteria decision making. *Soft Comput.* **2020**, *24*, 4667–4674. [CrossRef]
44. Lin, L.; Yuan, X.-H.; Xia, Z.-Q. Multicriteria fuzzy decision-making methods based on intuitionistic fuzzy sets. *J. Comput. Syst. Sci.* **2007**, *73*, 84–88. [CrossRef]
45. Wang, J.; Zhang, J.; Liu, S.Y. A New Score Function for Fuzzy MCDM Based on Vague Set Theory. *Int. J. Comput. Cogn.* **2006**, *4*, 44–48.
46. Ye, J. Using an improved measure function of vague sets for multicriteria fuzzy decision-making. *Expert Syst. Appl.* **2010**, *37*, 4706–4709. [CrossRef]

axioms

MDPI

Article

Decision Rules for Renewable Energy Utilization Using Rough Set Theory

Chuying Huang [1], Chun-Che Huang [2,*], Din-Nan Chen [2] and Yuju Wang [3]

1 Institute of Technoogyl Management, National Chung Hsing University, 145 Xingda Rd., South Dist., Taichung City 40227, Taiwan; ireneh@smail.nchu.edu.tw
2 Department of Information Management, National Chi Nan University, 1 University Road, Puli Township, Nantou 545301, Taiwan; s107213031@mail1.ncnu.edu.tw
3 International Business Administration Program, Providence University, 200, Sec.7, Taiwan Boulevard, Shalu Dist., Taichung City 43301, Taiwan; tammywang2022@pu.edu.tw
* Correspondence: cchuang@ncnu.edu.tw

Abstract: Rough Set (RS) theory is used for data analysis and decision making where decision-making rules can be derived through attribute reduction and feature selection. Energy shortage is an issue for governments, and solar energy systems have become an important source of renewable energy. Rough sets may be used to summarize and compare rule sets for different periods. In this study, the analysis of rules is an element of decision support that allows organizations to make better informed decisions. However, changes to decision rules require adjustment and analysis, and analysis is inhibited by changes in rules. With this consideration, a solution approach is proposed. The results show that not only can decision costs be reduced, but policymakers can also make it easier for the public to understand the incentives of green energy programs and the use of solar panels. The application process is simplified for the implementation of sustainable energy policies.

Keywords: Rough Set Theory; decision making; atribute reduction; decision support; sustainable; feature selection

Citation: Huang, C.; Huang, C.-C.; Chen, D.-N.; Wang, Y. Decision Rules for Renewable Energy Utilization Using Rough Set Theory. *Axioms* **2023**, *12*, 811. https://doi.org/10.3390/axioms12090811

Academic Editor: Darjan Karabašević

Received: 27 May 2023
Revised: 10 August 2023
Accepted: 15 August 2023
Published: 23 August 2023

1. Introduction

Energy shortages are an urgent issue for governments worldwide, and sustainable development is a common global goal. Governments attach great importance to the development of renewable energy to create better global energy policies. Energy conservation and carbon reduction are also indispensable elements of energy policies, so the Green Energy Roofs Project encourages households to install solar panels and actively participate in the sustainable development of renewable energy by promoting energy conservation and carbon reduction. Roof solar systems are a pollution-free and renewable energy source and represent a sustainable and green source of energy.

Climate change has exacerbated energy shortages since environmental sustainability has become more difficult due to economic and social development. In recent years, the Russian–Ukrainian war, the COVID-19 pandemic and extreme droughts in the Northern Hemisphere have been the principal climate change factors and have created a reduction in global economic growth. Governments use subsidies to address this problem, but their effectiveness is often limited, and the expected results are often not achieved [1]. Decision support systems must be used to assist decision makers in designing measures more quickly and accurately.

Rapid developments in information technology have allowed companies and government agencies to collect vast amounts of data, but these data must be extracted, processed and organized to be transformed into useful information that supports decision making. This transformation involves formulation, analysis, improvement and prediction.

Rough set knowledge is the most important pertinent element of decision making [2]. Rough set technology is used for complex decision-making problems that feature uncertainty to induce decision rules and to support decision-making formulation. Therefore, decision support systems must transform data into knowledge to increase the efficiency and accuracy of decision making.

The motivation of this study is as follows: decision rule sets may require adjustment over time and the relationship between rule sets can change, so analysis is difficult. Differences between rule sets may require flexible strategic modifications in response to business trends. Previous studies quickly established applicable algorithms and verified those models to create disposable solutions, but this process did not involve a comprehensive analysis of rules. Yen, Huang, Wen and Wang [3] showed that advanced rule sets are not accurately or thoroughly analyzed, although rule analysis is an indispensable element of decision support that responds to the challenges of changes in rule sets to allow institutions to make better-informed decisions.

Previous studies have determined methods to generate rule sets [4–6], but these studies do not pertain to correlation-based rule evolution and renewable energy exploration. Previous studies also did not induce decision rules or consider changes in energy usage for different time periods or the effect of environmental factors on energy use. This study compares rule sets to provide more relevant information and to support decision making in terms of sustainable energy. The results of this study are more useful than those for simple data analysis and address the need for rapid decision making with regard to sustainable energy.

Due to the shortage of energy resources, renewable energy has become a development trend in the future, and solar panels are one of the renewable energy sources that has attracted people's attention. This study uses the "Green Energy Roofs Project" that was proposed by the Taiwanese government as a case study of a nationwide participation policy. It monitors the encouragement of public participation and the sharing of benefits from the project to provide more useful information for decision making with regard to sustainable energy. This study compares rule sets for different periods for case studies and for empirical research. Rough sets are used to induce rule sets and to determine the differences and changes in the public's willingness to participate in energy incentives, with respect to rule evolution and feed-in tariffs.

This study is an in-depth case study of the analysis of participation in renewable energy programs before and after the COVID-19 pandemic. It provides specific policy recommendations for decision makers that allow governments to develop more effective sustainable energy policies. The contribution of this study is that not only will this study reduce the cost of decision-making, but policymakers can also make it easier for the public to understand green energy programs and incentives for solar panel use.

This study determines the evolution and application of decision-making rules for sustainable renewable energy in order to provide decision makers with relevant information to formulate more effective policy strategies. This study features five sections. Section 1 details the collection of information that is related to decision making and sustainable renewable energy. A literature review that shows the evolution and application of decision rules is presented in Section 2. Nine types of decision rules are compared in Section 3. The condition attributes and the results for the nine types of decision-making rules are summarized in Section 4. Section 5 draws conclusions and presents recommendations for attribute weight measurement and with regard to the evaluation criteria that are required to allow decision makers to make wiser decisions.

2. Literature Review

2.1. Rough Sets

Rough Set (RS) theory is a mathematical method and decision classification rule theory that was proposed by Pawlak in 1982. It is used for data that feature uncertainty, incompleteness, and fuzziness. Decision rules are presented in IF-THEN form to represent knowledge that is used for reasoning and classification. Rough sets are combined with decision rules to solve decision problems that involve uncertainty.

This theory divides a database into condition attributes and decision attributes. The information about objects is allocated to subsets based on its attributes. The approximate relationship between subsets of condition attributes and decision attributes is then determined, in order to generate decision rules. These decision rules provide the most direct result and are used to mine contextual rules in the database. These scenarios are used to demonstrate the effect of changes in conditional patterns and outcome attributes on decision rules to better understand the relationships between rule sets.

The results of this study show that rough set theory can be used to mine contextual rules in a database to better predict various phenomena and behaviors. The following details some basic concepts of rough set theory.

Rough set theory is a method of classifying knowledge that is used for decision making by determining the effect of changes in the condition type and the outcome attributes on decision rules. A finite set of objects that is described by a finite set of attributes is used to mine knowledge rules in the database, in order to categorize the information system to allow decision making using all available information about the set of objects in the information table [7]. An information system S is defined as

$$S = (U, A, V, p),\ A = C \cup D, \tag{1}$$

where

U: the set of objects

A: the set of attributes,

$V = \cup_{Va}$, Va, the set of values of attributes $a \in A$,

p: $U \times A \longrightarrow V$, an information function, $p_x : A \rightarrow V$,

$x \in U$ the information about x in S,

where

$$p_x(a) = p(x, a),$$

for every $x \in U$ and $a \in A$.

It is assumed that the empty set is fundamental in every S. The pair $S = (U, A)$ is called an approximation space. If U is the set $U = \{x_1, x_2, \cdots, x_n\}$, which is called the full domain, and attribute set A is an equivalence relation on U, then C is the condition attribute subset and D is the decision attribute subset.

An equivalence relationship features indiscernibility. If the set of attributes $B \subseteq A$ is a subset of the indiscernible relationship $ind(B)$ on the universe U, which is expressed as

$$ind(B) = (x, y)\Big|(x,\ y) \in U^2,\ \forall_{b \in B}(b(x) = b(y)), \tag{2}$$

where x objects are defined by the equivalence class $[x]_{ind(B)}$, $[x]_B$ or $[x]$, then $(U, [x]_{ind(B)})$ is called the approximation space.

For an information system $S = (U, A, V, p)$, for a subset $X \subseteq U$. The upper and lower approximation sets are defined as

$$\begin{aligned} \overline{apr}(X) &= x \in U|[x] \cap X \neq \varnothing, \\ \underline{apr}(X) &= \{x \in U|[x] \subseteq X\} \end{aligned} \tag{3}$$

where $[x]$ is the equivalence class for x.

Therefore, (U, X) forms an approximation space. The universe is divided into three disjointed regions, as shown in Figure 1, which shows the upper and lower approximation sets i.e., the positive, boundary and negative regions. X is bounded by the red circle line. The boundary region is defined as the difference between the upper and lower approximation sets [8]. As the area of the boundary increases, the degree of uncertainty increases. The approximate values and the three regions are expressed as,

$$
\begin{aligned}
POS(X) &= \underline{apr}(X), \\
BND(X) &= \overline{apr}(X) - \underline{apr}(X), \\
NEG(X) &= U - \overline{apr}(X),
\end{aligned}
$$

and

$$
\begin{aligned}
\overline{apr}(X) &= POS(X) \cup BND(X), \\
\underline{apr}(X) &= POS(X).
\end{aligned} \tag{4}
$$

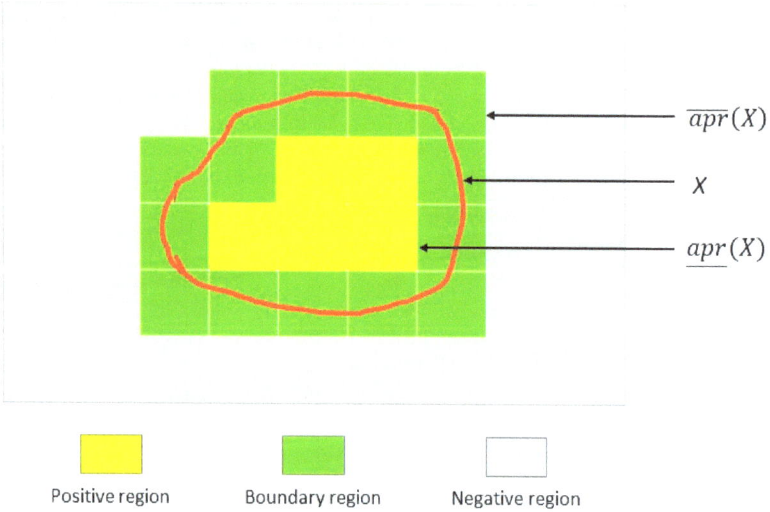

Figure 1. An approximation space.

If object $x \in POS\ (X)$, then it must belong to the target set X;

If object $x \in BND\ (X)$, then x definitely does not belong to the target set X;

If object $x \in NEG\ (X)$, it may or may not belong to the target set X, and therefore it is not possible to determine whether object x belongs to the target set X.

For any target subset $X \subseteq U$ and attribute subset $B \subseteq A$, X is called a rough set with respect to B if and only if $\overline{apr}(X) \neq \underline{apr}(X)$. The roughness of set X with respect to B is defined as

$$
P_B(X) = 1 - \frac{\left|\underline{apr}(X)\right|}{\left|\overline{apr}(X)\right|} \tag{5}
$$

where $X \neq \varnothing$(IF $X = \varnothing$, then $P_B(X) = 0$), and $|\cdot|$ denotes the base of a finite set.

If there is a large amount of information and many uncertainties, it is very difficult to make scientific and rational decisions. The condition attribute and the decision attribute in rough set theory are used to determine the similarities and differences in the approximation space, in order to classify the data messages into different equivalence categories and conduct a meaningful structural analysis. Decision support systems allow decision making using IF THEN rules. Decision support systems that use rough sets allow better decisions than common decision-making methods. Rough set theory determines these rules by analyzing a large amount of historical empirical data. Oblique and less complete attribute factors in the decision object are used to achieve essentially positive conclusions.

Pawlak [7] proposed that approximation spaces give rise to topological spaces. For each $X \subset U$, and for each approximation space, $S = (U, R)$.

$$\begin{gathered}\overline{apr}(X) \supset X \supset \underline{apr}(X), \\ \underline{apr}(U) = \overline{apr}(U) = U, \\ \underline{apr}(\varnothing) = \overline{apr}(\varnothing) = \varnothing, \\ \overline{apr}(\overline{apr}(X)) = \underline{apr}(\overline{apr}(X)) = \overline{apr}(X), \\ \underline{apr}\left(\underline{apr}(X)\right) = \overline{apr}\left(\underline{apr}(X)\right) = \underline{apr}(X). \end{gathered} \tag{6}$$

The notions of $\underline{apr}(X)$ and $\overline{apr}(X)$ can be understood as the interior and closure of the set in the associated topological space, respectively. Within the rough set theory of topological spaces, a considerable body of literature exists, encompassing various aspects such as fundamental concepts [9]; the study of generalized rough neighborhood systems involving rough approximations (lower and upper) and topological operators (interior and closed) [10]; the exploration of approximation spaces inspired by subset rough neighborhoods and their practical applications [11,12]; and the proposal of a novel rough approximation operator in the form of an abstract structure known as a "supra-topology" [13].

Al-shami [11] proposed neighborhood space of lower and upper approximations. Let (U, Ω, λ_j) be a *j-NS* such that is an equivalence relation and $X \subseteq U$. The pair $(\mathcal{F}_{Cj}\ (X),\ \mathcal{F}^{Cj}(X))$ represents the lower and upper approximations of a set X based on C_j-neighborhoods, respectively. The C_j is refer to C_j-neighborhoods (as containment neighborhoods), *j-NS* refers to *j*-neighborhood space and $j \in \{r,\ l,\ \langle r \rangle,\ \langle l \rangle,\ i,\ u,\ \langle i \rangle,\ \langle u \rangle\}$.

The C_j-neighborhoods defined as follows:

$$\begin{gathered}\mathcal{F}_{Cj}(X) = \{x \in U :\ C_j(x) \subseteq X\}, \\ \mathcal{F}^{Cj}(X) = \{x \in U :\ C_j(x) \cap X \neq \varnothing\}. \end{gathered}$$

The three regions and accuracy measure are expressed as:

$$\begin{gathered}\mathcal{B}_{Cj}(X) = \mathcal{F}^{Cj}(X) \backslash \mathcal{F}_{Cj}(X), \\ POS_{Cj}(X) = \mathcal{F}_{Cj}(X), \\ NEG_{Cj}(X) = U \backslash \mathcal{F}^{Cj}(X), \\ \mathcal{M}_{Cj}(X) = \frac{|\mathcal{F}_{Cj}(X)|}{|\mathcal{F}^{Cj}(X)|}, \end{gathered} \tag{7}$$

where $X \neq \varnothing$ for any j and each nonempty subset X of U. The best accuracy measures obtained with $j = i, \langle i \rangle$.

Decision-making problems often involve multiple conditions, goals or subjects. By integrating the use of topological space and rough set theory, the structure within the data can be understood in greater depth, thus reducing the boundary regions. This enables decision makers to make accurate decisions and obtain solutions quickly.

For two finite, non-empty sets U and A, where U is the universe of objects and cases and A is *a* set of attributes and features [14], the pair IS = (U, A) is called an information table. For each attribute, $a \in A$ is a set *Va* of its values, which is referred to as the domain of a. The object x and attributes $A = a_1, \ldots, a_n$ are defined as $a(x)$ with a data pattern $(a_1(x), \ldots, a_n(x))$.

The IS information table (U, A) divides A into two types of attributes: $C,\ D \subseteq A$. These are, respectively, called conditional and decision (action) attributes. Each decision table describes decisions (actions, results, etc.) that are determined when some conditions are satisfied, so each row of the decision table specifies a decision rule that determines the decisions for a specific set of conditions. DT $= (U, C, D)$ is called a decision table, in which $U = \{x_1, \ldots, x_N\}$, $C = \{a_1, \ldots, a_n\}$ and $D = d_1, \ldots, d_k$ are represented by means of a data sequence (also called a data set) of data patterns $((v_1, \text{target}_1), \ldots, (v_N, \text{target}_N))$, where $v_i = C(x_i)$, $\text{target}_i = D(x_i)$, and $C_i = (a_1(x_i), \ldots,\ a_n(x_i))$, $D_i = (d_1(x_i), \ldots,\ d_k(x_i))$, for

$i = 1, \ldots, N$. A data sequence also defines a decision table. The equivalence classes of $I(D)$ are the decision classes.

Using the attribute selection method for rough sets, the attribute subsets are subdivided into positive domains, boundaries and negative regions to identify significant and non-significant features and increase accuracy and efficiency. Classification, categorization, analysis and evaluation are used to determine the decision rules between data. After analysis of the data of different attributes and corresponding decisions, if the attributes and decisions are the same, a positive, certain, and non-conflicting rule is generated, which is called a consistent decision rule. If the attributes and decisions are different, a non-deterministic and conflicting rule is generated, which is an inconsistent decision rule.

Conflict analysis is a mathematical formal model that uses rough set theory to determine the relationship between the degree of conflict between subjects. In the context of conflict analysis, rough set theory is used to analyze and classify data that are related to a conflict. This process can be used to solve governmental, political, and business strategy formulation models [7]. There is no universal theory of conflict analysis using mathematical models. The domain has the greatest effect.

For a finite, non-empty set U that is called the *universe*, the elements of U are called agents. A function $v{:}U \to \{-1, 0, 1\}$, or $\{-, 0, +\}$ assigns a value -1, 0 or 1 to every agent, representing an opinion, view or voting result for an issue that, respectively, corresponds to against, neutral and favorable.

The pair $S = (U, v)$ is denoted as a conflict situation.

Three basic binary relationships are defined in the universe to express the relationship between subjects: conflict, neutrality and alliance. The auxiliary function is defined as

$$\varnothing_v(x,\ y) = \begin{cases} 1, \text{ if } v(x)v(y) = 1 \text{ or } x = y, \\ 0, \text{ if } v(x)v(y) = 0 \text{ and } x \neq y, \\ -1, \text{ if } v(x)v(y) = -1. \end{cases} \tag{8}$$

Therefore, if $\varnothing_v(x,\ y) = 1$, agents x and y share the same view on issue v (allied on v); if $\varnothing_v(x,\ y) = 0$, at least one agent, x or y, is neutral with regard to a (neutral on a); and if $\varnothing_v(x,\ y) = -1$, both agents differ on issue v (in conflict on v).

The three basic relationships R_v^+, R_v^0 and R_v^- on U^2, respectively, refer to alliance, neutrality and conflict relationships and are defined as

$$\begin{aligned} R_v^+(x,\ y) &\text{ if } \varnothing_v(x,\ y) = 1, \\ R_v^0(x,\ y) &\text{ if } \varnothing_v(x,\ y) = 0, \\ R_v^-(x,\ y) &\text{ if } \varnothing_v(x,\ y) = -1. \end{aligned} \tag{9}$$

Conflict analysis is used to resolve conflicts of interest and value in a complex decision-making environment. Rough sets are used to compare options in terms of benefit and cost, in order to determine the option that minimizes conflict and contradiction. Conflict analysis identifies conflicts and their root causes, and rough set theory is used to analyze complex and uncertain data that are related to conflicts and to extract useful knowledge to inform a conflict resolution process.

Pawlak [15] proposed a method of data analysis for a specific type of data table: a decision table. Zhang and Yao [16] and Zhang and Miao [17] proposed a three-level granularity structure for decision tables, which provides a framework for granularity calculations, data processing and attribute simplification. The decision table is represented by

$$\mathrm{DT} = (OB, AT = C \cup D, \{V_a \mid a \in AT\}, \{I_a \mid a \in AT\}), \tag{10}$$

where OB is a universe with finite objects; AT is the finite set of attributes, which includes the set of condition attributes C and the set of decision attributes D; V_a is the value domain for $a \in AT$; I_a: $OB \to V_a$ is an information function; $x \in OB$ has a value $I_a(x)$ on attribute a; and $DT = (OB, C \cup D)$ represents a simplified decision table.

Three-way decisions are used in RSs to reduce the cost of a decision. Positive certainty rules that are derived from the positive domain indicate acceptance of a concept, negative certainty rules that are derived from the negative domain indicate rejection of a concept and uncertainty rules that are derived from the boundary domain are used to delay decision making (deferment).

Solutions that quickly support decision making are necessary in an organization that is changing from traditional management to a more flexible and adaptable form [18]. This faster, more flexible change is beneficial to global organizations, but data change dynamically over time, so it is complex and time-consuming to obtain relevant and consistent up-to-date information across large organizations [19]. These organizations must use decision support to allow decision makers to make faster and more accurate judgments in response to changing environments. The process of extracting and transforming rules (in IF-THEN form) using expert knowledge is a knowledge management process, technique and methodology [2]. These rules are used for the reasoning process for decision support. Extracting useful information by analyzing rules and identifying the evolution of rules by comparing rule sets to make decisions increases efficiency and innovation in decision support.

In terms of rough sets, many studies propose effective multi-criteria decision-making methods. Wang and Zhang [20] used rough sets and fuzzy measures to propose a multi-criteria decision-making method that allows decision makers to deal with complex multi-criteria decision-making problems. Ayub et al. [21] proposed the linear Diophantine fuzzy rough set model, which is used for multi-stage decision analysis. Many studies also show that rough sets and rule evolution are practical in a real-world context [22–24]. These methods allow decision makers to make better decisions in a complex decision-making context.

2.2. Renewable Energy

There is an energy shortage crisis, so renewable energy sources must be identified for contemporary society. Sharma et al. [6] showed that solar energy is a renewable energy management system that is efficient and reduces energy wastage. Jafari and Malekjamshidi [25] proposed the use of rule control and rule optimization methods to manage sustainable energy sources.

In terms of renewable energy, Gung et al. [5] proposed a hybrid analytical approach that uses quantitative and qualitative analysis to determine the factors that affect household energy consumption. Alzahrany et al. [4] used rough set theory to determine the barriers and drivers in the use of solar energy in Saudi Arabia and showed that technical, financial and policy factors are the main barriers. However, Saudi Arabia has abundant solar resources, and there is an impetus for energy transition. These studies show that effective energy management strategies that promote sustainable development are possible.

In terms of environmental protection and sustainable development, rooftop solar systems offer many advantages [26,27]. To promote the use of renewable energy, the Taiwanese government launched the "Green Energy Roofs Project" in 2019. This was modeled on Germany's Renewable Energy Act of 2000, which used a feed-in tariff policy for solar power generation. Li, Wang, Dai, and Wu [28] showed that the Chinese government's support for the solar energy industry has given a strong impetus to the development of solar power generation.

The US government supports the development of the solar energy industry by implementing policy measures such as tax credits and by simplifying the application process for solar projects [29]. This incentive has greatly increased the number of solar panels installed. Incentives encourage individuals to install solar panels on their roofs to provide clean renewable energy for households and to generate additional income [26].

Costa, Ng, and Su [30] showed that if there are no incentives to stimulate consumers, few solar systems are installed. Barnes, Krishen, and Chan [31] showed that residents who had already installed solar panels in their communities had a positive impact on those who did not, and the use of solar panels increased. However, the adoption of solar panels is influenced by factors such as roof size, climate and equipment costs, so in promoting green roof programs, it is necessary to account for the needs and resources of different households and to develop suitable solar panel installation plans in order to establish effective rules for renewable energy use decisions.

Energy consumers are an essential part of the energy system and are the target audience for governments promoting renewable energy policies. As renewable energy production and markets expand globally, consumers have become involved in small-scale energy production, with policymakers playing a critical role [32]. In the past, consumers were passive buyers of energy and used traditional sources to meet their needs, but increasing numbers of consumers are becoming small energy producers by installing solar systems and actively participating in energy production [25]. This transition allows consumers to use renewable energy more effectively and reduces their dependence on traditional sources, which contributes to the goal of energy transformation.

Increasing energy demands from households have led to a continuous increase in energy consumption, which has increased environmental damage. To achieve sustainable development goals, interventions and guidance are required at multiple levels. The decision environment to increase participation in energy-saving programs is changed by macro- and micro-interventions, such as cash incentives and legal or energy-saving factors [33,34]. The Taiwan Power Company offers cash incentives to households to encourage participation in energy-saving plans with fixed electricity prices.

It is also possible to change individual behavior. Solar panels on household roofs conserve energy and reduce carbon dioxide emissions, and previous studies determined the factors that affect household energy consumption. Rausch and Kopplin [35] studied the effect of environmental awareness, values, beliefs and perceived knowledge on the willingness of consumers to purchase sustainable products. Namazkhan et al. [33] determined the effect of factors such as building characteristics, social demographics and psychological factors on natural gas consumption from the perspective of household natural gas consumption using a decision tree. This method offers a more comprehensive analysis and better guides policy development and energy management.

Studies show that environmental, economic, technological and policy factors have a significant effect on household energy consumption [34,36]. Voluntary behavior can be changed through behavioral intervention measures and individual factors, such as perception, preferences and abilities. Sustainable development goals can only be achieved by promoting energy conservation and establishing policies at various levels to meet the energy needs of households.

Few studies have determined the evolution of rules through comparative correlations, or the effect of rule changes on renewable energy incentives. In terms of the renewable energy industry as a future development trend, the effect of rule evolution on energy incentive issues and trends in the willingness of households to accept renewable energy incentives is important. Target households, as users and producers, are playing an increasingly important role in government policy. The goal is to provide valuable information to decision makers to allow the formulation of more effective policies. This study makes a substantive contribution to the formulation and implementation of renewable energy incentive policies.

3. Methodology and Conceptual Framework

3.1. Conceptual Framework

This study determines the factors that affect households in terms of installing solar panels on rooftops and determines the public's needs and expectations for solar panel systems. Using rough set theory, rules are induced for each object through attribute reduction, and these rules are saved in a rule set. Rule induction is used to generate a rule set, and then object classification and prediction are performed.

The recognition process for each type of rule set involves inputting two different rule sets for time t and $t + 1$. The process then evolves into nine output forms, based on changes in a condition attribute addition, deletion or modification or changes in result attributes. For each rule set, if there are changes in the condition attribute values, the rule set is of the seventh, eighth or ninth type. The remainder are assigned to the first through sixth types.

The evaluation process depends on whether the result attribute changes. In terms of the seventh, eighth and ninth types, if a condition attribute does not change, the rule set is assigned to the seventh type; if the result changes, the rule set is assigned to the eighth type; and if the result changes significantly, the rule set is assigned to the ninth type. If there are changes in the result attribute, the rule set is assigned to the second, fourth or sixth type.

The rule set is then classified based on the direction of changes in the condition attribute. If there is no change, it is assigned to the second type; if there is an increase, it is assigned to the fourth type; and if there is a decrease, it is assigned to the sixth type. If there are no changes in the result attribute, the rule set is of the first, third or fifth type. This is further classified based on the direction of changes in the condition attribute. If there is no change, it is assigned to the first type; if there is an increase, it is assigned to the third type; and if there is a decrease, it is assigned to the fifth type. The process is shown in Figure 2.

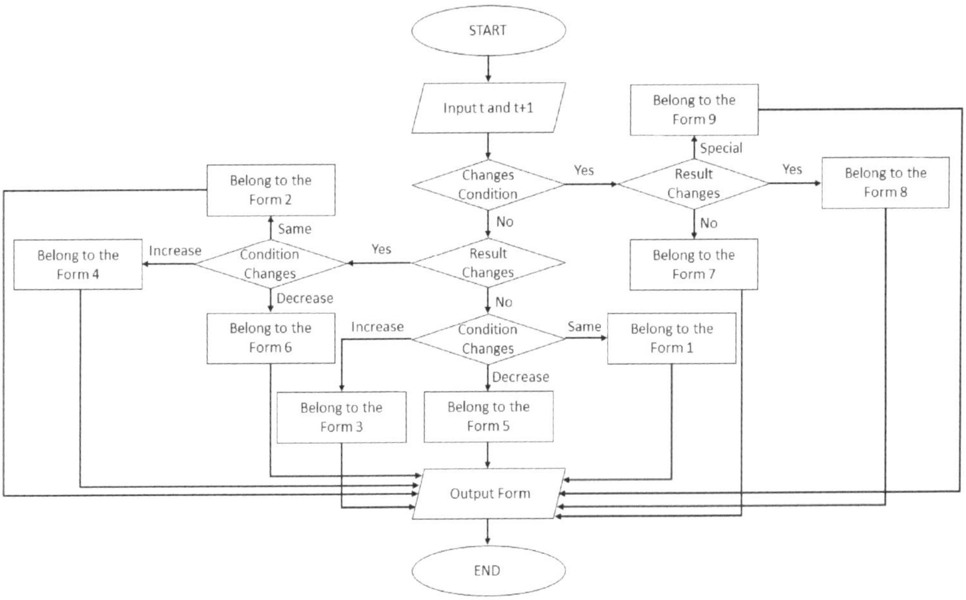

Figure 2. Flowchart for the pattern recognition process.

The classification rules are expressed in terms of pseudocode judgments, whose algorithm is expressed as follows: Algorithm 1.

This study uses RS decision categorization to create a summary. In terms of the differences in installation figures before and after the COVID-19 pandemic, changes in household energy use are determined to collect data on public willingness and the factors that affect the installation of solar panels. This allows decision-making recommendations to be derived that promote green energy plans for nationwide participation. Taiwan's "Green Energy Roofs Project" is used as a case study.

Algorithm 1.

1. Input t and $t + 1$
2. if condition changed
3.   if result changed
4.   if result change is significant
5.   output "form 9"
6.   else
7.   output "form 8"
8.   else
9.   output "form 7"
10. else
11.   if result changed
12.   if condition increased
13.   output "form 4"
14.   elif condition decreased
15.   output "form 6"
16.   else
17.   output "form 2"
18.   else
19.   if condition increased
20.   output "form 3"
21.   elif condition decreased
22.   output "form 5"
23.   else
24.   output "form 1"
25. end

3.2. Methodology

Rough set decision classification rules are used to compare intervals that feature different conditions for various periods and to determine the relationship between two rule sets r^t and r^{t+1}. Three different types of conditions, including addition, deletion and modification, are considered, and these are categorized into nine different forms, based on whether the outcome attribute changes (see Table 1). Attributes are either condition attributes or decision attributes, and all object information is classified into subsets according to the feature selection. The approximate relationship between the subset of condition attributes and the decision attributes is then determined, and decision rules are generated.

Table 1. Classification of rule set forms.

	Result Unchanged	Result Changes
Condition attribute remains unchanged	Form 1	Form 2
Addition of a condition attribute	Form 3	Form 4
Removal of a condition attribute	Form 5	Form 6
Change in a condition attribute value	Form 4	Form 8
Special case		Form 9

Notations:

- t: time interval.
- r_{ij}: rule set.
- A: condition attribute.
- O: result (decision attribute).
- v: variable of condition attribute.
- w: variable of result attribute.
- i: rule index, $i = 1 - r^t, i \in 1, \ldots, n$.

- j: rule index, $j = 1 - r^{t+1}, j \in 1, \ldots, n$.
- n: number of condition attributes
- m: number of result attributes.

The nine forms of condition changes are described in the following.

Form 1: Maintaining Equality

For rule sets r^t and r^{t+1}, the condition and result attributes are consistent.

If $r_{ij}^t = r_{ij}^{t+1}$, i.e., $(A1_i = v1_i) \cap \ldots \cap (An_i = vn_i) \rightarrow (O1_j = w1_j) \cap \ldots \cap (Om_j = wm_j)$, where $v_i^t = v_i^{t+1}$, $w_j^t = w_j^{t+1}$, then r_{ij}^t and r_{ij}^{t+1} pertain to maintaining equality.

Form 2: Result Change

Two rule sets have the same condition attributes in r^t and r^{t+1}, but the results are different.

If $r_{ij}^t = r_{ij}^{t+1}$, i.e., $(A1_i = v1_i) \cap \ldots \cap (An_i = vn_i) \rightarrow (O1_j = w1_j) \cap \ldots \cap (Om_j = wm_j)$, where $v_i^t = v_i^{t+1}$, $w_j^t \neq w_j^{t+1}$, then r_{ij}^t and r_{ij}^{t+1} pertain to a change in the result.

Form 3: Condition Attribute Addition (Increased Restriction)

Two rule sets are different in terms of one condition attribute for r^t and r^{t+1}, where t + 1 has an additional attribute, but the results are the same.

If $r_{ij}^t = r_{ij}^{t+1}$, i.e., $(A1_i = v1_i) \cap \ldots \cap (An_i = vn_i) \rightarrow (O1_j = w1_j) \cap \ldots \cap (Om_j = wm_j)$, where $v_i^t = v_{i+1}^{t+1}$, $w_j^t = w_j^{t+1}$, then r_{ij}^t and r_{ij}^{t+1} pertain to an additional condition attribute.

This represents the addition of an attribute between time t and t + 1 that produces no change in the results.

Form 4: Increase in Condition Attributes and Change in Results

There is a difference in the condition attributes for rule sets in r^t and r^{t+1}, where t + 1 is later than t, and the results are different.

If $r_{ij}^t = r_{ij}^{t+1}$, i.e., $(A1_i = v1_i) \cap \ldots \cap (An_i = vn_i) \rightarrow (O1_j = w1_j) \cap \ldots \cap (Om_j = wm_j)$, where $v_i^t = v_{i+1}^{t+1}$, $w_j^t \neq w_j^{t+1}$, then r_{ij}^t and r_{ij}^{t+1} pertain to this form.

There is a change in the results when a condition attribute is added between times t and t + 1.

Form 5: Reduction in Condition Attributes (Reduced Condition)

There is a difference in the condition attributes for rule sets r^t and r^{t+1}, where t + 1 is less than t, but the results are the same.

If $r_{ij}^t = r_{ij}^{t+1}$, i.e., $(A1_i = v1_i) \cap \ldots \cap (An_i = vn_i) \rightarrow (O1_j = w1_j) \cap \ldots \cap (Om_j = wm_j)$, $v_{i+1}^t = v_i^{t+1}$, $w_j^t = w_j^{t+1}$, then r_{ij}^t and r_{ij}^{t+1}

A condition attribute is removed between time t and t + 1, and there is no change in the results.

Form 6: Reduction in Condition Attributes and Change in Results (Reduced Condition and Change in Results)

There is a difference in the condition attributes for rule sets r^t and r^{t+1}, where t + 1 is less than t, and the results are different.

If $r_{ij}^t = r_{ij}^{t+1}$, i.e., $(A1_i = v1_i) \cap \ldots \cap (An_i = vn_i) \rightarrow (O1_j = w1_j) \cap \ldots \cap (Om_j = wm_j)$, where $v_{i+1}^t = v_i^{t+1}$, $w_j^t \neq w_j^{t+1}$, then r_{ij}^t and r_{ij}^{t+1} belong to this form.

The result changes because the target audience has one condition attribute removed at t + 1.

Form 7: Change in Condition Attribute Values (Adjusted Condition)

The majority of the condition attributes in the two rule sets are the same, but there is a change in the value of one condition attribute between r^t and r^{t+1}.

If $r_{ij}^t = r_{ij}^{t+1}$, i.e., $(A1_i = v1_i) \cap \ldots \cap (Ak_i = vk_i) \rightarrow (O1_j = w1_j) \cap \ldots \cap (Om_j = wm_j)$, where $v_i^t = v_i^{t+1}$, $w_j^t = w_j^{t+1}$, then r_{ij}^t and r_{ij}^{t+1} pertain to a change in the condition attribute values, where k means the condition of adjustment.

There is a change in the condition attribute values from time t to t + 1, but the results remain the same.

Form 8: Change in Result Values (Adjusted Condition and Change in Results)

The majority of the condition attributes in the two rule sets are the same, but there is no change in the condition attribute values between r^t and r^{t+1}, and the result values change.

If $r^t_{ij} = r^{t+1}_{ij}$, i.e., $(A1_i = v1_i) \cap \ldots \cap (Ak_i = vk_i) \rightarrow (O1_j = w1_j) \cap \ldots \cap (Ok_j = wk_j)$, where $v^t_i = v^{t+1}_i$, $w^t_j \neq w^{t+1}_j$, but $k^t = k^{t+1}$, then r^t_{ij} and r^{t+1}_{ij} pertain to a change in the result value.

This is similar to Form 7, but the result value changes.

Form 9: Special Case

There are significant changes in both the condition attributes and the results between time t and $t + 1$.

If $r^t_{ij} = r^{t+1}_{ij}$, i.e., $(A1_i = v1_i) \cap \ldots \cap (Ak_i = vk_i) \cap \ldots \cap (An_i = vn_i) \rightarrow (O1_j = w1_j) \cap \ldots \cap (Ok_j = wk_j) \cap \ldots \cap (Om_j = wm_j)$, where $v^t_i \neq v^{t+1}_i$, $w^t_j \neq w^{t+1}_j$, but $k^t = k^{t+1}$, then r^t_{ij} and r^{t+1}_{ij} pertain to this type.

In practical applications, it is difficult to perform analysis or support a decision because there are significant changes in the premise and in the conclusion.

These decision rules allow for a direct interpretation of the results and show contextual rules in the database. The effects of changes in the condition type and outcome attributes on decision rules are determined. The results of this study show the relationships between rule sets. Rough set theory is used to mine contextual rules in the database, in order to predict phenomena and behaviors.

3.3. *Case Study*

The global economy was significantly affected by the COVID-19 pandemic, which was the greatest economic shock since the Second World War, and governments around the world faced enormous challenges. Individuals who were forced to work and study from home due to social distancing measures spent almost 24 h a day at home, so there was a significant increase in household electricity consumption. Therefore, green and energy-efficient practices must be promoted to households, and policymakers must establish decision-making rules for household energy use. This study determines the changes in attitudes to rooftop solar panel installations before (PBe) and after (PAf) the COVID-19 pandemic to establish effective policies.

This study determines the willingness of households to accept incentives for renewable energy. This study uses the "Green Energy Roofs Project" of the Taiwanese government between 2019 and 2022 as a case study and focuses on the two different rule sets before and after the outbreak of the COVID-19 pandemic. Two decision rules are used to determine the changes in public willingness to accept feed-in tariffs for installing solar panels on rooftops before and after the COVID-19 pandemic. Rule analysis is used to categorize the influencing factors as either basic data factors, public background factors or energy incentive factors.

To encourage the public to install solar panels, the government implemented an energy incentive that includes an additional percentage (1 + %) on top of the feed-in tariff. The basic data factor is based on the area where the equipment is installed, such as the northern region, offshore islands or other remote areas, and can be changed through policies such as incentives or subsidies. The public background factors include family living conditions, building environmental data, household income and self-perceived value. These can be changed by interventions such as media advocacy, legal restrictions and social expectations. This study determines the characteristics and contours of households.

There is an additional bonus solar green billing rate (see Table 2) to the 15% bonus for installations in Taipei, New Taipei, Taoyuan, Hsinchu, Miaoli, Yilan and Hualien (northern region). Starting in 2020, participants in indigenous or remote areas received a 1% bonus. In terms of the capacity tier, there is a distinction between those who pay a grid connection fee (GF) and those who do not pay a grid connection fee (NGF). Those who participated in the Green Energy Roofs Project received a 3% bonus, with the initial 15% bonus remaining if no submarine cable is connected to the island (expressed as OT1), being reduced to 4% if one is connected to the island (expressed as OT2).

Table 2. Feed-in tariff for solar photovoltaic power generation equipment.

									(Announced Upper Limit, Unit: NTD/kWh)
		2022		**2021**		**2020**		**2019**	
		First Quarter	**Second Quarter**	**First Quarter**	**Second Quarter**	**First Quarter**	**Second Quarter**	**First Quarter**	**Second Quarter**
Capacity ranges = x (unit: kWh)									
1 < x < 20		5.8952	5.7848	5.6707	5.6281	5.7132	5.7132	5.7983	5.7983
20 < x < 100	NGF	4.5549	4.4538	4.3304	4.2906	4.4366	4.3701	4.5925	4.5083
	GF	4.4861	4.3864						
100 < x < 500		4.0970	3.9666	3.9975	3.9227	4.1372	4.0722	4.3175	4.2355
<500	NGF	4.1122	3.9727	3.9449	3.8980	4.0571	3.9917	4.2313	4.1579
	GF			4.4191	4.3722	4.5245	4.4591	4.6902	4.6168
Rate increase items and percentages:									
Green Energy Roofs Project		3%		3%		3%		3%	
Indigenous or remote areas		none	1%	none	1%	1%		none	none
Northern Taiwan		15%		15%		15%		15%	
OT1		15%		15%		15%		15%	
OT2		4%		4%		4%		4%	

Source: Bureau of Energy, Ministry of Economic Affairs.

Table 3 shows the types of variables that are used for the case study and whether they are subject to interference. Household income is defined as the net amount of comprehensive income after deducting any tax exemption and other relevant deductions. The maximum useful life of a building is defined as 50 years in the fixed asset depreciation table of the Ministry of Finance, but this study defines the age of a house as the number of years since completion. Common classification categories include new houses (completed less than 5 years ago), second-hand houses (completed between 6 and 20 years ago) and old houses (completed more than 20 years ago).

The effect of the basic data factor on the target household in terms of the decision rule sheet is changed by changing the incentive method and the percentage bonus. The effect of the background factor on decision making is determined by changing the feature selection, and determining whether the outcome attribute changes. Three different condition types are proposed—add, delete, and modify—in order to allow more accurate decision making. The decision table for the various attributes is shown in Table 4.

This study changes a variety of interventions, such as cash incentives and legal and energy conservation factors, to change the decision-making environment. Taiwanese power companies offer cash incentives for household electricity consumption at a fixed tariff to encourage greater participation in the program. This study determines the relationship between the household sector, green energy and government regulatory interventions to develop recommendations for decisions on energy consumption and feed-in tariff rates that promote sustainable development and carbon reduction.

Table 3. Definitions of variables.

Type	Attribute	Definition	Intervention
Basic information factors	Region	Offshore islands, northern region, others	Yes
	Remote area	Remote area: no, yes	Yes
Background factors	Family type	Single family, childless family, single-parent family, grandparent family, extended family, others	No
	Gender	Gender of the head of household: F = female, M = male	No
	Age	Age of the head of household (years, adults over 20 years old): 20~29, 30~39, 40~49, 50~59, 60~69, >69	No
	Education	Head of household education level: 0 = elementary and under, 1 = junior, 2 = senior, 3 = college, 4 = graduate and above	No
	Population (pop.)	Number of persons: ≤1, 2~3, 4~5, >5	Yes
	Income	Household annual net income (USD): 0~18,000; 18,001~40,000; 40,001~80,000; 80,001~150,000; >150,001	Yes
	Age of house	Age of house (years): ≤5, 6~20, >20	Yes
	Number of stories	Number of stories: 1~2, 3~4, >5	Yes
	Capacity (cap)	Device capacity (kilowatts): <20, 20~100, 100~500, >500	Yes
	Perceived value (PV)	Perceived value/benefit or no benefit: no, yes	Yes
Reward factors	Reward	Incentive bonus for household installation of rooftop solar panels: 3%, 4%, 7%, 8%, 18%, 19%	Yes

Table 4. Decision table for condition attributes.

	Basic Attributes		Background Attributes						Outcome Attribute	
No	A1 Region	A2 Remote	B1 pop.	B2 Income	B3 House	B4 Stories	B5 cap	B6 PV	O1 Ratio	O2 Accept
0	Offshore islands	No	≤1	0~18,000	≤5	1~2	<20	No	3%	Low
1	Northern region	Yes	2~3	18,001 ~40,000	6~20	3~4	20~100	Yes	4%	Sustain
2	Others		4~5	40,001 ~80,000	>20	>5	100~500		7%	High
3			>5	80,001 ~150,000			>500		8%	
4				>150,001					18%	
5									19%	

4. Results and Discussion

For Case 3.3, the results are as follows.

4.1. Result

Form 1: Maintenance

PBe: if A1 = 2, A2 = 0, B1 = 1, B2 = 0, B3 = 2, B4 = 0, B5 = 0 and B6 = 1, then O1 = 2 and O2 = 1.

PAf: if A1 = 2, A2 = 0, B1 = 1, B2 = 0, B3 = 2, B4 = 0, B5 = 0 and B6 = 1, then O1 = 2 and O2 = 1.

The COVID-19 outbreak has changed the global political and economic environment, and there are new challenges to green energy development and policy making for sustainable energy. In the post-pandemic period, government policymakers must determine how to continue to promote existing green energy programs and initiatives in a new environment. The government must work closely with the public to ensure that targeted households continue to support these programs.

The government must also conduct regular, rolling tracking and evaluation to ensure the sustainability and effectiveness of policy implementation. The pandemic provides an opportunity for the government to strengthen its support for green energy development and to promote sustainable socio-economic development. Encouraging the installation of renewable energy facilities, such as solar panels on the roofs of homes, will reduce carbon emissions and reliance on traditional energy sources, which will help to achieve the goal of sustainable energy development.

Form 2: Result Change

PBe: if A1 = 2, A2 = 0, B1 = 0, B2 = 0, B3 = 2, B4 = 0, B5 = 0 and B6 = 1, then O1 = 1 and O2 = 1.

PAf: if A1 = 2, A2 = 0, B1 = 0, B2 = 0, B3 = 2, B4 = 0, B5 = 0 and B6 = 1, then O1 = 1 and O2 = 0.

To avoid the risk of cluster infections of COVID-19, governments encouraged people to restrict travel and to work or study at home. This change in lifestyle due to the pandemic has forced governments to reduce economic activity, so there has been a significant increase in the demand for electricity from the household sector [37], but electricity consumption by the commercial and industrial sectors has decreased.

The pandemic has changed lifestyles. Citizens work from home for extended periods and are more concerned about their home environment, energy consumption and increased household expenses. Therefore, target households no longer feel as incentivized and do not expect rewards or subsidies from government policies that are as high as those that were previously offered. The pandemic has changed the environment, but the government must ensure the continuity of existing plans and programs and sustain support for target households in terms of these initiatives. The government must conduct regular, rolling tracking and evaluation to ensure the continuity and effectiveness of policy implementation and outcomes. If there is no sense of urgency, target households will prioritize maintaining their existing standard of living over considering additional policy plans, so the government must develop policies to meet the various demands of target households by ensuring that policies are feasible and effective.

Form 3: Adding Restrictions

PBe: if A1 = 2, A2 = 0, B2 = 0, B3 = 2, B4 = 1 and B5 = 0, then O1 = 1 and O2 = 1.

PAf: if A1 = 2, A2 = 0, B1 = 2, B2 = 0, B3 = 2, B4 = 1 and B5 = 0, then O1 = 1 and O2 = 1.

In recent years, reducing the carbon footprint has become an important environmental policy goal for many governments. Wiedenhofer, Smetschka, Akenji, Jalas and Haberl [38] showed that household population is a crucial factor in reducing the carbon footprint since household appliances are shared in larger households, meaning the individual carbon footprint is reduced. Reducing the carbon footprint and increasing the supply of green electricity are important elements of a renewable energy policy.

During the COVID-19 pandemic, the government encouraged working from home, so there was an increase in daytime electricity consumption. Therefore, adding the restriction of "household population size" may not have a significant effect on the target households. Decision makers must reconsider the appropriateness and importance of the "household population size" attribute to ensure policy effectiveness. Target households underwent changes during the pandemic, and the government must re-evaluate existing policies to account for new lifestyles and environments.

Form 4: Adding Restrictions with Result Change

PBe: if A1 = 2, A2 = 0, B1 = 1, B2 = 0, B3 = 2 and B5 = 0, then O1 = 1 and O2 = 1.

PAf: if A1 = 2, A2 = 0, B1 = 1, B2 = 0, B3 = 2, B4 = 3 and B5 = 0, then O1 = 1 and O2 = 0.

Alrwashdeh [39] compared the energy output from two different heights of solar towers and showed that energy output is proportional to tower height. In the same year, the Energy Bureau of the Ministry of Economic Affairs proposed a solar photovoltaic strategy of "rooftops first, ground later". The government aims to ensure a stable energy supply, improve energy storage efficiency for rooftop solar panels and enhance lighting conditions. During the COVID-19 pandemic, the "number of floors" was added as an incentive condition to encourage households to participate in the program. However, the addition of this attribute is directly related to the existing building, so some households are less willing to participate since they do not meet these conditions.

Policymakers must consider the relevance of the number of floors in target households. Attributes are a significant consideration for target households, so the green energy plan must be evaluated to determine whether it meets their needs. If the plan is unable to achieve the original objectives, it requires review and adjustment.

Form 5: Reduction in Conditionality

PBe: if A1 = 2, A2 = 0, B1 = 1, B2 = 0, B3 = 2, B4 = 0 and B5 = 0, then O1 = 1 and O2 = 1.

PAf: if A1 = 2, A2 = 0, B1 = 1, B2 = 0, B4 = 0 and B5 = 0, then O1 = 1 and O2 = 1.

Aksoezen et al. [1] found a correlation between building age and energy consumption. The report by the Bureau of Energy, MOEA, in 2019 showed that current solar energy projects require a contract of at least 20 years and evaluation by professional installers and public agencies to determine the suitability of solar equipment installation. Therefore, building age has little impact on the installation of solar equipment: the structure and safety of the house are more important. During the COVID-19 pandemic, the age of a house had a less significant impact on the power generation efficiency for rooftop solar panels, and the transitional period during the pandemic was shorter than the average age of houses. Therefore, deleting the "age of the house" attribute has no significant impact on the policy, so the policy has little effect on target households, and this is an unnecessary attribute.

Form 6: Reduction in Conditionality and Result Changes

PBe: if A1 = 2, A2 = 0, B1 = 1, B2 = 0, B3 = 4, B4 = 0 and B5 = 0, then O1 = 1 and O2 = 1.

PAf: if A1 = 2, B1 = 1, B2 = 0, B3 = 4, B4 = 0 and B5 = 0, then O1 = 1 and O2 = 0.

Residents in remote areas have a lower willingness than urban dwellers to participate in the Green Energy Roofs Project. They tend to lead a more natural lifestyle with less reliance on modern technology, so their electricity demand is lower. In terms of the economy, there are fewer job opportunities and lower salaries. However, during the COVID-19 pandemic, economic activities significantly decreased, so more difficulties were experienced in remote areas, where individuals are less willing to participate in such projects.

If the "remote area" attribute is deleted, there may be significant changes in the results since a lack of incentives may lead to a significant reduction in rural residents' willingness to participate. This attribute has a significant impact on target households, and policymakers must account for this attribute. Remote areas have fewer high-rise buildings and longer sunshine hours, and they are well suited to the Green Energy Roofs Project. Therefore, it is necessary to listen more to the needs of residents in remote areas and to formulate appropriate policy adjustments.

Form 7: Adjustment Conditions

PBe: if A1 = 1, A2 = 0, B1 = 1, B2 = 0, B3 = 2, B4 = 0 and B5 = 0, then O1 = 3 and O2 = 1.

PAf: if A1 = 1, A2 = 0, B1 = 1, B2 = 0, B3 = 2, B4 = 0 and B5 = 2, then O1 = 3 and O2 = 1.

The concept of capacity may be unfamiliar to the public since the installation of capacity equipment must comply with the local electricity demand and building restrictions. The capacity for households is closely related to the rooftop space. The variability of capacity in households is much less significant than that for large buildings, and its impact on energy consumption is relatively small. Funkhouser et al. [29] showed that residential solar energy accounts for an increasing proportion of installed capacity, surpassing large-scale capacities in public utilities. Therefore, adjusting the attribute value for "nameplate capacity" may result in very similar before-and-after scenarios, with a smaller impact on the judgment of the target households.

Form 8: Adjustment Conditions and Result Changes

PBe: if A1 = 2, A2 = 0, B1 = 1, B2 = 0, B3 = 6, B4 = 0 and B5 = 1, then O1 = 1 and O2 = 0.

PAf: if A1 = 2, A2 = 0, B1 = 1, B2 = 1, B3 = 6, B4 = 0 and B5 = 1, then O1 = 1 and O2 = 1.

The income of households with solar panels installed is usually greater than that of those that do not have solar panels [40]. The government actively intervened in the housing market policy by adjusting the Central Bank's Regulations on Financial Institutions' Real Estate Mortgage Loan Business at the end of 2020 [41], with the intention of discouraging fewer investment buyers from purchasing homes. However, this did not affect the willingness of the investing public to buy houses, and the phenomenon of "you do your thing, I'll do mine" appeared, possibly due to the excessive amount of capital in the market and the shortage of labor and materials due to the pandemic, so the economy became stagnant, and real estate remains a popular investment. The government also adjusted the tax base and rate for property holding tax in 2014, which directly reduced income and willingness to purchase homes. Two different examples of housing market policies show that the government's policy adjustments have a direct impact on people's income attributes and are of concern to them.

The COVID-19 pandemic had a significant impact on the global power sector and a direct effect on oil and natural gas prices [33]. In this environment, households have experienced a decrease in disposable income, so policymakers must increase the willingness of target households to participate by adjusting the "income" attribute value to provide more achievable subsidy conditions. Adjustment of the "income" attribute is crucial since it has a direct and significant impact on the outcome. Policymakers must determine whether the adjustment of this attribute aligns with the original planning goals.

Form 9: Special Circumstances

PBe: if A1 = 2, A2 = 0, B1 = 1, B2 = 0, B3 = 2, B4 = 0 and B5 = 0, then O1 = 1 and O2 = 1.

PAf: if A1 = 1, A2 = 0, B1 = 3, B2 = 0, B3 = 0, B4 = 3 and B5 = 3, then O1 = 0 and O2 = 0.

In special circumstances, analysis is challenging and prone to errors, so new decision rules must be formulated. Comparing rule sets within a short period of time from t to $t + 1$ is not feasible.

Table 5 summarizes the nine types of evolutionary implications of case studies using different time differences and RSs for decision rule analysis. The results show the changes in the evolutionary process.

Table 5. Integration of evolutionary implications.

Condition Attributes	Outcome	Type	Evolutionary Implications
Maintenance	Result does not change	1	Maintain original attitude
	Result changes	2	No longer feel policy relevance or have higher expectations
Increasing restrictions	Result does not change	3	Attributes are not noticed by the target households or have little impact on participation conditions
	Result changes	4	Attributes are more valued by the target households, and the program's original goals may need to be reassessed
Reducing conditions	Result does not change	5	Deleted attributes are less necessary
	Result changes	6	Attributes have a significant impact on target households
Adjusting conditions	Result does not change	7	Low changes before and after adjustment, with little impact on willingness to participate
	Result changes	8	Changes in attributes have a significant impact on target households, and subtle adjustments may have a significant impact
Special circumstances		9	New decision rules are generated

4.2. Case Study Summary

This study shows that the majority of the public support the Green Energy Roofs Project, particularly since the outbreak of the COVID-19 pandemic, which has led to a decrease in income. This program provides a passive income and serves as a channel for household revenue. However, some members of the public have concerns about the safety of the installation process and the ongoing maintenance that is required. It is important that the incentives for the Green Energy Roofs Project and knowledge related to solar panels are made more easily comprehensible to the public. The results of these studies allow the government to make decisions on the development of green energy programs to increase the installation of solar panels on household roofs.

However, due to the changing policies and the differences in the installation of solar energy equipment due to environmental conditions, the amount of electricity that is generated and the reduction in the service life, current rates may not be suitable for all users. This study shows that different rates or incentives are required to meet the needs of different households. This study recommends that the feed-in tariff be adjusted to meet the needs of different households, so as to avoid any burden on users, policy changes or cost issues that may prevent the long-term stability of renewable energy generation. There is a current shortage of resources, so the use of solar power reduces the dependence on traditional energy sources, and carbon emissions. Solar power is a widely used green energy source that is renewable and features low maintenance costs, so the government must formulate policies that feature more comprehensive considerations in order to implement green energy policies.

From a decision-making perspective, the RS rule set evolves over two different periods. During the pandemic, public concern about energy increased, so the government had an opportunity to increase the implementation of green energy policies. Before and after the COVID-19 outbreak, the factors that influence the decision to install solar panels may have changed. The negative impact of the pandemic on income and expenses may make economic cost more important. The pandemic may also have an impact on energy markets and policies, so policymakers must promote green energy and environmental conservation and adopt a more holistic attitude to issues related to energy policy. Policies must be flexible to mitigate the impact of the pandemic on household energy costs. Governments and businesses are reducing carbon emissions, achieving carbon neutrality and increasing the use of green energy.

5. Conclusions

This study compares two rule sets to determine the weighting and assessment criteria for each attribute in different temporal and spatial contexts. Past experience can provide strategies for decision making if there are variations between conditions and outcomes and can be applied to decision making in different industries. This approach selects different features and generates more favorable decisions. There is a global energy shortage, so in terms of energy saving and achieving the carbon reduction goals, solar panels are an ideal green energy source. This study determines the willingness of the public to participate in the Green Energy Roofs Project and the effectiveness of adjustment strategies developed by the government in its implementation.

Individuals assess their suitability to participate in the program based on their own conditions, such as renting unused roofs to install solar panels or joining the program under appropriate conditions. The government adjusts to the response from target families and achieves its policy goals through intervention and other means. Appropriate policy support must also be provided to the public in terms of systems and infrastructure. Governments are promoting rooftop solar projects to improve energy efficiency and reduce environmental impact. During the pandemic, increased working and studying from home increased household energy use.

This study determines the changes in willingness to install solar panels on roofs and the considerations before and after the outbreak. The results show that after the pandemic, people's interest in green energy and energy reduction increased, which increased their willingness to install solar panels on roofs. Despite the positive attitude to installing solar panels on rooftops, the negative impact of the pandemic on income and expenses also had an impact on the financial cost of this decision. The government and industry must increase their efforts in terms of policy advocacy and marketing.

Rooftop solar power systems are simpler to implement than ground-based systems. They do not require a large area of land and people can easily participate in the program by providing their roofs. It is important to ensure that households have access to information and to streamline the application process. A level of awareness and confidence in solar panels will naturally lead to increased participation. The government and the industry must also continue to promote the implementation of sustainable energy policies through advocacy and effective funding to provide the public with information about the Green Energy Roofs Project incentives and information on solar panels.

This study determined the impact of rule evolution and feed-in tariff rates on households' willingness to participate in energy incentive programs, using Taiwan's Green Energy Roofs Project as a case study. The research scope was limited to the energy incentive for the household sector and excluded other sectors, such as energy development and industrial rooftop solar installations. Therefore, the generalizability of the research results was limited. The decision-making model for this study must also be improved by accounting for more decision factors to increase the accuracy and reliability of the decision that is made.

The proposed approach is based on the rough set theory. The limitation of classical rough sets requires large amounts of labeled data. Computation is time-consuming when dealing with large amounts of labeled datasets based on rough sets. In addition, when faced with real-valued continuous data, the rough set theory has difficulties in dealing with it, since it is more suitable for discrete data information systems [42].

To better achieve the goal of sustainable development, the following steps will be take: (1) a future study will expand the research scope to other sectors to determine different sectors' views and willingness toward energy subsidies and incentives; (2) future research will determine the innovation and development trends of solar panel technology that further reduce the cost of solar panels, improve their competitiveness and increase the scope of application; (3) the research results will also be extended to other fields to improve the study's usability and strengthen communication with other research areas; (4) a future study will propose more extensive and detailed levels to promote the application of renewable energy and achieve sustainable development; and (5) to improve computation efficiency, discretization methods and different types of neighborhoods such as containment neighborhoods may be considered in a future study to reduce the boundary region and improve the accuracy measure [12].

Author Contributions: Conceptualization, C.H. and C.-C.H.; Methodology, D.-N.C.; Software, D.-N.C.; Validation, C.H.; Formal analysis, C.H. and D.-N.C.; Investigation, C.H.; Resources, C.H.; Data curation, D.-N.C.; Writing—original draft, C.-C.H.; Writing—review & editing, Y.W.; Visualization, Y.W.; Supervision, C.-C.H. All authors have read and agreed to the published version of the manuscript.

Funding: This research was funded by the National Science Foundation—Ministry of Science and Technology of Taiwan, grant number MOST 110-2410-H-260 -016 -MY2, 110-2813-C-260-015-H, and NSTC 112-2410-H-260-010-MY3.

Data Availability Statement: Not applicable.

Conflicts of Interest: The authors declare no conflict of interest.

References

1. Aksoezen, M.; Daniel, M.; Hassler, U.; Kohler, N. Building age as an indicator for energy consumption. *Energy Build.* **2015**, *87*, 74–86. [CrossRef]
2. Satyadas, A.; Harigopal, U.; Cassaigne, N.P. Knowledge management tutorial: An editorial overview. *IEEE Trans. Syst. Man Cybern. Part C-Appl. Rev.* **2001**, *31*, 429–437. [CrossRef]
3. Yen, C.E.; Huang, C.C.; Wen, D.W.; Wang, Y.P. Decision support to customer decrement detection at the early stage for theme parks. *Decis. Support Syst.* **2017**, *102*, 82–90. [CrossRef]
4. Alzahrany, A.; Kabir, G.; Al Zohbi, G. Evaluation of the barriers to and drivers of the implementation of solar energy in Saudi Arabia. *Int. J. Sustain. Dev. World Ecol.* **2022**, *29*, 543–558. [CrossRef]
5. Gung, R.R.; Huang, C.C.; Hung, W.I.; Fang, Y.J. The use of hybrid analytics to establish effective strategies for household energy conservation. *Renew. Sustain. Energy Rev.* **2020**, *133*, 10. [CrossRef]
6. Sharma, S.; Dua, A.; Singh, M.; Kumar, N.; Prakash, S. Fuzzy rough set based energy management system for self-sustainable smart city. *Renew. Sustain. Energy Rev.* **2018**, *82*, 3633–3644. [CrossRef]
7. Pawlak, Z. Rough sets. *Int. J. Comput. Inf. Sci.* **1982**, *11*, 341–356. [CrossRef]
8. Yao, Y.; Greco, S.; Słowiński, R. Probabilistic rough sets. In *Springer Handbook of Computational Intelligence*; Springer: Berlin/Heidelberg, Germany, 2015; pp. 387–411. [CrossRef]
9. Lashin, E.F.; Kozae, A.M.; Khadra, A.A.A.; Medhat, T. Rough set theory for topological spaces. *Int. J. Approx. Reason.* **2005**, *40*, 35–43. [CrossRef]
10. Azzam, A.; Al-shami, T.M. Five Generalized Rough Approximation Spaces Produced by Maximal Rough Neighborhoods. *Symmetry* **2023**, *15*, 751. [CrossRef]
11. Al-shami, T.M. An improvement of rough sets' accuracy measure using containment neighborhoods with a medical application. *Inf. Sci.* **2021**, *569*, 110–124. [CrossRef]
12. Al-shami, T.M.; Mhemdi, A. Approximation spaces inspired by subset rough neighborhoods with applications. *Demonstr. Math.* **2023**, *56*, 24. [CrossRef]
13. Al-shami, T.M.; Alshammari, I. Rough sets models inspired by supra-topology structures. *Artif. Intell. Rev.* **2023**, *56*, 6855–6883. [CrossRef] [PubMed]
14. Swiniarski, R.W.; Skowron, A. Rough set methods in feature selection and recognition. *Pattern Recognit. Lett.* **2003**, *24*, 833–849. [CrossRef]
15. Pawlak, Z. *Rough Sets: Theoretical Aspects of Reasoning about Data*; Springer Science & Business Media: Berlin/Heidelberg, Germany, 1991; Volume 9.
16. Zhang, X.; Yao, Y. Tri-level attribute reduction in rough set theory. *Expert Syst. Appl.* **2022**, *190*, 116187. [CrossRef]
17. Zhang, X.; Miao, D. Three-layer granular structures and three-way informational measures of a decision table. *Inf. Sci.* **2017**, *412*, 67–86. [CrossRef]
18. Murray, L.W. Agile manufacturing: Forging new frontiers—Kidd, PT. *J. Prod. Innov. Manag.* **1996**, *13*, 181–182. [CrossRef]
19. Tseng, T.L.; Huang, C.C. Sustainable service and energy provision based on agile rule induction. *Int. J. Prod. Econ.* **2016**, *181*, 273–288. [CrossRef]
20. Wang, J.Q.; Zhang, X.H. A Novel Multi-Criteria Decision-Making Method Based on Rough Sets and Fuzzy Measures. *Axioms* **2022**, *11*, 15. [CrossRef]
21. Ayub, S.; Shabir, M.; Riaz, M.; Karaaslan, F.; Marinkovic, D.; Vranjes, D. Linear Diophantine Fuzzy Rough Sets on Paired Universes with Multi Stage Decision Analysis. *Axioms* **2022**, *11*, 18. [CrossRef]
22. Jia, Q.S.; Ho, Y.C.; Zhao, Q.C. Comparison of selection rules for ordinal optimization. *Math. Comput. Model.* **2006**, *43*, 1150–1171. [CrossRef]
23. Li, R.; Xu, M.; Chen, Z.; Gao, B.; Cai, J.; Shen, F.; He, X.; Zhuang, Y.; Chen, D. Phenology-based classification of crop species and rotation types using fused MODIS and Landsat data: The comparison of a random-forest-based model and a decision-rule-based model. *Soil Tillage Res.* **2021**, *206*, 12. [CrossRef]
24. Lim, G.G.; Kang, J.Y.; Lee, J.K.; Lee, D.C. Rule-based personalized comparison shopping including delivery cost. *Electron. Commer. Res. Appl.* **2011**, *10*, 637–649. [CrossRef]
25. Jafari, M.; Malekjamshidi, Z. Optimal energy management of a residential-based hybrid renewable energy system using rule-based real-time control and 2D dynamic programming optimization method. *Renew. Energy* **2020**, *146*, 254–266. [CrossRef]
26. Crago, C.L.; Grazier, E.; Breger, D. Income and racial disparities in financial returns from solar PV deployment. *Energy Econ.* **2023**, *117*, 12. [CrossRef]
27. Lau, L.S.; Senadjki, A.; Ching, S.L.; Choong, C.K.; Seow, A.; Choong, Y.O.; Wei, C.Y. Solar photovoltaic as a means to sustainable energy consumption in Malaysia: The role of knowledge and price value. *Energy Sources Part B-Econ. Plan. Policy* **2021**, *16*, 303–323. [CrossRef]
28. Li, Y.H.; Wang, S.Y.; Dai, W.; Wu, L.S. Prediction of the Share of Solar Power in China Based on FGM (1,1) Model. *Axioms* **2022**, *11*, 10. [CrossRef]
29. Funkhouser, E.; Blackburn, G.; Magee, C.; Rai, V. Business model innovations for deploying distributed generation: The emerging landscape of community solar in the U.S. *Energy Res. Soc. Sci.* **2015**, *10*, 90–101. [CrossRef]

30. Costa, A.; Ng, T.S.; Su, B. Long-term solar PV planning: An economic-driven robust optimization approach. *Appl. Energy* **2023**, *335*, 16. [CrossRef]
31. Barnes, J.L.; Krishen, A.S.; Chan, A. Passive and active peer effects in the spatial diffusion of residential solar panels: A case study of the Las Vegas Valley. *J. Clean. Prod.* **2022**, *363*, 11. [CrossRef]
32. Varho, V.; Rikkonen, P.; Rasi, S. Futures of distributed small-scale renewable energy in Finland—A Delphi study of the opportunities and obstacles up to 2025. *Technol. Forecast. Soc. Chang.* **2016**, *104*, 30–37. [CrossRef]
33. Namazkhan, M.; Albers, C.; Steg, L. A decision tree method for explaining household gas consumption: The role of building characteristics, socio-demographic variables, psychological factors and household behaviour. *Renew. Sustain. Energy Rev.* **2020**, *119*, 11. [CrossRef]
34. Vlek, C.; Skolnik, M.; Gatersleben, B. Sustainable development and quality of life: Expected effects of prospective changes in economic and environmental conditions. *Z. Fur Exp. Psychol.* **1998**, *45*, 319–333.
35. Rausch, T.M.; Kopplin, C.S. Bridge the gap: Consumers' purchase intention and behavior regarding sustainable clothing. *J. Clean. Prod.* **2021**, *278*, 15. [CrossRef]
36. Gatersleben, B.; Vlek, C. Household consumption, quality of life, and environmental impacts: A psychological perspective and empirical study. In *Green Households*; Routledge: London, UK, 2014; pp. 141–183.
37. Liu, S.-J. COVID-19 Impact Analysis and Recommendations for the Power Industry. 2020. Available online: https://km.twenergy.org.tw/Knowledge/knowledge_more?id=8407 (accessed on 1 October 2020).
38. Wiedenhofer, D.; Smetschka, B.; Akenji, L.; Jalas, M.; Haberl, H. Household time use, carbon footprints, and urban form: A review of the potential contributions of everyday living to the 1.5 degrees C climate target. *Curr. Opin. Environ. Sustain.* **2018**, *30*, 7–17. [CrossRef]
39. Alrwashdeh, S.S. The effect of solar tower height on its energy output at Ma'an-Jordan. *AIMS Energy* **2018**, *6*, 959–966. [CrossRef]
40. Rai, V.; McAndrews, K. Decision-making and behavior change in residential adopters of solar PV. In Proceedings of the World Renewable Energy Forum, Denver, CO, USA, 13–17 May 2012.
41. Central Bank of the Republic of China (Taiwan). Press Release on the Resolution of the Joint Conference of the Central Bank Supervisors. 2021. Available online: https://www.cbc.gov.tw/tw/cp-302-141562-49221-1.html (accessed on 23 September 2020).
42. Fan, Y.N.; Tseng, T.L.; Chern, C.C.; Huang, C.C. Rule induction based on an incremental rough set. *Expert Syst. Appl.* **2009**, *36*, 11439–11450. [CrossRef]

axioms

MDPI

Article

Sugeno Integral Based on Overlap Function and Its Application to Fuzzy Quantifiers and Multi-Attribute Decision-Making

Xiaoyan Mao [1,2,*], Chaolu Temuer [3] and Huijie Zhou [2]

1 College of Information Engineering, Shanghai Maritime University, Shanghai 200135, China
2 College of Science and Technology, Ningbo University, Ningbo 315211, China; zhouhuijie@nbu.edu.cn
3 School of Sciences, Shanghai Maritime University, Shanghai 200135, China; tmchaolu@shmtu.edu.cn
* Correspondence: maoxiaoyan@nbu.edu.cn; Tel.: +86-057463712651

Abstract: The overlap function is an important class of aggregation function that is closely related to the continuous triangular norm. It has important applications in information fusion, image processing, information classification, intelligent decision-making, etc. The usual multi-attribute decision-making (MADM) is to select the decision object that performs well on all attributes (indicators), which is quite demanding. The MADM based on fuzzy quantifiers is to select the decision object that performs well on a certain proportion or quantification (such as most, many, more than half, etc.) of attributes. Therefore, it is necessary to study how to express and calculate fuzzy quantifiers such as most, many, etc. In this paper, the Sugeno integral based on the overlap function (called the O-Sugeno integral) is used as a new information fusion tool, and some related properties are studied. Then, the truth value of a linguistic quantified proposition can be estimated by using the O-Sugeno integral, and the O-Sugeno integral semantics of fuzzy quantifiers is proposed. Finally, the MADM method based on the O-Sugeno integral semantics of fuzzy quantifiers is proposed and the feasibility of our method is verified by several illustrative examples such as the logistics park location problem.

Keywords: overlap function; sugeno integral; fuzzy quantifier; multi-attribute decision-making

MSC: 03B52; 03E72; 90B50

Citation: Mao, X.; Temuer, C.; Zhou, H. Sugeno Integral Based on Overlap Function and Its Application to Fuzzy Quantifiers and Multi-Attribute Decision-Making. *Axioms* **2023**, *12*, 734. https://doi.org/10.3390/axioms12080734

Academic Editors: Ta-Chung Chu and Wei-Chang Yeh

Received: 6 June 2023
Revised: 20 July 2023
Accepted: 25 July 2023
Published: 27 July 2023

1. Introduction

The triangular norm (t-norm) first appeared in Menger's paper "Statistical metrics" in 1942, which proposed t-norm as a natural generalization of triangular inequalities in classical metric spaces [1]. A t-norm is an aggregation operator that satisfies commutativity, monotonicity, and associativity, and has the identity element 1. From the mathematical structure, t-norms and t-conorms are pairs of dual operators. T-norms play an important role as the general fuzzy "and" operator in the fuzzy logic community. In order to apply more widely, researchers proposed many generalized forms of t-norms and t-conorms, such as t-seminorms [2], pseudo-t-norms [3], t-operators [4], uninorms [5], semiuninorms [6], etc. Recently, many scholars still studied the extension structure of t-norms. For example, Dan proposed a universal way to study t-semi(co)norms and semiuninorms in terms of behavior operations [7]. For partially defined binary operations in practical problems, Borzooei et al. introduced a partial t-norm on a bounded lattice [8]. Zhang et al. further investigated the partial residual implications of partial t-norms and partial residuated lattices [9].

The measures discussed in classical measure theory are additive, and they are abstractions of real-world concepts such as length, area, volume and weight. However, additivity cannot be satisfied in many practical situations, e.g., the work efficiency of two people in cooperation is often greater or less than the combined work efficiency of two people. In 1971, Shilkret introduced the maxitive measure in place of the usual additive measure, and then proposed the integral with respect to a maxitive measure in [10]. The concepts of fuzzy measures and fuzzy integrals (also called Sugeno integrals) were first introduced by Sugeno in his doctoral thesis in 1974 [11]. Fuzzy measure is a class of set functions using weaker monotonicity instead of additivity. Fuzzy measures have been widely used in different scenarios and can be described as similar concepts such as importance, reliability, and satisfaction. Sugeno integrals replace the addition and multiplication of Lebesgue integrals with the maximum and minimum operators, respectively. These operations have limitations. Subsequently, many scholars further generalized Sugeno's integral theory based on other operators. For example, Garcia and Alvarez defined semi-normed fuzzy integrals and semiconorm fuzzy integrals and pointed out that Sugeno integrals are a special case [2]. Dudois et al. introduced Sugeno-like qualitative integrals and qualitative co-integrals defined in terms of fuzzy conjunctions and implications, respectively [12]. In 2010, Klement et al. Proposed the framework covering generalizations of Sugeno integrals, in which the role of multiplication is played by semicopulas [13]. Note that the multiplication of Shilkret integrals is still the standard product. Jin et al. introduced the concept of weak universal integrals based on semicopulas, which are generalizations of Sugeno integrals and Shilkret integrals [14]. Mihailovi and Pap defined Sugeno integrals based on set functions that have the properties of absolute monotony and sign stability [15]. In recent years, many scholars have been interested in Sugeno integrals and their generalizations, such as [16–20].

Eslami et al. pointed out that t-norms are not suitable for solving natural interpretations of language words [21]. In addition, Fodor and Keresztfalvi proposed non-associative conjunctions are very effective in generalized inference patterns [22]. Bustince et al., in 2010, introduced the concept of overlap functions as a special class of bivariate continuous aggregation functions, which are closely related to continuous t-norms [23]. Subsequently, scholars deeply studied the theory of overlap functions and their application, such as [24–29]. Overlap functions are mainly used for image processing, classification problems, decision analysis, and intelligent information fusion, in which the associative law is not strongly required. In order to be applied in more fields, overlap functions were generalized in various ways, including general overlap functions [30], Archimedean overlap functions [31], quasi-overlap functions [32], pseudo-overlap functions [33], semi-overlap functions [34], and interval-valued pseudo-overlap functions [35].

The MADM requires comprehensively considering multiple attributes through the aggregation function, and gives the optimal choice or sorts the schemes. The usual MADM is to select the decision object that performs well on all attributes (indicators), which is quite demanding. Attribute weight values can reflect the importance of attributes. In the MADM problem with known attribute weights, the decision-maker considers all the attributes together by means of an aggregation function and gives the optimal choice or ranking of decision objects. However, different attribute weights will affect the decision results and it is difficult to obtain the optimal attribute weights. The MADM problem with unknown attribute weights has been studied by many researchers from different perspectives. For example, a new MADM method based on rough sets and fuzzy measures was proposed by Wang et al. [36]. Because decision objects that perform well on all attributes are difficult to be selected out, many scholars have begun to study the MADM based on fuzzy quantifiers. The MADM problem based on fuzzy quantifiers selects the decision object that performs well on a certain proportion or quantification (such as most, many, more than half, etc.) of attributes. Therefore, it is necessary to study how to express and calculate fuzzy quantifiers such as most, many, etc.

In 1983, Zadeh first used the term "fuzzy quantifier" and described a method for quantifying fuzzy sets [37]. Zadeh treated fuzzy quantifiers as fuzzy numbers, and linguistic quantified propositions correspond to fuzzy sets defined by linguistic predicates. Zadeh obtained the truth value of a quantification proposition by calculating the cardinality of the fuzzy set. In 1988, Yager, an American scholar, proposed the method of evaluating a linguistic quantified proposition based on the ordered weighted average (OWA) operators [38]. Recently, Dvorak et al. proposed the notion of fuzzy quantifiers over fuzzy domains and investigated relevant semantic properties [39]. Medina et al. further investigated the properties of generalized quantifiers and defined the semantics of multi-adjoint logic programs [40]. In 2006, Ying, a Chinese scholar, proposed a method for modeling linguistic statements involving fuzzy quantifiers in natural language, in which fuzzy measures can be used to represent fuzzy quantifiers and Sugeno integrals can be used to calculate the truth value of a quantified statement [41]. Zhang et al. studied fuzzy quantifiers and their integral semantics based on the Sugeno integral with t-norm, and successfully applied them to the problem [42].

For wider application, we generalized the t-norm-based Sugeno integral in [42] by replacing t-norms with overlap function, which is non-associative and continuous. The Sugeno integral based on overlap functions (O-Sugeno integral) is proposed as a new information fusion tool, and its related properties are studied. Then, the O-Sugeno integral is used to deal with fuzzy quantifiers and the O-Sugeno integral semantics of fuzzy quantifiers is proposed. In fuzzy quantifier integral semantics, fuzzy measures are usually used to represent fuzzy quantifiers, and O-Sugeno integrals are used to calculate the truth value of a quantified proposition. Finally, a novel MADM method is proposed based on the O-Sugeno integral semantics of fuzzy quantifiers. The method is used to solve the fuzzy quantifiers-based MADM problems.

2. Preliminaries

We briefly review the basic definitions and conclusions that are used in our discussion of overlap functions, fuzzy quantifiers, and Sugeno integrals.

Definition 1. *Binary mapping O: $[0, 1]^2 \to [0, 1]$ is called an overlap function if it satisfies the following requirements: for any $x, y \in [0, 1]$:*

(*i*) *O is commutative, that is, $O(x, y) = O(y, x)$;*
(*ii*) *$O(x, y) = 0$ if and only if $x\,y = 0$;*
(*iii*) *$O(x, y) = 1$ if and only if $x\,y = 1$;*
(*iv*) *O is non-decreasing; and*
(*v*) *O is an continuous function [23].*

Definition 2. *Overlap function O is inflationary if it satisfies the condition $O(x, 1) \geq x$, and is deflationary if it satisfies $O(x, 1) \leq x$; and has unit element 1 if $O(x, 1) = x$ holds for each $x \in [0, 1]$ [43].*

Example 1.

(1) *The binary function is defined by*

$$O(x,y) = xy\frac{x+y}{2}$$

where x, y are two arbitrary element on the unit interval. Then it is an overlap function that does not have associativity and 1 is not a unit element, therefore, it is not a continuous t-norm.

(2) *The binary function is defined by*

$$O(x, y) = min(x^p, y^p)$$

for every $x, y \in [0, 1]$ and $p > 0$. Then it is an overlap function and is deflationary if $p > 1$ and inflationary if $0 < p < 1$, and has neutral element 1 if $p = 1$.

In natural languages, many "vague" words are used to express quantity, such as "several", "a few", "quite a few", "most", "many", "very many", "not many", "not very many", "approximately eight", "frequently", etc. These linguistic components used to represent inexact amounts are called fuzzy quantifiers [37].

Definition 3. *A fuzzy quantifier includes two items:*
For arbitrary non-empty set X, a Borel field $\wp_X$ over X; and

a selection function

$$Q\colon (X, \wp_X) \text{ a } Q(X, \wp_X) \in M(X, \wp X)$$

of the truth class {M(X, ℘ X): (X, ℘ X) is a measurable space} [41].

For convenience, the selection function $Q(X, \wp_X)$ is usually abbreviated as Q_X when the Borel field does not need to be specifically indicated. Given X as a discourse domain, if *E* represents individuals in X that have a specific attribute *A*, then $Q_X(E)$ is seen as the truth value of the linguistic quantified statement "*Q* Xs are *A*s".

Example 2.
The quantifier "at least five" is defined as follows:

$$at\ least\ five_X(E) = \begin{cases} 1,\ if\ |E| \geq 5, \\ 0,\ otherwise. \end{cases}$$

where the domain X is any nonempty set, and E is any subset of X. Then the quantifier "at least five" is a crisp quantifier because at least $five_X(E) \in \{0, 1\}$.

As is well known, ∀ and ∃ are also crisp quantifiers. The following example gives three typical fuzzy quantifiers.

Example 3.
The terms "many", "most", and "almost all" are often used to indicate inexact amounts in natural language, and are defined based on the following fuzzy measures [41]:

$$many_X(E) = \frac{|E|}{|X|}, most_X(E) = \left(\frac{|E|}{|X|}\right)^{3/2}, almost_X(E) = \left(\frac{|E|}{|X|}\right)^{2},$$

for every non-empty set X and any subset E of X, where | E | *represents the cardinality of E.*

Definition 4: Suppose *(X,* 2^X*, m) is a fuzzy measure space. If h: X → [0, 1] is a measurable function, then the Sugeno integral of h over A ∈ ℘ is defined as follows [11]:*

$$\int_A h \circ m = \sup_{F \in 2^X} \min[\inf_{x \in F} h(x), m(A \cap F)]$$

Theorem 1. *Given (X, ℘, m) as a fuzzy measure space, for any ℘ measurable function h: X → [0, 1], we have*

$$\int_A h \circ m = \sup_{\lambda \in [0,1]} \min[\lambda, m(A \cap h_\lambda)]$$

where $h_\lambda = \{x \in X\colon h(x) \geq \lambda\}$ *for every* $\lambda \in [0, 1]$ *[11].*

In particular, $\int_A h \circ m$ will be abbreviated as $\int h \circ m$ whenever *A* = X.

3. Sugeno Integrals Based on Overlap Functions

Definition 5. *Given (X, ℘, m) as a fuzzy measure space and O:* $[0, 1]^2 \to [0, 1]$ *as an overlap function, if h: X → [0, 1] is a ℘ measurable function, then the Sugeno integral based on the overlap function (O-Sugeno integral) of h over A ∈ ℘ is defined by*

$$\int_A^{(OS)} h \circ m = \sup_{\lambda \in [0,1]} O[\lambda, m(A \cap h_\lambda)]$$

where $h_\lambda = \{x \in X\colon h(x) \geq \lambda\}$ *for every* $\lambda \in [0, 1]$.

When the Borel field in measurable space is the power set of the underlying set, the O-Sugeno integral can be simplified.

Theorem 2. *Assume* $(X, \wp, m)$ *is a fuzzy measure space and* $O: [0, 1]^2 \rightarrow [0, 1]$ *is an overlap function. If* $\wp = 2^X$, *then for any* $\wp$ *measurable function* $h: X \rightarrow [0, 1]$ *and any subset* A *of* X, *we have*

$$\int_A^{(OS)} h \circ m = \sup_{F \subseteq X} O[\inf_{x \in F} h(x), m(A \cap F)]$$

where $h_\lambda = \{x \in X: h(x) \geq \lambda\}$ *for every* $\lambda \in [0, 1]$.

Proof of Theorem 2.

(1) $\forall F \subseteq X$, Let $\lambda' = \inf_{x \in F} h(x)$.

If $\lambda' = 0$, then $h_{\lambda'} = X$, so $F \subseteq h_{\lambda'}$;
If $\lambda' > 0$, then $\forall x \in F$, $h(x) \geq \lambda'$, so $F \subseteq h_{\lambda'}$.
Hence, we have

$$O[\lambda', m(A \cap h_{\lambda'})] \geq O[\inf_{x \in F} h(x), m(A \cap F)]$$

Further, we obtain

$$\begin{aligned} \int_A^{(TS)} h \circ m &= \sup_{\lambda \in [0,1]} O[\lambda, m(A \cap h_\lambda)] \\ &\geq \sup_{F \subseteq X} O[\lambda', m(A \cap h_{\lambda'})], \ (\lambda' = \inf_{x \in F} h(x)) \\ &\geq \sup_{F \subseteq X} O[\inf_{x \in F} h(x), m(A \cap F)] \end{aligned}$$

(2) $\forall \lambda \in [0, 1]$, Let $F' = h_\lambda$, then $\forall x \in F'$, $h(x) \geq \lambda$, so $\inf_{x \in F'} h(x) \geq \lambda$.

Hence, we have

$$O[\inf_{x \in F'} h(x), m(A \cap h_\lambda)] \geq O[\lambda, m(A \cap h_\lambda)]$$

Further, we obtain

$$\begin{aligned} \int_A^{(OS)} h \circ m &= \sup_{\lambda \in [0,1]} O[\lambda, m(A \cap h_\lambda)] \\ &\leq \sup_{\lambda \in [0,1]} O[\inf_{x \in F'} h(x), m(A \cap F')], \ (F' = h_\lambda) \\ &\leq \sup_{F \subseteq X} O[\inf_{x \in F} h(x), m(A \cap F)] \end{aligned}$$

In summary, we can get

$$\int_A^{(OS)} h \circ m = \sup_{F \subseteq X} O[\inf_{x \in F} h(x), m(A \cap F)]$$

□.

In the case where the domain is finite, the O-Sugeno integral over it can be further simplified.

Theorem 3. *Given domain* $X = \{x_1, \ldots, x_n\}$ *as a finite set, and* $\wp = 2^X$, $(X, \wp, m)$ *as a fuzzy measure space, and* $O: [0, 1]^2 \rightarrow [0, 1]$ *as an overlap function, if a* $\wp$ *measurable function* $h: X \rightarrow [0, 1]$ *such* *that $h(x_i) \leq h(x_{i+1})$, for $1 \leq i \leq n-1$ (if not, rearrange $h(x_i)$, $1 \leq i \leq n$). Then the O-Sugeno integral of h over A is further simplified as follows:*

$$\int_A^{(OS)} h \circ m = \max_{i=1}^{n} O[h(x_i), m(A \cap X_i)]$$

where A is any subset of X and $X_i = \{x_j : i \leq j \leq n\}$, $1 \leq i \leq n$.

Proof of Theorem 3. For any $\lambda \in [0, 1]$, it holds that

$$h_\lambda = \begin{cases} X_1 = X, 0 \leq \lambda \leq h(x_1) \\ X_2, h(x_1) < \lambda \leq h(x_2) \\ X_3, h(x_2) < \lambda \leq h(x_3) \\ \cdots \\ X_n, h(x_{n-1}) < \lambda \leq h(x_n) \\ \Phi, h(x_n) < \lambda \end{cases}$$

So, we obtain

$$\begin{aligned} \int_A^{(OS)} h \circ m &= \sup_{\lambda \in [0,1]} O[\lambda, m(A \cap h_\lambda)] \\ &= \max_{i=1}^{n+1} \left\{ \sup_{\lambda \in [h(x_{i-1}), h(x_i)]} O[\lambda, m(A \cap h_\lambda)] \right\} (h(x_0) = 0, h(x_{n+1}) = 1) \\ &= \max_{i=1}^{n+1} \left\{ \sup_{\lambda \in [h(x_i), h(x_{i+1})]} O[\lambda, m(A \cap X_i)] \right\} (X_{n+1} = \Phi) \\ &= \max_{i=1}^{n+1} \{O[h(x_i), m(A \cap X_i)]\} \\ &= \max_{i=1}^{n} \{O[h(x_{i+1}), m(A \cap X_{i+1})]\} \vee O[1, 0] \\ &= \max_{i=1}^{n} \{O[h(x_{i+1}), m(A \cap X_{i+1})]\} \end{aligned}$$

□.

Theorem 4. *Given (X, ℘, m) as a fuzzy measure space and ℘ = 2^X and O: $[0, 1]^2 \to [0, 1]$ as an overlap function, for arbitrary ℘ measurable functions h, h_1, and h_2 and arbitrary subset A of X, the following conclusion is established:*

(1) *If $h_1 \leq h_2$ (i.e., for any $x \in X$, $h_1(x) \leq h_2(x)$), then it holds that*

$$\int_A^{(OS)} h_1 \circ m \leq \int_A^{(OS)} h_2 \circ m$$

(2) *If m(A) = 0, then it holds that*

$$\int_A^{(OS)} h \circ m = 0$$

(3) *If a constant $c \in [0, 1]$, then it holds that*

$$\int_A^{(OS)} c \circ m = O[c,\ m(A)]$$

(4) *If a constant $c \in [0, 1]$, and for any $x \in X$, max(c, h)(x) = max{c, h(x)}, then it holds that*

$$\int_A^{(OS)} \max(c, h) \circ m = \max\left(\int_A^{(OS)} c \circ m,\ \int_A^{(OS)} h \circ m\right)$$

(5) *If $A_1 \subseteq A_2$, then it holds that*

$$\int_{A_1}^{(OS)} h \circ m \leq \int_{A_2}^{(OS)} h \circ m$$

(6)

$$\int_{A}^{(OS)} \max(h_1, h_2) \circ m \geq \max\left(\int_{A}^{(OS)} h_1 \circ m, \int_{A}^{(OS)} h_2 \circ m\right)$$

Proof of Theorem 4.

(1) For any $\lambda \in [0, 1]$, it holds that for any $x \in X$, $\lambda \leq h_1(x) \leq h_2(x)$.

Then,

$$h_{1\lambda} = \{x \in X: h_1(x) \geq \lambda\} \subseteq \{x \in X: h_2(x) \geq \lambda\} = h_{2\lambda}$$

So,

$$m(A \cap h_{1\lambda}) \leq m(A \cap h_{2\lambda}).$$

Then, we have

$$O(\lambda, m(A \cap h_{1\lambda})) \leq O(\lambda, m(A \cap h_{2\lambda}))$$

Furthermore, we obtain

$$\sup_{\lambda \in [0,1]} O(\lambda, m(A \cap h_{1\lambda})) \leq \sup_{\lambda \in [0,1]} O(\lambda, m(A \cap h_{2\lambda}))$$

that is,

$$\int_{A}^{(OS)} h_1 \circ m \leq \int_{A}^{(OS)} h_2 \circ m$$

(2) From $m(A) = 0$, we have $m(A \cap h_\lambda) = 0$.

Thus, it holds that

$$\int_{A}^{(OS)} h \circ m = \sup_{\lambda \in [0,1]} O[\lambda, m(A \cap h_\lambda)] = \sup_{\lambda \in [0,1]} O[\lambda, 0] = 0$$

(3) For any $x \in X$, we define $h(x) = c$.

If $\lambda \leq c$, then $h_\lambda = X$. So,

$$O[\lambda, m(A \cap h_\lambda)] = O[\lambda, m(A \cap X)] = O[\lambda, m(A)].$$

If $\lambda > c$, then $h_\lambda = \Phi$. So,

$$O[\lambda, m(A \cap \Phi)] = O[\lambda, m(\Phi)] = O[\lambda, 0] = 0$$

Hence, we can obtain

$$\begin{aligned} \int_{A}^{(OS)} h \circ m &= \sup_{\lambda \in [0,1]} O[\lambda, m(A \cap h_\lambda)] \\ &= \max\left(\sup_{\lambda \in [0,c]} O[\lambda, m(A \cap h_\lambda)], \sup_{\lambda \in [c,1]} O[\lambda, m(A \cap h_\lambda)]\right) \\ &= \max\left(\sup_{\lambda \in [0,c]} O[\lambda, m(A)], \sup_{\lambda \in [c,1]} O[\lambda, m(\Phi)]\right) \\ &= \max(O[c, m(A)], 0) \\ &= O[c, m(A)] \end{aligned}$$

(4) If $\lambda \leq c$, then for every $x \in X$,

$$\max(c, h)(x) = \max(c, h(x)) \geq c \geq \lambda,$$

that is, $\max(c, h)_\lambda = X$.

Hence,

$$O[\lambda, m(A \cap \max(c, h)_\lambda)] = O[\lambda, m(A \cap X)] = O[\lambda, m(A)].$$

If $\lambda > c$, then

$$\{x \in X: \max(c, h(x)) \geq \lambda\} = \{x \in X: h(x) \geq \lambda\},$$

that is, $\max(c, h)_\lambda = h_\lambda$.

Hence,

$$O[\lambda, m(A \cap \max(c, h)_\lambda)] = O[\lambda, m(A \cap h_\lambda)].$$

Furthermore, we obtain

$$\begin{aligned} &\sup_{\lambda\in[0,1]} O[\lambda, m(A\cap\max(c,h)_\lambda)] \\ &= \max\left(\sup_{\lambda\in[0,c]} O[\lambda, m(A\cap\max(c,h)_\lambda)],\ \sup_{\lambda\in[c,1]} O[\lambda, m(A\cap\max(c,h)_\lambda)]\right) \\ &= \max\left(O[c, m(A)],\ \sup_{\lambda\in[c,1]} O[\lambda, m(A\cap h_\lambda)]\right) \end{aligned}$$

And because

$$\sup_{\lambda\in[0,c]} O[\lambda, m(A\cap h_\lambda)] \leq \sup_{\lambda\in[0,c]} O[\lambda, m(A)] \leq O[c, m(A)]$$

we obtain

$$\begin{aligned} \int_A^{(OS)} \max(c,h)\circ m &= \sup_{\lambda\in[0,1]} O[\lambda, m(A\cap\max(c,h)_\lambda)] \\ &= \max\left(O[c, m(A)],\ \sup_{\lambda\in[0,1]} O[\lambda, m(A\cap h_\lambda)]\right) \\ &= \max\left(\int_A^{(OS)} c\circ m,\ \int_A^{(OS)} h\circ m\right) \end{aligned}$$

(5) For any $\lambda \in [0, 1]$ and $A_1 \subseteq A_2$, we have

$$(A_1 \cap h_\lambda) \subseteq (A_2 \cap h_\lambda), \text{ then } m(A_1 \cap h_\lambda) \leq m(A_2 \cap h_\lambda).$$

Thus, we can obtain

$$\sup_{\lambda\in[0,1]} O[\lambda, m(A_1\cap h_\lambda)] \leq \sup_{\lambda\in[0,1]} O[\lambda, m(A_2\cap h_\lambda)]$$

that is,

$$\int_{A_1}^{(OS)} h\circ m \leq \int_{A_2}^{(OS)} h\circ m$$

(6) For any $x \in X$, it holds that

$$\max(h_1, h_2)(x) = \max(h_1(x), h_2(x)).$$

So, for any $\lambda \in [0, 1]$, we have

$$\max(h_1, h_2)(x) \geq h_1(x) \geq \lambda \text{ and } \max(h_1, h_2)(x) \geq h_2(x) \geq \lambda.$$

Then,

$$h_{1\lambda} \subseteq \max(h_1, h_2)_\lambda \text{ and } h_{2\lambda} \subseteq \max(h_1, h_2)_\lambda.$$

Therefore, we get

$$m(A \cap h_{1\lambda}) \leq m(A \cap \max(h_1, h_2)_\lambda) \text{ and } m(A \cap h_{2\lambda}) \leq m(A \cap \max(h_1, h_2)_\lambda).$$

Furthermore, we obtain

$$\int_A^{(OS)} \max(h_1, h_2) \circ m \geq \int_A^{(OS)} h_1 \circ m, \int_A^{(OS)} \max(h_1, h_2) \circ m \geq \int_A^{(OS)} h_2 \circ m$$

that is,

$$\int_A^{(Os)} \max(h_1, h_2) \circ m \geq \max\left(\int_A^{(OS)} h_1 \circ m, \int_A^{(OS)} h_2 \circ m\right)$$

□.

Example 4. *Consider the decision-making problem of a commercial housing purchase. After preliminary screening, the buyer needs to choose one of two properties. Assume that the evaluation of the property mainly considers the attributes geographical location, floor, and orientation, which are recorded as s_1, s_2, and s_3. Let the attribute set $X = \{s_1, s_2, s_3\}$. The importance of each attribute is determined by experts and house buyers as follows:*

$$m(\Phi) = 0, m(\{s_1\}) = 0.7, m(\{s_2\}) = 0.5, m(\{s_3\}) = 0.4, m(\{s_1, s_2\}) = 0.9, m(\{s_1, s_3\}) = 0.6, m(\{s_2, s_3\}) = 0.8, m(\{s_1, s_2, s_3\}) = 1$$

The buyer rates the two properties and the three attributes as follows:

First property: $h_1(\{s_1\}) = 0.9$, $h_1(\{s_2\}) = 0.8$, $h_1(\{s_3\}) = 0.5$; second property: $h_2(\{s_1\}) = 0.6$, $h_2(\{s_2\}) = 0.9$, $h_2(\{s_3\}) = 0.7$.

Taking the importance of the attribute in the property evaluation as a measure of the attribute set, it easy to see that it is non-additive. The h_1 and h_2 scores of the two properties are regarded as functions of the property set X. The overlap function is defined by

$$O(x, y) = x^2y^2, \text{ for any } x, y \in [0, 1].$$

Then the buyer's composite score for the first property can be calculated by the O-Sugeno integral of h_1 over X, as follows:

$$\begin{aligned} \int^{(OS)} h_1 \circ m &= \max_{i=1}^{3} O[h_1(x_i), m(X_i)] \\ &= \max\{O(h_1(x_3), m(X)), O(h_1(x_2), m(\{x_1, x_2\})), O(h_1(x_1), m(\{x_1\}))\} \\ &= \max\{0.5^2 \times 1^2, 0.8^2 \times 0.9^2, 0.9^2 \times 0.7^2\} \\ &= 0.5184 \end{aligned}$$

The buyer's composite score for the second property can be calculated by the O-Sugeno integral of h_2 over X, as follows:

$$\begin{aligned} \int^{(OS)} h_2 \circ m &= \max_{i=1}^{3} O[h_2(x_i), m(X_i)] \\ &= \max\{O(h_2(x_1), m(X)), O(h_2(x_3), m(\{x_2, x_3\})), O(h_2(x_2), m(\{x_2\}))\} \\ &= \max\{0.6^2 \times 1^2, 0.7^2 \times 0.6^2, 0.9^2 \times 0.5^2\} \\ &= 0.36 \end{aligned}$$

This shows that the buyer has a higher comprehensive score for the first property, so he should buy the first property.

4. O-Sugeno Integral Semantics of Fuzzy Quantifiers

For the sake of completeness, we recall several concepts of a first-order logical language L_q with fuzzy quantifiers.

Definition 6. *A first order logical language L_q contains the following:*

(i) An enumerable set of individual variables: x_0, x_1, x_2;
(ii) A set of predicate symbols: $F = \cup_{n=0}^{\infty} F_n$, where F_n indicates the set of all n-place predicate symbols for every $n \geq 0$, assuming that $\cup_{n=0}^{\infty} F_n \neq \Phi$;
(iii) Propositional connectors: ~ and ∧; and
(iv) Parentheses: ( ) [41].

The following definition gives the syntax of language L_q:

Definition 7. *The minimum set of symbol strings is called set Wff of well-formed formula if the following conditions are satisfied:*

(i) For every $n \geq 0$, if F is an n-place predicate symbol and $y_1, \ldots, y_n$ are individual variables, then $F(y_1, \ldots, y_n)$ is a well-formed formula;
(ii) If Q is a quantifier, x is an individual variable, and φ is a well-formed formula, then $(Qx)\ \varphi$ is also a well-formed formula; and
(iii) If φ, φ_1, and φ_2 are all well-formed formulas, then $\sim\varphi$, φ_1, and $\wedge\varphi_2$ are also well-formed formulas [41].

The following definitions give the semantics of language L_q:

Definition 8. *The following items comprise an interpretation I of the logic language:*

(i) A measurable space $(X, \wp)$, which is called the domain of the interpretation;
(ii) For every $n \geq 0$, there exists an element x_i^I in X corresponding to the individual variable x_i; and
(iii) For every $n \geq 0$ and any $F \in F_n$, there exists a $\wp^n$-measurable function F^I: $X^n \to [0, 1]$ [41].

Definition 9. *Assume that I is an interpretation. Then the truth value $T_I(\varphi)$ of formula φ under I based on O-Sugeno integrals is defined recursively as follows:*

(i) If $\varphi = F(y_1, \ldots, y_n)$, then

$$T^I(\varphi) = F({y_1}^I, \ldots, {y_n}^I).$$

(ii) If $\varphi = (Qx)\ y$, then

$$T_I(\varphi) = \int^{(OS)} T_{I\{\cdot/x\}}(\psi) \circ Q_X$$

where X is the domain of I, $T_{I\{.\ /x\}}$: $X \to [0, 1]$ is a mapping such that

$$T_{I\{.\ /\ x\}}(\varphi)(u) = T_{I\{u/x\}}(\varphi), \text{for every } u \in X,$$

and I {u/x} is the interpretation which is different from I only in the assignment of the individual variable x, that is,

$$y^{I\{u/x\}} = y^I \text{ and } x^{I\{u/x\}} = u, \text{for every } x, y \in X \text{ and } x \neq y.$$

(iii) If $\varphi = \sim \psi$, then

$$T_I(\varphi) = 1 - T_I(\psi),$$

and if $\varphi = \varphi_1 \wedge \varphi_2$, *then*

$$T_I(\varphi) = \min\{T_I(\varphi_1), T_I(\varphi_2)\} = \int^{(OS)} T_{I\{\cdot/x\}}(\varphi) \circ Q_X$$

Proposition 1. *For any quantifier Q and for any formula* $\varphi \in Wff$, *if O is an overlap function with unit element 1 and I is an interpretation with the domain of a single point set* $X = \{u\}$, *then*

$$T_I((Qx)\varphi) = T_I(\varphi).$$

Proof of Proposition 1.

$$\begin{aligned} T_I((Qx)\varphi) &= \int^{(OS)} T_{I\{\cdot/x\}}(\varphi) \circ Q_X \\ &= \sup_{F\subseteq X} O[\inf_{u\in F} T_{I\{u/x\}}(\varphi), Q_X(F)] \\ &= O[\inf_{u\in\varphi} T_{I\{u/x\}}(\varphi), Q_X(\varphi)] \vee O[T_{I\{u/x\}}(\varphi), Q_X(\{u\})] \\ &= O[\inf_{u\in\varphi} T_{I\{u/x\}}(\varphi),\ 0)] \vee O[T_{I\{u/x\}}(\varphi),\ 1)] \\ &= O[T_{I\{u/x\}}(\varphi),\ 1)] \\ &= T_{I\{u/x\}}(\varphi) \\ &= T_I(\varphi) \end{aligned}$$

□.

The above proposition states that quantification degenerates in a single point domain.

Proposition 2. *For any quantifier Q and for any formula* $\varphi \in Wff$, *if O is an overlap function with unit element 1, then for any interpretation I with domain X*

(1) $T_I((\exists x)\varphi) = \sup_{u\in X} T_{I\{u/x\}}(\varphi)$ *and*

(2) $T_I((\forall x)\varphi) = \inf_{u\in X} T_{I\{u/x\}}(\varphi)$.

Proof of Proposition 2.

(1)

$$\begin{aligned} T_I((\exists x)\varphi) &= \int^{OS} T_{I\{u/x\}}(\varphi) \circ \exists_X \\ &= \sup_{F\subseteq X} O[\inf_{u\in F} T_{I\{u/x\}}(\varphi), \exists_X(F)] \\ &= O[\inf_{u\in\varphi} T_{I\{u/x\}}(\varphi), \exists_X(\varphi)] \vee \sup_{\varphi\neq F\subseteq X} O[\inf_{u\in F} T_{I\{u/x\}}(\varphi), \exists_X(F)] \\ &= O[\inf_{u\in\varphi} T_{I\{u/x\}}(\varphi), 0)] \vee \sup_{\varphi\neq F\subseteq X} O[\inf_{u\in F} T_{I\{u/x\}}(\varphi),\ 1)] \\ &= \sup_{\varphi\neq F\subseteq X} O[\inf_{u\in F} T_{I\{u/x\}}(\varphi),\ 1] \\ &= \sup_{\varphi\neq F\subseteq X} \inf_{u\in F} T_{I\{u/x\}}(\varphi) \\ &= \sup_{u\in X} T_{I\{u/x\}}(\varphi) \end{aligned}$$

(2)

$$\begin{aligned} T_I((\forall x)\varphi) &= \int^{OS} T_{I\{u/x\}}(\varphi) \circ \exists_X \\ &= \sup_{F\subseteq X} O[\inf_{u\in F} T_{I\{u/x\}}(\varphi), \forall_X(F)] \\ &= O[\inf_{u\in X} T_{I\{u/x\}}(\varphi), \forall_X(X)] \vee \sup_{F\subset X} O[\inf_{u\in F} T_{I\{u/x\}}(\varphi), \forall_X(F)] \\ &= O[\inf_{u\in X} T_{I\{u/x\}}(\varphi),\ 1] \vee \sup_{F\subset X} O[\inf_{u\in F} T_{I\{u/x\}}(\varphi),\ 0] \\ &= O[\inf_{u\in X} T_{I\{u/x\}}(\varphi),\ 1] \\ &= \inf_{u\in X} T_{I\{u/x\}}(\varphi) \end{aligned}$$

□.

The above proposition shows that for the two quantifiers ∀ and ∃, the method of calculating the truth value of a quantified proposition based on O-Sugeno integrals corresponds to the standard method, which shows that the O-Sugeno integral semantics of fuzzy quantifiers is reasonable.

Example 5. *We consider a comprehensive evaluation of students' health status (see Example 43 in [41]). Assuming X is a set consisting of 10 students, X =* $\{s_1, s_2, \ldots, s_{10}\}$*, and H is a linguistic predicate, "to be healthy". The health evaluation of these students is shown in Table 1.*

Table 1. Health condition of 10 students.

	s_1	s_2	s_3	s_4	s_5	s_6	s_7	s_8	s_9	s_{10}
$H(x)$	0.95	0.1	0.73	1	0.84	0.7	0.67	0.9	1	0.81

Next, we choose the fuzzy quantifier Q = "most" to describe the overall health status of this group of students.

Let $QxH(x)$ be the proposition "Most students are healthy" and I be the interpretation given in Table 1, then $T_I((Qx)H(x))$ represents the truth value of $QxH(x)$ under I calculated by the O-Sugeno integral. According to Theorem 2 in Section 3 and Definition 4 in Section 4, we have

$$T_I((Qx)H(x)) = \int^{(OS)} T_{I\{\cdot/x\}}(H(x)) \circ Q_X = \max_{i=1}^{10} O[h(x_i), Q_X(X_i)]$$

where $h(x_i)$ is the result of rearranging the possible values of $H(x)$ in non-decreasing order, and $X_i = \{x_j: i \leq j \leq 10\}$ for $1 \leq i \leq 10$. Table 2 presents the rearranged truth values $h(x_i)$ for $1 \leq i \leq 10$.

Table 2. Rearranged truth values.

	1	**2**	**3**	**4**	**5**	**6**	**7**	**8**	**9**	**10**
$h(x_i)$	0.1	0.67	0.7	0.73	0.81	0.84	0.9	0.95	1	1

According to the definition of the quantifier "most" in Example 3, we calculate the fuzzy measures of X_i as follows:

$$Q_X(X_i) = (|X_i| / |X|)^{3/2} = [(11 - i)/10]^{3/2}, \text{ for } 1 \leq i \leq 10$$

Using the O-Sugeno integral in which the overlap function $O(x, y) = x^2y^2$ for every $x, y \in [0, 1]$, the truth value of $QxH(x)$ is calculated as follows:

$$
\begin{aligned}
T_I((Qx)H(x)) &= \max_{i=1}^{10} O[h(x_i), Q_X(X_i)] \\
&= 0.01 \vee 0.327 \vee 0.251 \vee 0.183 \vee 0.142 \vee 0.088 \vee 0.052 \vee 0.024 \vee 0.008 \vee 0.001 \\
&= 0.327
\end{aligned}
$$

If the overlap function by $O(x, y) = \min(\sqrt{x}, \sqrt{y})$ for any $x, y \in [0, 1]$, the truth value of $QxH(x)$ is calculated as follows:

$$
\begin{aligned}
T_I((Qx)H(x)) &= \max_{i=1}^{10} O[h(x_i), Q_X(X_i)] \\
&= 0.316 \vee 0.819 \vee 0.837 \vee 0.765 \vee 0.682 \vee 0.595 \vee 0.503 \vee 0.405 \vee 0.299 \vee 0.178 \\
&= 0.837
\end{aligned}
$$

The above example shows that choosing different overlap functions to calculate the true value of the proposition "Most students are healthy" under the interpretation will lead to different results. In decision-making problems based on preference relationships, different overlap functions can reflect different fuzzy preferences, which provide multiple choices for decision-makers (they can manifest the preference relationship by choosing different overlap functions).

Example 6. *We consider a comprehensive evaluation of the weather conditions for a week (see Example 42 in [41]). Let X be a set consisting of 7 days, X = {Sunday, Monday, Tuesday, Wednesday, Thursday, Friday, Saturday}. And let P_1 and P_2 represent respectively the linguistic predicates "to be cloudy" and "to be cold". The respective weather conditions of the week are indicated in Table 3. Suppose Q is a fuzzy quantifier, "most", then the formula $\varphi = (Qx)\psi = (Qx)\ (P_1(x) \wedge\sim P_2(x))$ represents "many days (in this week) are cloudy but not cold".*

Table 3. Truth values of linguistic predicates P_1 and P_2.

	Sunday	Monday	Tuesday	Wednesday	Thursday	Friday	Saturday
$P_1{}^I$	0.1	0	0.5	0.8	0.6	1	0.2
$P_2{}^I$	1	0.9	0.4	0.7	0.3	0.4	0

With interpretation I and truth values P_1 and P_2 given in Table 3, then $T_I(\varphi) = T_I((Qx)\ (P_1(x) \wedge\sim P_2(x)))$ represents the truth value of $\varphi = (Qx)\ (P_1(x) \wedge\sim P_2(x))$ under interpretation I about the O-Sugeno integral. According to Theorem 2 in Section 3 and Definition 4 in Section 4, we have

$$
T_I(\varphi) = \int_A^{(OS)} T_{I\{\cdot/x\}}[P_1^I(x) \wedge \sim P_2^I(x)] \circ Q_X = \max_{i1}^{7} O[h(x_i), Q_X(X_i)]
$$

where $h(x_i)$ is the result of the rearranged possible values of $T_I\ (P_1(x) \wedge\sim P_2(x))$ in non-decreasing order, and $X_i = \{x_j : i \le j \le 7\}$ for $1 \le i \le 7$. Table 4 presents the rearranged truth values $h(x_i)$ for $1 \le i \le 7$.

Table 4. Rearranged truth values.

	1	2	3	4	5	6	7
$h(x_i)$	0	0	0.2	0.3	0.5	0.6	0.6

According to the definition of the quantifier "most" in Example 3, fuzzy measures about the fuzzy quantifier of X_i are calculated as follows:

$Q_X(X_1) = 1$, $Q_X(X_2) = 6/7$, $Q_X(X_3) = 5/7$, $Q_X(X_4) = 4/7$, $Q_X(X_5) = 3/7$, $Q_X(X_6) = 2/7$, $Q_X(X_7) = 1/7$.

The overlap function be defined as $O(x, y) = \min(\sqrt{x}, \sqrt{y})$ for any $x, y \in [0, 1]$, and by using the O-Sugeno integral, the truth value of $(Qx)(P_1(x) \wedge \sim P_2(x))$ is calculated as follows:

$$\begin{aligned} T_I(\varphi) &= \max_{i=1}^{7} O[h(x_i), Q_X(X_i)] \\ &= 0 \vee 0 \vee 0.447 \vee 0.548 \vee 0.655 \vee 0.535 \vee 0.378 \\ &= 0.655 \end{aligned}$$

5. Applying Integral Semantics of Fuzzy Quantifiers to MADM

The MADM based on fuzzy quantifiers is to select the decision object that performs well on a certain proportion or quantification (such as most, many, more that half, etc.) of attributes. In this section, we propose a MADM method based on O-Sugeno integral semantics of fuzzy quantifiers to solve the MADM problem involving fuzzy quantifiers. The specific process is described as follows.

The basic representations are as follows: $S = \{s_1, s_2, \ldots, s_m\}$ is a set of m decision objects (also known as feasible alternatives), $G = \{g_1, g_2, \ldots, g_n\}$ is a set of n evaluation indicators (also called attributes), and Q represents the fuzzy quantifiers such as most, many, more than half, etc.

Step 1: Calculate the truth values of linguistic predicates under the interpretations and rearrange them to obtain the standardized truth values.

The performance of each decision-making object on the attributes is regarded as an interpretation I. For any $x \in G$, $\varphi(x)$ means that the predicate meets the requirements of attribute x. For each decision-making object $s \in S$, we compute the truth value of the linguistic predicate $\varphi(x)$ under interpretation I, and then rearrange all of them to get $h(x_i)$, where for $1 \leq i \leq n-1$, $h(x_i) \leq h(x_{i+1})$.

Step 2: Calculate fuzzy measures about the fuzzy quantifier.

According to the semantic analysis of the fuzzy quantifier, we can calculate a family of fuzzy measures $Q_X(X_i)$, where for $1 \leq i \leq n$, $X_i = \{x_j : i \leq j \leq n\}$.

Step 3: Calculate the truth value of the proposition for each decision object based on O-Sugeno integral semantics.

We consider the proposition $(Qx)\varphi(x)$ = "A decision object meets the requirements of attributes with the fuzzy proportion Q". Based on the O-Sugeno integral semantics of fuzzy quantifiers, we calculate the truth values $D(s_i)$ of proposition $(Qx)\varphi(x)$ under its interpretation for each decision-making object $s_i \in S$:

$$D(s) = T_I((Qx)\varphi(x)) = \int^{(OS)} s \circ Q_X = \max_{i=1}^{n} O[h(x_i), Q_X(X_i)]$$

for any $s \in G$

Step 4: Obtain the optimal object by ranking the truth values of decision objects.

Example 7. *Decision-making problem for selecting excellent students. The best of three high school students will be recommended to enter a well-known university based on their mathematics, physics, biology, chemistry, and literature grades. The relevant data in Table 5 show the grades for each student in each course.*

Table 5. Performance of three students.

	Mathematics	Physics	Biology	Chemistry	Literature
s_1	0.75	0.85	0.95	0.90	0.86
s_2	0.85	0.92	0.91	0.95	0.86
s_3	0.92	0.87	0.90	0.89	0.91

In order to obtain a comprehensive evaluation of each student, we consider the proposition "(student) has performed well in almost all courses". The domain is indicated as X = {mathematics, physics, biology, chemistry, literature}, the fuzzy quantifier is Q = "almost all", and the predicate is $\varphi(x)$ = "(student) performed well on x" for each student, then the proposition "(student) performs well in almost all courses" is expressed as the logic formula $(Qx)\varphi(x)$.

Each student's performance in the five courses is considered as an interpretation I. For each student, we rearrange the truth values of the linguistic predicate $\varphi(x)$ under its interpretation I to get $h(x_i)$ for $1 \leq i \leq 5$. Table 6 presents the rearranged truth values.

Table 6. Rearranged truth values.

	1	2	3	4	5
s_1 $(h_1(x_i))$	0.75	0.85	0.86	0.9	0.95
s_2 $(h_2(x_i))$	0.85	0.86	0.91	0.92	0.95
s_3 $(h_3(x_i))$	0.87	0.89	0.90	0.91	0.92

According to the definition of "almost all" in Example 3, we can calculate fuzzy measures of $X_i = \{x_j: i \leq j \leq 5\}$ for $1 \leq i \leq 5$ about the fuzzy quantifier as follows:

$$Q_X(X_1) = 1,\ Q_X(X_2) = 0.64,\ Q_X(X_3) = 0.36,\ Q_X(X_4) = 0.16,\ Q_X(X_5) = 0.04.$$

The overlap function is defined by $O(x, y) = x\,y\,(x + y)/2$ for any $x, y \in [0, 1]$, then the true value $(Qx)\varphi(x)$ of each student under interpretation I based on the O-Sugeno integral is calculated as follows:

$$\begin{aligned}
&D(s_1) = T_I((Qx)\varphi(x)) = \int^{(OS)} s_1 \circ Q_X = \max_{i=1}^{5} O[h_1(x_i), Q_X(X_i)] \\
&= 0.75 \times 1 \times (0.75 + 1)/2 \vee 0.85 \times 0.64 \times (0.85 + 0.64)/2 \vee 0.86 \times 0.36 \times (0.86 + 0.36)/2 \\
&\vee 0.9 \times 0.16 \times (0.9 + 0.16)/2 \vee 0.95 \times 0.04 \times (0.95 + 0.04)/2 \\
&= 0.656 \vee 0.405 \vee 0.189 \vee 0.076 \vee 0.019 \\
&= 0.656.
\end{aligned}$$

$$\begin{aligned}
&D(s_2) = T_I((Qx)\varphi(x)) = \int^{(OS)} s_2 \circ Q_X = \max_{i=1}^{5} O[h_2(x_i), Q_X(X_i)] \\
&= 0.85 \times 1 \times (0.85 + 1)/2 \vee 0.86 \times 0.64 \times (0.86 + 0.64)/2 \vee 0.91 \times 0.36 \times (0.91 + 0.36)/2 \\
&\vee 0.92 \times 0.16 \times (0.92 + 0.16)/2 \vee 0.95 \times 0.04 \times (0.95 + 0.04)/2 \\
&= 0.786 \vee 0.413 \vee 0.208 \vee 0.079 \vee 0.019 \\
&= 0.786.
\end{aligned}$$

$$\begin{aligned}
&D(s_3) = T_I((Qx)\varphi(x)) = \int^{(OS)} s_3 \circ Q_X = \max_{i=1}^{5} O[h_3(x_i), Q_X(X_i)] \\
&= 0.87 \times 1 \times (0.87 + 1)/2 \vee 0.89 \times 0.64 \times (0.89 + 0.64)/2 \vee 0.9 \times 0.36 \times (0.9 + 0.36)/2 \\
&\vee 0.91 \times 0.16 \times (0.91 + 0.16)/2 \vee 0.92 \times 0.04 \times (0.92 + 0.04)/2 \\
&= 0.813 \vee 0.436 \vee 0.204 \vee 0.078 \vee 0.018 \\
&= 0.813.
\end{aligned}$$

The maximum value $D(s_3)$ can be obtained by ranking these true values, so student s_3 is the best student.

Example 8. *Decision-making problem about supplier selection. A factory needs to choose a supplier for an important raw material, and the decision-maker intends to select from four alternative suppliers, which are represented as s_1, s_2, s_3, and s_4. The decision-maker evaluates these suppliers in four aspects, which are called decision attributes: product price, product quality, service level, and reputation. The specific data are given in Table 7.*

Table 7. Attribute values of four suppliers.

	Product Price	Product Quality	Service Level	Reputation
s_1	0.95	0.71	0.85	0.8
s_2	0.8	0.76	0.92	0.83
s_3	0.85	0.81	0.7	0.86
s_4	0.76	0.9	0.75	0.84

In order to obtain a comprehensive evaluation of each supplier, we consider the proposition "(supplier) meets the requirements for most attributes". Domain X is indicated as X = {product price, product quality, service level, reputation}, the fuzzy quantifier is Q = "most", and the predicate is $\varphi\ (x)$ = "(supplier) meets the requirement of x" for each $x \in X$, then the proposition "(supplier) meets the requirements for most attributes" is expressed as the logic formula $(Qx)\varphi(x)$.

Each supplier's performance on four attributes is considered as an interpretation I. For each supplier, we rearrange the truth values of the linguistic predicate $\varphi(x)$ under its interpretation I to get $h(x_i)$ for $1 \leq i \leq 4$. Table 8 presents the rearranged truth values.

Table 8. Rearranged truth values.

	1	2	3	4
$s_1\ (h_1(x_i))$	0.71	0.8	0.85	0.95
$s_2\ (h_2(x_i))$	0.76	0.8	0.83	0.92
$s_3\ (h_3(x_i))$	0.7	0.81	0.85	0.86
$s_4\ (h_4(x_i))$	0.75	0.76	0.84	0.9

According to the definition of "most" in Example 3, we can calculate fuzzy measures of $X_i = \{x_j: i \leq j \leq 4\}$ for $1 \leq i \leq 4$ about the fuzzy quantifier as follows:

$$Q_X(X_1) = 1,\ Q_X(X_2) = (3/4)^{3/2} \approx 0.650,\ Q_X(X_3) = (1/2)^{3/2} \approx 0.354,\ Q_X(X_4) = (1/4)^{3/2} \approx 0.125.$$

Let the overlap function O: $[0, 1]^2 \to [0, 1]$ be defined as

$$O(x, y) = \frac{1}{1 - xy + 1/xy}$$

for any $x, y \in [0, 1]$, then the true value $(Qx)\varphi(x)$ of each supplier under its interpretation I based on the O-Sugeno integral is calculated as follows:

$$\begin{aligned}
D(s_1) &= T_I((Qx)\varphi(x)) = \int^{(OS)} s_1 \circ Q_X = \max_{i=1}^{4} O[h_1(x_i), Q_X(X_i)] \\
&= \frac{1}{1-0.71+1/0.71} \vee \frac{1}{1-0.8\times 0.65+1/(0.8\times 0.65)} \\
&\vee \frac{1}{1-0.85\times 0.354+1/(0.85\times 0.354)} \vee \frac{1}{1-0.95\times 0.125+1/(0.95\times 0.125)} \\
&= 0.589 \vee 0.416 \vee 0.248 \vee 0.108 \\
&= 0.589.
\end{aligned}$$

$$\begin{aligned}
D(s_2) &= T_I((Qx)\varphi(x)) = \int^{(OS)} s_2 \circ Q_X = \max_{i=1}^{4} O[h_2(x_i), Q_X(X_i)] \\
&= \frac{1}{1-0.76+1/0.76} \vee \frac{1}{1-0.8\times 0.65+1/(0.8\times 0.65)} \\
&\vee \frac{1}{1-0.83\times 0.354+1/(0.83\times 0.354)} \vee \frac{1}{1-0.92\times 0.125+1/(0.92\times 0.125)} \\
&= 0.643 \vee 0.416 \vee 0.243 \vee 0.104 \\
&= 0.643.
\end{aligned}$$

$$D(s_3) = T_I(Qx)\varphi(x)) = \int^{(OS)} s_3 \circ Q_X = \max_{i=1}^{4} O[h_3(x_i), Q_X(X_i)]$$
$$= \frac{1}{1-0.7+1/0.7} \vee \frac{1}{1-0.81\times0.65+1/(0.81\times0.65)}$$
$$\vee \frac{1}{1-0.85\times0.354+1/(0.85\times0.354)} \vee \frac{1}{1-0.86\times0.125+1/(0.86\times0.125)}$$
$$= 0.579 \vee 0.421 \vee 0.248 \vee 0.098$$
$$= 0.579.$$

$$D(s_4) = T_I(Qx)\varphi(x)) = \int^{(OS)} s_4 \circ Q_X = \max_{i=1}^{4} O[h_4(x_i), Q_X(X_i)]$$
$$= \frac{1}{1-0.75+1/0.75} \vee \frac{1}{1-0.76\times0.65+1/(0.76\times0.65)}$$
$$\vee \frac{1}{1-0.84\times0.354+1/(0.84\times0.354)} \vee \frac{1}{1-0.9\times0.125+1/(0.9\times0.125)}$$
$$= 0.632 \vee 0.395 \vee 0.246 \vee 0.102$$
$$= 0.632.$$

Therefore, the evaluation shows that supplier s_2 has the highest score, thus supplier s_2 should be selected.

Example 9. *Decision-making problem about logistics park location. A city wants to build a logistics park, and the decision-maker plans to choose from eight alternatives, which are represented as s_i, for $1 \le i \le 8$. The decision-maker evaluates these alternatives in 12 aspects, which are called decision attributes: urban support, traffic conditions, geological environment, land price, urban traffic improvement, convenient delivery, surrounding facilities, neighboring enterprises, talent attraction, logistics development space, prospect of environmental development, and predicted economic development. The specific data are given in Table 9.*

Table 9. Evaluation of 12 attributes.

	s_1	s_2	s_3	s_4	s_5	s_6	s_7	s_8
Urban support	0.912	0.97	0.824	0.706	0.964	0.556	0.656	0.734
Traffic conditions	0.9	0.846	0.786	0.93	0.824	0.972	0.738	0.892
Geological environment	0.89	0.876	0.93	0.824	0.772	0.932	0.936	0.814
Land price	0.69	0.574	0.856	0.712	0.592	0.93	0.726	0.794
Urban traffic improvement	0.7	0.624	0.858	0.89	0.652	0.978	0.972	0.904
Convenient delivery	0.85	0.864	0.904	0.774	0.902	0.606	0.596	0.912
Surrounding facilities	0.648	0.774	0.912	0.842	0.804	0.67	0.806	0.796
Neighboring enterprises	0.806	0.828	0.912	0.774	0.812	0.604	0.772	0.804
Talent attraction	0.846	0.972	0.826	0.774	0.962	0.604	0.796	0.806
Logistics development space	0.796	0.712	0.912	0.804	0.806	0.608	0.778	0.952
Prospect of environmental development	0.792	0.774	0.956	0.796	0.846	0.734	0.752	0.846
Predicted economic development	0.808	0.808	0.816	0.842	0.792	0.774	0.804	0.912

In order to obtain a comprehensive evaluation of each alternative, we consider the proposition "(alternative) meets the requirements of most attributes". Domain X is indicated as X = {urban support, traffic conditions, geological environment, land price, urban traffic improvement, convenient delivery, surrounding facilities, neighboring enterprises, talent attraction, logistics development space, prospect of environmental development, predicted economic development}, the fuzzy quantifier is Q = "most", and the predicate is φ (x) = "(alternative) meets the requirement of x" for each $x \in X$, then the proposition "(alternative) meets the requirements of most attributes" is expressed as the logic formula $(Qx)\varphi(x)$.

The performance of each alternative on 12 attributes is considered as an interpretation I. For each alternative, we rearrange the truth values of the linguistic predicate $\varphi(x)$ under its interpretation I to get $h(x_i)$ for $1 \le i \le 12$. Table 10 presents the rearranged truth values.

Table 10. Rearranged truth values.

	s_1 $(h_1(x_i))$	s_2 $(h_2(x_i))$	s_3 $(h_3(x_i))$	s_4 $(h_4(x_i))$	s_5 $(h_5(x_i))$	s_6 $(h_6(x_i))$	s_7 $(h_7(x_i))$	s_8 $(h_8(x_i))$
1	0.648	0.574	0.786	0.706	0.592	0.556	0.596	0.734
2	0.69	0.624	0.816	0.712	0.652	0.604	0.656	0.794
3	0.7	0.712	0.824	0.774	0.772	0.604	0.726	0.796
4	0.792	0.774	0.826	0.774	0.792	0.606	0.738	0.804
5	0.796	0.774	0.856	0.774	0.804	0.608	0.752	0.806
6	0.806	0.808	0.858	0.796	0.806	0.67	0.772	0.814
7	0.808	0.828	0.904	0.804	0.812	0.734	0.778	0.846
8	0.846	0.846	0.912	0.824	0.824	0.774	0.796	0.892
9	0.85	0.864	0.912	0.842	0.846	0.93	0.804	0.904
10	0.89	0.876	0.912	0.842	0.902	0.932	0.806	0.912
11	0.9	0.97	0.93	0.89	0.962	0.972	0.936	0.912
12	0.912	0.972	0.956	0.93	0.964	0.978	0.972	0.952

According to the definition of "most" in Example 3, we can calculate fuzzy measures of $X_i = \{x_j: i \le j \le 12\}$ for $1 \le i \le 12$ about the fuzzy quantifier as follows:

$$Q_X(X_i) = (\mid X_i \mid / \mid X \mid)^{3/2} = [(13 - i)/12]^{3/2}, \text{ for } 1 \le i \le 12.$$

After the calculation, we can obtain:

$$Q_X(X_1) = 1, Q_X(X_2) \approx 0.878, Q_X(X_3) \approx 0.761, Q_X(X_4) \approx 0.650, Q_X(X_5) \approx 0.544, Q_X(X_6) \approx 0.446,$$

$$Q_X(X_7) \approx 0.354, Q_X(X_8) \approx 0.269, Q_X(X_9) \approx 0.192, Q_X(X_{10}) = 0.125, Q_X(X_{11}) \approx 0.068, Q_X(X_{12}) \approx 0.024.$$

The overlap function is defined as

$$O(x,y) = \min(\sqrt{x}, \sqrt{y})$$

for any $x, y \in [0, 1]$, then the true value $(Qx)\varphi(x)$ of each supplier under its interpretation I based on the O-Sugeno integral is calculated as follows:

$$D(s_j) = T_I(Qx)\varphi(x)) = \int^{(OS)} s_j \circ Q_X = \max_{i=1}^{12} O[h_j(x_i), Q_X(X_i)]$$

for $1 \le j \le 8$.

After the calculation, we obtain the comprehensive evaluation values of all alternatives, as shown in Table 11.

Table 11. Comprehensive evaluation values of eight alternatives.

	s_1	s_2	s_3	s_4	s_5	s_6	s_7	s_8
Evaluation value	0.837	0.844	0.903	0.872	0.872	0.778	0.852	0.891

Therefore, the evaluation shows that alternative s_3 has the highest score, thus alternative s_3 should be selected.

Example 10. *Decision-making problem about the purchase of a new energy car. A customer is going to buy a new energy car. After preliminary screening, the customer has four alternatives. These alternatives are represented as s_i, for $1 \le i \le 4$. In order to purchase a satisfactory car, the customer browsed the comments of each alternative on various network platforms and evaluated them from seven aspects (attributes): appearance, interior, space, comfort, power, operation difficulty,* *and cost performance. Through text sentiment analysis, all evaluation information is converted into specific data, as shown in Table 12.*

Table 12. Evaluation of 7 attributes.

	s_1	s_2	s_3	s_4
Appearance	0.8149	0.7320	0.8352	0.6786
Interior	0.6890	0.7302	0.7056	0.6810
Space	0.5969	0.3858	0.2555	0.3183
Comfort	0.7058	0.6030	0.7398	0.6429
Power	0.5708	0.5227	0.6259	0.4488
Operation difficulty	0.6632	0.6041	0.4893	0.4579
Cost performance	0.6765	0.4597	0.6123	0.4090

In order to obtain a comprehensive evaluation of each alternative, we consider the proposition of "(alternative) meets the requirements for almost all attributes". Domain X is indicated as X = {appearance, interior, space, comfort, power, operation difficulty, cost performance}, the fuzzy quantifier is Q = "almost all", and the predicate is $\phi(x)$ = "(alternative) meets the requirement of x" for each $x \in X$, then the proposition "(alternative) meets the requirements of almost all attributes" is expressed as the logic formula $(Qx)\phi(x)$.

The performance of each alternative on seven attributes is considered as an interpretation I. For each alternative, we rearrange the truth values of the linguistic predicate $\phi(x)$ under its interpretation I to get $h(x_i)$ for $1 \leq i \leq 7$. Table 13 presents the rearranged truth values.

Table 13. Rearranged truth values.

	s_1 ($h_1(x_i)$)	s_2 ($h_2(x_i)$)	s_3 ($h_3(x_i)$)	s_4 ($h_4(x_i)$)
1	0.5708	0.3858	0.2555	0.3183
2	0.5969	0.4597	0.4893	0.4090
3	0.6632	0.5227	0.6123	0.4488
4	0.6765	0.6030	0.6259	0.4579
5	0.6890	0.6041	0.7056	0.6429
6	0.7058	0.7302	0.7398	0.6786
7	0.8149	0.7320	0.8352	0.6810

According to the definition of "almost all" in Example 3, we can calculate fuzzy measures of $X_i = \{x_j: 1 \leq i \leq 7\}$ for $1 \leq i \leq 7$ about fuzzy quantifier as follows:

$$Q_X(X_i) = (|X_i| / |X|)^2 = [(8 - i)/7]^2, \text{ for } 1 \leq i \leq 7.$$

After the calculation, we can obtain:

$$Q_X(X_1) = 1, Q_X(X_2) = 0.735, Q_X(X_3) = 0.510, Q_X(X_4) = 0.327, Q_X(X_5) = 0.184, Q_X(X_6) = 0.082, Q_X(X_7) = 0.020$$

The overlap function is defined as

$$O(x, y) = \min(\sqrt{x}, \sqrt{y}),$$

for any $x, y \in [0, 1]$, then the truth value $(Qx)\phi(x)$ of each alternative under its interpretation I based on the O-Sugeno integral is calculated as follows:

$$D(s_j) = T_I((Qx)\varphi(x)) = \int^{OS} s_j \circ Q_X = \max_{i=1}^{7} O[h_j(x_i), Q_X(X_i)]$$

for $1 \leq j \leq 4$.

After the calculation, we obtain the comprehensive evaluation values of all alternatives, as shown in Table 14.

Table 14. Comprehensive evaluation values of four alternatives.

	s_1	s_2	s_3	s_4
Evaluation value	0.773	0.714	0.714	0.670

Therefore, the evaluation shows that alternatives s_1 has the highest score, thus alternatives s_1 should be selected.

Example 11. *Decision-making problem about red wine selection. There are currently four types of red wines. In order to select the optimal one, the components (attributes) of each wine needs to be measured and evaluated, including fixed acidity, volatile acidity, citric acid, residual sugar, chlorides, free sulfur dioxide, total sulfur dioxide, sulfate and alcohol. The specific data are revealed in Table 15. (data from open source datasets website).*

Table 15. Evaluation of nine components.

	s_1	s_2	s_3	s_4
Fixed acidity	0.9689	0.9222	0.8210	0.8327
Volatile acidity	0.4055	0.9843	0.7323	0.7323
Citric acid	0.6648	0.5369	0.8750	0.7983
Residual sugar	0.8147	0.6207	0.8922	0.9914
Chlorides	0.7727	0.4513	0.8312	0.7581
Free sulfur dioxide	0.8285	0.7531	0.3430	0.9456
Total sulfur dioxide	0.7143	0.8937	0.6246	0.9867
Sulfate	0.6903	0.8148	0.8726	0.9580
Alcohol	0.7607	0.7855	0.8020	0.8682

In order to obtain a comprehensive evaluation of each alternative, we consider the proposition of "(alternative) meets the requirements for almost all attributes". Domain X is indicated as X = {fixed acidity, volatile acidity, citric acid, residual sugar, chlorides, free sulfur dioxide, total sulfur dioxide, sulfate, alcohol}, the fuzzy quantifier is Q = "almost all", and the predicate is $\phi(x)$ = "(alternative) meets the requirement of x" for each $x \in X$, then the proposition "(alternative) meets the requirements for almost all attributes" is expressed as the logic formula $(Qx)\phi(x)$.

The performance of each alternative on nine attributes is considered as an interpretation I. For each alternative, we rearrange the truth values of the linguistic predicate $\phi(x)$ under its interpretation I to get $h(x_i)$ for $1 \leq i \leq 4$. Table 16 presents the rearranged truth values.

Table 16. Rearranged truth values.

	s_1 $(h_1(x_i))$	s_2 $(h_2(x_i))$	s_3 $(h_3(x_i))$	s_4 $(h_4(x_i))$
Fixed acidity	0.4055	0.4513	0.3430	0.7323
Volatile acidity	0.6648	0.5369	0.6246	0.7581
Citric acid	0.6903	0.6207	0.7323	0.7983
Residual sugar	0.7143	0.7531	0.8020	0.8327
Chlorides	0.7607	0.7855	0.8210	0.8682
Free sulfur dioxide	0.7727	0.8148	0.8312	0.9456
Total sulfur dioxide	0.8147	0.8937	0.8726	0.9580
Sulfate	0.8285	0.9222	0.8750	0.9867
Alcohol	0.9689	0.9843	0.8922	0.9914

According to the definition of "almost all" in Example 3, we can calculate fuzzy measures of $X_i = \{x_j: 1 \leq i \leq 9\}$ for $1 \leq i \leq 9$ about the fuzzy quantifier as follows:

$$Q_X(X_i) = (|X_i|/|X|)^2 = [(9-i)/9]^2, \text{ for } 1 \leq i \leq 9$$

After the calculation, we can obtain:

$$Q_X(X_1) = 1, Q_X(X_2) = 64/81, Q_X(X_3) = 49/81, Q_X(X_4) = 4/9, Q_X(X_5) = 25/81, Q_X(X_6) = 16/81, Q_X(X_7) = 1/9, Q_X(X_8) = 4/81, Q_X(X_9) = 1/81.$$

The overlap function is defined as

$$O(x,y) = \min(\sqrt{x}, \sqrt{y}),$$

for any $x, y \in [0, 1]$, then the truth value $(Qx)\varphi(x)$ of each supplier under its interpretation I based on the O-Sugeno integral is calculated as follows:

$$D(s_j) = T_I((Qx)\varphi(x)) = \int^{OS} s_j \circ Q_X = \max_{i=1}^{9} O[h_j(x_i), Q_X(X_i)]$$

for $1 \leq j \leq 4$.

After the calculation, we obtain the comprehensive evaluation values of all alternatives, as shown in Table 17.

Table 17. Comprehensive evaluation values of nine alternatives.

	s_1	s_2	s_3	s_4
Evaluation value	0.815	0.778	0.790	0.871

Therefore, the evaluation shows that alternatives s_4 has the highest score, and the alternatives s_4 should be selected.

6. Conclusions

In this study, we proposed O-Sugeno integrals and studied their basic properties. Since overlap functions can be non-associative, the range of applications of O-Sugeno integrals is greatly expanded. Fuzzy quantifiers can be quantified by fuzzy measures, and linguistic quantifier propositions containing fuzzy quantifiers can be calculated their truth values using O-Sugeno integrals. Then, we researched the O-Sugeno integral semantics of fuzzy quantifiers. Finally, we proposed a MADM method based on O-Sugeno integral semantics of fuzzy quantifiers to solve the MADM problem involving fuzzy quantifier-based.

In future work, we will introduce Choquet integrals based on overlap functions and apply them to MADM problems involving fuzzy quantifiers.

Author Contributions: Writing—original draft preparation, X.M.; writing—review and editing, C.T.; writing—inspection and modification, H.Z. All authors have read and agreed to the published version of the manuscript.

Funding: This research was funded by the Natural Science Foundation of Zhejiang Province, grant number LY20A010012, and by the Zhejiang Provincial Soft Science Research Project, grant number 2022C35101.

Data Availability Statement: Not applicable.

Conflicts of Interest: The authors declare no conflict of interest.

References

1. Menger, K. Statistical Metrics. *Proc. Natl. Acad. Sci. USA* **1942**, *28*, 535–537.
2. García, F.S.; Álvarez, P.G. Two families of fuzzy integrals. *Fuzzy Sets Syst.* **1986**, *18*, 67–81.
3. Flondor, P.; Georgescu, G.; Iorgulescum, A. Pseudo-t-morms and pseudo-BL algebras. *Soft Comput.* **2001**, *5*, 355–371.
4. Mas, M.; Mayor, G.; Torrens, J. t-operators. *Int. J. Uncertain. Fuzziness Knowl.-Based Syst.* **1999**, *7*, 31–50.
5. Yager, R.R.; Rybalov, A. Uninorm aggregation operators. *Fuzzy Sets Syst.* **1996**, *80*, 111–120.
6. Liu, H. Semi-uninorms and implications on a complete lattice. *Fuzzy Sets Syst.* **2012**, *191*, 72–82.
7. Dan, Y. A unified way to studies of t-seminorms, t-semiconorms and semi-uninorms on a complete lattice in terms of behaviour operations. *Int. J. Approx. Reason.* **2023**, *156*, 61–76.

8. Borzooei, R.A.; Dvurecenskij, A.; Sharafi, A.H. Material implications in lattice effect algebras. *Information Sciences* **2018**, *433–434*, 233–240.
9. Zhang, X.; Sheng, N.; Borzooei, R.A. Partial residuated implications induced by partial triangular norms and partial residuated lattices. *Axioms* **2023**, *12*, 63.
10. Shilkret, N. Maxitive measure and integration. *Indag. Math.* **1971**, *33*, 109–116.
11. Sugeno, M. Theory of Fuzzy Integral and Its Application. Ph.D. Thesis, Tokyo Institute of Technology, Tokyo, Japan, 1974.
12. Dubois, D.; Prade, H.; Rico, A.; Teheux, B. Generalized qualitative Sugeno integrals. *Inf. Sci.* **2017**, *415–416*, 429–445. [CrossRef]
13. Klement, E.P.; Mesiar, R.; Rap, E. A universal integral as common frame for Choquet and Sugeno integral. *IEEE Trans. Fuzzy Syst.* **2010**, *18*, 178–187. [CrossRef]
14. Jin, L.; Kolesarova, A.; Mesiar, R. Semicopula based integrals. *Fuzzy Sets Syst.* **2021**, *412*, 106–119.
15. Mihailovi, B.; Pap, E. Sugeno integral based on absolutely monotone real set functions. *Fuzzy Sets Syst.* **2010**, *161*, 2857–2869.
16. Liu, X. Further discussion on convergence theorems for seminormed fuzzy integrals and semiconormed fuzzy integrals. *Fuzzy Sets Syst.* **1993**, *55*, 219–226.
17. Luan, T.N.; Hoang, D.H.; Thuyet, T.M.; Phuoc, H.N.; Dung, K.H. A note on the smallest semicopula-based universal integral and an application. *Fuzzy Sets Syst.* **2022**, *430*, 88–101.
18. Hoang, D.H.; Son, P.T.; Duc, H.Q.; Duong, D.V.; Luan, T.N. On a convergence in measure theorem for the seminormed and semiconormed fuzzy integrals. *Fuzzy Sets Syst.* **2023**, *457*, 156–168. [CrossRef]
19. Luan, T.N.; Hoang, D.H.; Thuyet, T.M. On the coincide of lower and upper generalized Sugeno integrals. *Fuzzy Sets Syst.* **2023**, *457*, 169–179. [CrossRef]
20. Tao, Y.; Sun, G.; Wang, G. Generalized K-Sugeno integrals and their equivalent. *Comput. Appl. Math.* **2022**, *41*, 52.
21. Eslami, E.; Khosravi, H.; Sadeghi, F. Very and more or less in non-commutative fuzzy logic. *Soft Comput.* **2008**, *12*, 275–279.
22. Fodor, J.C.; Keresztfalvi, T. Nonstandard conjunctions and implications in fuzzy logic. *Int. J. Approx. Reason.* **1995**, *12*, 69–84. [CrossRef]
23. Bustince, H.; Fernandez, J.; Mesiar, R.; Orduna, R. Overlap function. *Nonlinear Anal.* **2010**, *72*, 1488–1499.
24. Bedregal, B.C.; Dimuro, G.P.; Bustince, H.; Barrenechea, E. New results on overlap and grouping function. *Inf. Sci.* **2013**, *249*, 148–170.
25. Dimuro, G.P.; Bedregal, B. On residual implications derived from overlap functions. *Inf. Sci.* **2015**, *312*, 78–88.
26. Qiao, J.; Hu, B. On generalized migrativity property for overlap functions. *Fuzzy Sets Syst.* **2019**, *357*, 91–116.
27. Jurio, A.; Bustince, H.; Pagola, M.; Pradera, A.; Yager, R.R. Some properties of overlap and grouping function and their application to image thresholding. *Fuzzy Sets Syst.* **2013**, *229*, 69–90.
28. Bustince, H.; Pagola, M.; Mesiar, R.; Hullermeier, E.; Herrera, F. Grouping, overlap, and generalized bientropic functions for fuzzy modeling of pairwise comparisons. *IEEE Trans. Fuzzy Syst.* **2012**, *20*, 405–415.
29. Gomez, D.; Rodriguez, J.T.; Yanez, J.; Montero, J. A new modularity measure for fuzzy community problems based on overlap and grouping function. *Int. J. Approx. Reason.* **2016**, *74*, 88–107.
30. Miguel, L.D.; Gomez, D.; Rodriguez, J.T.; Montero, J.; Bustince, H.; Dimuro, G.P.; Sanz, J.A. General overlap functions. *Fuzzy Sets Syst.* **2019**, *372*, 81–96. [CrossRef]
31. Dimuro, G.P.; Bedregal, B. Archinmedean overlap functions: The ordinal sum and the cancellation, idempotency and limiting properties. *Fuzzy Sets Syst.* **2014**, *252*, 39–54.
32. Paiva, R.; Santiago, R.; Bedregal, B.; Palmeira, E. Lattice-valued overlap and quasi-overlap functions. *Inf. Sci.* **2021**, *562*, 180–199. [CrossRef]
33. Zhang, X.; Liang, R.; Bustince, H.; Bedregal, B.; Fernandez, J.; Li, M.; Qu, Q. Pseudo overlap functions, fuzzy implications and pseudo grouping functions with applications. *Axioms* **2022**, *11*, 593. [CrossRef]
34. Zhang, X.; Wang, M.; Bedregal, B.; Li, M.; Liang, R. Semi-overlap functions and novel fuzzy reasoning algorithms with applications. *Inf. Sci.* **2022**, *614*, 104–122. [CrossRef]
35. Liang, R.; Zhang, X. Interval-valued pseudo overlap functions and application. *Axioms* **2022**, *11*, 216. [CrossRef]
36. Wang, J.; Zhang, X. A novel multi-criteria decision-making method based on rough sets and fuzzy measures. *Axioms* **2022**, *11*, 275. [CrossRef]
37. Zadeh, L.A. A computational approach to fuzzy quantifiers in natural languages. *Comput. Math. Appl. Spec. Issue Comput. Linguist.* **1983**, *9*, 149–184. [CrossRef]
38. Yager, R.R. On ordered weighted averaging aggregation operators in multicriteria decisionmaking. *IEEE Trans. Syst. Man Cybern.* **1988**, *18*, 183–190. [CrossRef]
39. Dvorak, A.; Holcapek, M. Fuzzy quantifiers defined over fuzzy domains. *Fuzzy Sets Syst.* **2022**, *431*, 39–69. [CrossRef]
40. Medina, J.; Torne-Zambrano, J.A. Immediate consequences operator on generalized quantifiers. *Fuzzy Sets Syst.* **2023**, *456*, 72–91. [CrossRef]
41. Ying, M. Linguistic quantifiers modeled by Sugeno integrals. *Artif. Intell.* **2006**, *170*, 581–606. [CrossRef]

42. Zhang, X.; She, Y. *Fuzzy Quantifiers and Their Integral Semantics*; Science Press: Beijing, China, 2017; pp. 104–123.
43. Paiva, R.; Santiago, R.; Bedregal, B.; Rivieccio, U. Inflationary BL-algebras obtained from 2-dimensional general overlap functions. *Fuzzy Sets Syst.* **2021**, *418*, 64–83. [CrossRef]

axioms

MDPI

Article

Ranking Startups Using DEMATEL-ANP-Based Fuzzy PROMETHEE II

Huyen Trang Nguyen [1] and Ta-Chung Chu [2,*]

[1] College of Business, Southern Taiwan University of Science and Technology, Tainan City 710301, Taiwan; da51g207@stust.edu.tw
[2] Department of Industrial Management and Information, Southern Taiwan University of Science and Technology, Tainan City 710301, Taiwan
* Correspondence: tcchu@stust.edu.tw

Abstract: In entrepreneurship management, the evaluation and selection of startups for acceleration programs, especially technology-based startups, are crucial. This process involves considering numerical and qualitative criteria such as sales, prior startup experience, demand validation, and product maturity. To effectively rank startups based on the varying importance of these criteria, a fuzzy multi-criteria decision-making (MCDM) approach is needed. Although MCDM methods have been successful in handling complex problems, their application in startup selection and evaluating criteria interrelationships from the accelerator perspective is underexplored. To address this gap, a hybrid DEMATEL-ANP-based fuzzy PROMETHEE II model is proposed in this study, facilitating startup ranking and examining interrelationships among factors. The resulting preference values are fuzzy numbers, necessitating a fuzzy ranking method for decision-making. An extension of ranking fuzzy numbers using a spread area-based relative maximizing and minimizing set is suggested to enhance the flexibility of existing ranking MCDM methods. Algorithms, formulas, and a comparative analysis validate the proposed method, while a numerical experiment verifies the viability of the hybrid model. The final ranking of four startup projects is $A_4 < A_1 < A_3 < A_2$ which indicates that startup project A_2 has the highest comprehensive potential, followed by startup project A_3.

Keywords: DEMATEL; ANP; PROMETHEE II; ranking fuzzy numbers; startups

MSC: 91B06

Citation: Nguyen, H.T.; Chu, T.-C. Ranking Startups Using DEMATEL-ANP-Based Fuzzy PROMETHEE II. *Axioms* **2023**, *12*, 528. https://doi.org/10.3390/axioms12060528

Academic Editor: Oscar Castillo

Received: 21 April 2023
Revised: 24 May 2023
Accepted: 25 May 2023
Published: 28 May 2023

Correction Statement: This article has been republished with a minor change. The change does not affect the scientific content of the article and further details are available within the backmatter of the website version of this article.

1. Introduction

Entrepreneurship has been recognized as a significant driver of economic growth, both directly and indirectly, and as well as a catalyst for more investments in knowledge creation and generation [1]. Notably, technology-based startups can transform the traditional economy into a digital economy through innovation [2]. The key determinants of entrepreneurial success encompass a range of factors, including entrepreneurs' networks, leadership skills, financial competency, aptitude, knowledge, and support services [3]. Stam [3] defined the entrepreneurial ecosystem as a complex network of interconnected actors and factors that collaborate to facilitate productive entrepreneurship. Among the factors, accelerators are the primary players in the entrepreneurial ecosystem and are actively engaged in fostering innovation and nurturing startups. They develop startup projects, including financing, services, networking, mentoring, and training [4]. Not only do accelerators support through networking services, mentorships, and educational endeavors, but they also play a crucial role in augmenting the financial capabilities of entrepreneurial firms. However, despite their critical role, exploring the selection process employed by accelerators in identifying and evaluating entrepreneurial firms and the underlying selection criteria remains relatively scarce [5].

The initial phase in the process of the entry-boost-exit process is to select a suitable startup. Accelerators whose financial gains are contingent upon the successful exit of the startups in which they invest must exercise discernment when evaluating prospective projects [5]. The selection process encompasses three distinct steps: soliciting startup submissions, conducting comprehensive examinations and evaluations of the projects, and, based on the input of key decision-makers (DMs), eliminating unpromising projects while investing in those that exhibit promise [6]. Lin et al. [7] used the hesitant fuzzy linguistic (HFL) multi-criteria decision-making (MCDM) method to evaluate startups from a technology business incubator perspective, taking into account DMs' psychology. The researchers developed a ratio of score value to deviation degree to compare HFL term sets and defined the HFL information envelopment efficiency, analysis, and preference model. Their numerical example showed the method's applicability, and they concluded that it is more flexible and general. Nonetheless, it should be noted that this method exclusively applies to HFL information environments with unrevealed criteria weight values. Furthermore, the authors acknowledged the limited extent of research on ranking startups within the existing literature.

The process of selecting startups for acceleration programs involves intricate consideration of both qualitative and quantitative criteria. Qualitative criteria encompass factors such as competitive advantage and demand validation, while quantitative criteria include investment costs and team size. Consequently, the ranking of startups poses an MCDM problem. MCDM, as a research field, contributes to the development and implementation of decision-support methodologies and tools [8]. Additionally, MCDM methods are valuable in resolving multiplex problems involving objectives, multiple criteria, and alternatives rated by DMs. It is important to note that the DMs' judgment through qualitative criteria is crucial to the decision-making process, despite its inherent subjectivity and vagueness. Fuzzy numbers (FNs) offer a more effective means of modeling human thought compared to their crisp counterparts.

However, the conventional MCDM method solely adheres to classical mathematical theory, and different methods must be improved or combined to adapt to actual MCDM [9]. Moreover, the amalgamation of DEMATEL-ANP-based fuzzy PROMETHEE II has not been previously applied. This study aims to bridge this gap by investigating the technology startup selection process from the perspective of accelerators, utilizing the DEMATEL-ANP-based fuzzy PROMETHEE II approach. To the best of our knowledge, no prior research has scrutinized this hybrid method in evaluating startups. Accordingly, our study explores its feasibility and effectiveness. A ranking method based on spread areas is proposed with formulas to support the decision-making process, and a comparison is conducted to demonstrate the method's advantages. Subsequently, a numerical example is presented to elucidate the complete process of the hybrid method.

The subsequent sections of this paper are structured as follows. Section 2 provides a literature review of the accelerator, selection criteria, and MCDM techniques. Section 3 introduces the classical concept of fuzzy set theory and outlines the hybrid DEMATEL-ANP-based fuzzy PROMETHEE II method. In Section 4, a comparative analysis is presented to underscore the advantages of the ranking technique. Section 5 presents a numerical example that illustrates the applicability and implementation of the hybrid approach in real-world problems. Finally, Section 6 concludes the work by summarizing key findings and suggesting potential avenues for future research.

2. Literature Review

2.1. Accelerators and the Startup Selection Approach

In the last 15 years, accelerators have boomed due to their effects on startup development, entrepreneurial ecosystem formation, and innovation support [10]. The Y-Combinator, the first accelerator founded by Paul Graham in 2005, was a milestone for the growth of startup accelerators worldwide. By April 2023, according to Seed-DB, 8153 companies were accelerated with funding of USD 88,874,580,633 [11]. Worldwide high-impact accelerators include Y-Combinator, with 1801 companies accelerated and USD 52,211,811,615 of funding, Techstars with 1336 companies accelerated and USD 12,690,624,018 of funding, and 500 startups with 1686 companies accelerated and USD 4,030,020,819 of funding. In the entrepreneurial ecosystem, many organizations support startups in their early stages with financial and nonfinancial investment, including incubators, accelerators, angel investors, venture capitalists, and governments. However, accelerators are the primary players with their mission of fostering innovative ecosystems and nurturing startups.

Accelerators provide mentoring and networking for selected startups in their intensive programs that develop startups' ability to seek investors. "Accelerators are organizations that serve as gatekeepers and validators of promising business innovations through their embeddedness in their respective ecosystems and, thus, play an active and salient role in socioeconomic and technological advancement" ([10], p. 2). Moreover, various accelerators require equity to counterbalance the support services. For example, the structured investment of one of the biggest accelerators, 500 startups, is USD 150,000 for 6% of their companies [12]. The primary return of profit-driven accelerators is from initial public offerings or acquisitions when a startup exits [13]. Therefore, accelerators must be selective when evaluating startup projects. The filtering process is crucial yet challenging for both accelerators and startups; however, research on the selection criteria and process is still lacking [5].

When investigating the Singapore-based Joyful Frog Digital Incubator (JFDI), Yin and Luo [5] adopted an RWW framework for innovation projects to apply to the accelerator program's assessment. Using a scoreboard of 30 criteria based on the RWW framework, they identified eight vital criteria in the initial screening process. Among these factors, market attractiveness factors explain the existing markets and potential customers, including "demand validation", "customer affordability", and "market demographics", and product feasibility factors include "concept maturity", "sales and distribution", and "product maturity". In addition, product advantage factors, such as "value proposition" and "sustainable advantage", and team competence factors, such as "technology expertise", "prior startup experience", and "feedback mechanism", were crucial. Furthermore, "growth strategy" was considered an essential criterion.

Mariño-Garrido et al. [14] used statistical methods on a Spanish accelerator case study analysis to determine the essential criteria for selecting an entrepreneurial project. Out of the nine criteria investigated, six were significant: speed of acceleration, the extent of innovation, the extent of investment ability, creativity, negotiation, and the extent of team consistency.

Learning about ranking startup methods is crucial for investors, incubators, accelerators, and other stakeholders as it facilitates effective decision-making, risk management, resource allocation, and benchmarking and ultimately increases the chances of success in the dynamic and competitive startup ecosystem. More recent studies about startups can be found at [15–17].

2.2. DEMATEL

MCDM methods assist in resolving complex problems that entail multiple objectives, criteria, and alternatives evaluated by decision-makers (DMs). A review of MCDM methods can be found in various studies [18–21].

DEMATEL [22,23] is a constructive method for identifying cause–effect-linked components of a multiplex system. Using a visual systemic model, the technique evaluates interrelationships among criteria and uncovers the critical interrelationships. Moraga et al. [24] used DEMATEL to create a quantitative strategy map identifying causal relationships. Using an MCDM method, the authors developed the final strategy map with qualitative and quantitative approaches that improve and assist managers' assessment process. Altuntas and Gok [25] applied DEMATEL to making correct quarantine decisions, aiming to reduce the burden of the COVID-19 pandemic on the hospitality industry. In 2023,

Wang et al. [26] suggested a new approach for group recommendation, named GroupRecD, which utilizes data mining and the DEMATEL technique to allocate user weights scientifically and rationally. Si et al. [27] conducted a systematic review of DEMATEL. They claimed that the DEMATEL has advantages, including effectively analyzing the direct and indirect effects among factors, visualizing the interdependent relationships between factors by network relation maps, and identifying critical criteria. However, the review also pointed out that DEMATEL cannot achieve the desired level of alternatives, as in Vise Kriterijumska Optimizacija I Kompromisno Resenje (VIKOR) method, or produce partial ranking sequences, as in the ELimination Et Choix Traduisant la REalite (ELECTRE) method. Hence, the DEMATEL was combined with different MCDM methods to obtain appropriate outcomes [27].

2.3. AHP

Saaty [28] introduced both the AHP and ANP methods. The AHP method [29] assumes criteria independence and analyzes decision-making problems in a hierarchical criteria structure. To overcome this limitation, Saaty [30,31] developed the ANP method, which considers dependencies and feedback among elements in a network structure to obtain criteria weights. A systematic review of both methods can be found in [32]. The ANP method has been applied to various fields of research. Galankashi et al. [33] amalgamated fuzzy logic and linguistic expression with ANP for investment portfolio selection. When sorting portfolios, multiple studies have focused on financial factors; however, the results indicated that other factors, such as risk, the market, and growth, are essential. The study demonstrated that ANP could present the internal relations between criteria, which is critical in decision-making. In 2023, Saputro et al. [34] utilized Multi-Dimensional Scaling (MDS) and ANP to examine the sustainability approach for developing rural tourism in Panjalu, Ciamis, Indonesia. Kadoić [35] noted that the ANP method effectively analyzes interconnections and consistency within a decision system. When the criteria are interdependent, only the ANP technique can be used [36]. When rating startups, the evaluation values may change over time, thus a network structure to express interdependencies is required, and the original weight of each criterion should be turned into the comprehensive weight. As a result, ANP is chosen over AHP to deal with this problem more effectively.

2.4. Fuzzy PROMETHEE II

The Preference Ranking Organization Method for Enrichment Evaluations (PROMETHEE), developed by Brans [37], is one of the most common MCDM methods. PROMETHEE was extended to decision-making in many studies, such as PROMETHEE I for partial ranking and PROMETHEE II for complete ranking [38]. The method has undergone many modifications and improvements to assist humans in decision-making [39]. Among them, PROMETHEE II is the most frequently used because it allows a DM to establish a full ranking [40]. Numerous studies have applied hybrid models combining the PROMETHEE method and other MCDM techniques. Khorasaninejad et al. [41] used a hybrid model to determine the best prime mover in a thermal power plant. The model combined fuzzy ANP-DEMATEL to assess criteria importance and relationships and PROMETHEE to rank alternatives. Govindan et al. [42] used an integrated Fuzzy Delphi, a DEMATEL-based ANP (DANP), and a PROMETHEE method to choose the best supplier based on corporate social responsibility practices and to identify the key factors. Seikh and Mandal [40] proposed an integrated approach, combining PROMETHEE II and SWARA within a fuzzy environment, to streamline the selection of the best bio-chemistry waste management organization. The effectiveness and practicality of their approach were demonstrated through a case study. Hua and Jing [43] extended the classical PROMETHEE method by incorporating the generalized Shapley value in interval-valued Pythagorean fuzzy sets to achieve a more rigorous ranking outcome. To verify the effectiveness of this approach, a case study is conducted to evaluate sustainable suppliers.

The comprehensive literature review conducted in this study revealed the effectiveness and reliability of combining DEMATEL, ANP, and PROMETHEE methods in assisting decision-making in various fields. However, despite the proven success of these individual methods, the amalgamation of DEMATEL-ANP-based fuzzy PROMETHEE II has not been previously applied. Given the intricate nature of rating startups, a robust hybrid approach is essential to effectively address the complexities involved. Considering the multitude of qualitative and quantitative criteria that need to be evaluated, the incorporation of DEMATEL in the initial stage becomes crucial. DEMATEL allows for the examination of cause–effect relationships among these criteria, facilitating the identification and elimination of nonsignificant factors. Subsequently, ANP emerges as the optimal choice for determining criterion weights, as it accounts for criterion interdependencies and provides a comprehensive weighting scheme. To establish a complete ranking, the fuzzy-based PROMETHEE II method is employed with utmost precision. This method accommodates the inherent uncertainties and subjectivity in decision-making processes, enabling a more robust assessment of the startups. By integrating DEMATEL, ANP, and PROMETHEE II within a fuzzy framework, this hybrid approach offers a novel and effective solution for the evaluation and ranking of startups, particularly in contexts where qualitative and quantitative criteria interact and require comprehensive analysis.

2.5. *Ranking Fuzzy Numbers*

Lofi Zadeh [44] introduced fuzzy sets to efficiently model human thought. Fuzzy sets have widely affected many areas of scientific research, including mathematics [45], engineering [46], business, and management [47]. A literature review of the historical evolutions of fuzzy sets, their application, and their frequencies was conducted by Kahraman et al. [48].

Ranking FNs became a critical problem in linguistic decision-making. Jain [49] proposed the first FN ranking method based on maximizing sets. Since then, various methods have been presented, such as the Pos index and its dual Nec index [50], maximizing set and minimizing set [51], area compensation [52], an area method using a radius of gyration [53], deviation degree [54], defuzzified values, heights and spreads [55] and mean of relative values [56].

Wang et al. [54] proposed a ranking method based on left and right deviation degrees derived from maximal and minimal reference sets. Additionally, Wang and Luo [57] introduced an area ranking method using positive and negative ideal points, which they claimed more effectively discriminated FNs than Chen's maximizing and minimizing sets [51]. Asady [58] pointed out that the methods of Wang et al. [54] could not correctly rank fuzzy images. Therefore, he proposed a revised method using parametric forms. Nejad and Mashinchi [59] developed a technique based on the left and right areas to improve the deviation degree method. Yu et al. [60] proposed an extension using an epsilon-deviation degree. Nevertheless, Chutia [61] observed that the approach of Yu et al. still presented limitations in discriminating FNs. Chutia suggested a modified method constituting the ill-defined magnitude value and the angle of the fuzzy set. However, this method cannot be used when FNs have non-linear left and right membership functions [61]. Ghasemi et al. [62] discovered a disadvantage in both the deviation degree method [54] and area ranking based on positive and negative ideal points [57]. The author accordingly introduced an improved approach that considers DMs' risk attitudes. Moreover, numerical examples that demonstrated the efficiency of ranking the proposed method's FNs were provided.

Chu and Nguyen [63] suggested a method to improve Chen's [51] maximizing and minimizing sets to rank FNs. In their study, comparative examples were provided. An experiment demonstrated that the relative maximizing and minimizing set (RMMS) could consistently and logically rank the final fuzzy values of alternatives. This study proposed a fuzzy ranking approach inspired by area ranking and using four spread areas. Based on the RMMS model, the areas were measured and integrated with a confidence level μ to assist the FN ranking procedure. The DMs provided confidence levels, which indicated their confidence toward alternatives.

3. Model Establishment

3.1. Fuzzy Set Theory

Fuzzy Sets

$A = \{(x, f_A(x))|x \in U\}$ where x is an element in the space of points U, A is a fuzzy set in U, $f_A(x)$ is the membership function of A at x [44]. The larger $f_A(x)$, the stronger the grade of membership for x in A.

Fuzzy Numbers

A real FN A is described as any fuzzy subset of the real line R with a membership function f_A that possesses the following properties [50]. f_A is a continuous mapping from R to [0, 1], $f_A(x) = 0$ for all $x \in (-\infty, a]$. f_A is strictly increasing on the left membership function $[a, b]$ and is strictly decreasing on the right membership function $[c, d]$. $f_A(x) = 1$ for all $x \in [b, c]$ and $f_A(x) = 0$ for all $x \in [d, \infty)$, where a, b, c, and d are real numbers.

We may let $a = -\infty$, or $a = b$, or $b = c$, or $c = d$, or $d = +\infty$. Unless elsewhere defined, A is assumed to be convex, normalized, and bounded, i.e., $-\infty < a, d < \infty$. A can be indicated as $[a, b, c, d]$, $a \leq b \leq c \leq d$. Let $f_A^L(x)$, $a \leq x \leq b$ represent and $f_A^R(x)$, $c \leq x \leq d$ represent the left and the right membership function of A, respectively, and $f_A(x) = 1$, $b \leq x \leq c$.

In this research, TFNs will be used. The FN A is a TFN if its membership function f_A is given as follows [51].

$$f_A(x) = \begin{cases} (x-a)/(b-a), & a \leq x \leq b, \\ (x-c)/(b-c), & b \leq x \leq c, \\ 0, & \text{otherwise,} \end{cases} \tag{1}$$

where a, b, and c are real numbers.

α-Cuts

The α-cuts of FN A can be determined as $A^\alpha = \{x|f_A(x) \geq \alpha\}$, $\alpha \in [0, 1]$, where A^α is a non-empty bounded closed interval is contained in R and can be denoted by $A^\alpha = [A_l^\alpha, A_u^\alpha]$, where A_l^α are lower bounds and A_u^α are upper bounds [64].

Arithmetic Operations on Fuzzy Numbers

Given FNs A and B, A, $B \in R^+$, $A^\alpha = [A_l^\alpha, A_u^\alpha]$ and $B^\alpha = [B_l^\alpha, B_u^\alpha]$. By the interval arithmetic, some primary operations of A and B can be described as follows [64].

$$(A \oplus B)^\alpha = [A_l^\alpha + B_l^\alpha,\ A_u^\alpha + B_u^\alpha] \tag{2}$$

$$(A \ominus B)^\alpha = [A_l^\alpha - B_u^\alpha,\ A_u^\alpha - B_l^\alpha] \tag{3}$$

$$(A \otimes B)^\alpha = [A_l^\alpha \cdot B_l^\alpha,\ A_u^\alpha \cdot B_u^\alpha] \tag{4}$$

$$r(A)^\alpha = [r \cdot A_l^\alpha,\ r \cdot A_u^\alpha],\ r \in R^+ \tag{5}$$

Linguistic Values

A linguistic variable is a variable whose values are represented in linguistic terms. It is advantageous for dealing with complicated matters or is ambiguous to be rationally described in traditional quantitative information [51,65]. DMs are assumed to have agreed to weight alternatives over criteria using linguistic values such as *Extremely Poor (EP)*, *Very Poor (VP)*, *Poor (P)*, *Moderate (M)*, *High (H)*, *Very High (VH)*, and *Extremely High (EH)* which can also be represented by TFNs such as *EP* = (0,0.1,0.25), *VP* = (0.1,0.2,0.35), *P* = (0.25,0.35,0.5), *M* = (0.35,0.5,0.65), *H* = (0.5,0.65,0.75), *VH* = (0.65,0.8,0.9), and *EH* = (0.75,0.9,1).

3.2. Relative Maximizing and Minimizing Sets

Chu and Nguyen [63] suggested a technique to improve Chen's [51] maximizing and minimizing set to rank FNs. In their study, numerical comparisons and examples were conducted to demonstrate that the RMMS can consistently and logically rank fuzzy values of alternatives. The RMMS [63] technique is introduced as follows.

Assume there are n FNs $A_i = (a_i, b_i, c_i)$, $i = 1, \ldots, n$, $n \geq 2$, $f_{A_i} \in R$. $x_{\min} = \inf S$, $x_{\max} = \sup S$, $S = U_{i=1}^{n} S_i$, $S_i = \{x | f_{A_i}(x) > 0\}$. FNs $A_g = (a_g, b_g, c_g)$ and $A_l = (a_l, b_l, c_l)$ are added to the right and left sides of the above n FNs $A_i = (a_i, b_i, c_i)$, $i = 1, \ldots, n$, respectively. Assume $x_{\min} = a_1$, $x_{\max} = c_n$, $c_g \geq x_{\max}$ and $a_l \leq x_{\min}$. Let $\delta_R = c_g - x_{\max}$ and $\delta_L = x_{\min} - a_l$, where $x_{\max} = c_n$, $x_{\min} = a_1$, $\delta_R \geq 0$, $\delta_L \geq 0$. The new supremum element is defined as $x'_{\max} = x_{\max} + \delta$ and the new infimum element is defined as $x'_{\min} = x_{\min} - \delta$, where $\delta = \max\{\delta_L, \delta_R\}$.

The relative maximizing set M' and the relative minimizing set N' are determined as:

$$f_{M'}(x) = \begin{cases} \left(\frac{x_{R_i} - (x_{\min} - \delta)}{(x_{\max} + \delta) - (x_{\min} - \delta)} \right)^k, & (x_{\min} - \delta) \leq x_{R_i} \leq (x_{\max} + \delta) \\ 0, \text{ otherwise} \end{cases} \tag{6}$$

$$f_{N'}(x) = \begin{cases} \left(\frac{x_{L_i} - (x_{\max} + \delta)}{(x_{\min} - \delta) - (x_{\max} + \delta)} \right)^k, & (x_{\min} - \delta) \leq x_{L_i} \leq (x_{\max} + \delta) \\ 0, \text{ otherwise} \end{cases} \tag{7}$$

Herein, k is set to 1. The value of k can be varied to suit the application. The total relative utility of each A_i is denoted as in Equation (8).

$$U_{T'}(A_i) = \frac{1}{4}[U_{R_{i1}}(A_i) + ((1 - U_{L_{i1}}(A_i)) + U_{L_{i2}}(A_i) + ((1 - U_{R_{i2}}(A_i))], i = 1, \ldots, (n+2) \tag{8}$$

where the first right relative utility $U_{R_{i1}}(A_i) = \sup\left(f_{M'}(x) \wedge f_{A_i}^R(x)\right)$, the first left relative utility $U_{L_{i1}}(A_i) = \sup\left(f_{N'}(x) \wedge f_{A_i}^L(x)\right)$, the second left relative utility $U_{L_{i2}}(A_i) = \sup\left(f_{M'}(x) \wedge f_{A_i}^L(x)\right)$ and the second right relative utility $U_{R_{i2}}(A_i) = \sup\left(f_{N'}(x) \wedge f_{A_i}^R(x)\right)$.

3.3. Spread Area-Based RMMS

In 2011, Nejad and Mashinchi [59] pointed out the shortcomings of Wang et al.'s [54] deviation degree method that when the values of the left area, the right area, the transfer coefficient λ_i or $1 - \lambda_i$ is zero, the ranking result is inaccurate. Hence, to prevent these problems from occurring, expanding $x_{\max}$ and $x_{\min}$ is needed when ranking. Chu and Nguyen [63] also found out that when adding a new FN, $x_{\max}$ and $x_{\min}$ must be modified by adding equal values to consider both sides of membership functions. Consequently, four utilities need to be accounted for to reduce the inconsistency of Chen's [51] maximizing and minimizing set. However, if a set of FNs with $x_{\min} = 3$, then a new FN $A_g = (3, 3, 3)$ is added, there is no extended value applicable in this situation. Therefore, this work suggests integrating confidence levels in ranking FNs to solve the mentioned problems.

Yeh and Kuo [66] in their research on evaluating passenger service quality of Asia-Pacific international airports, suggested incorporating a DM's confidence level α and a preference index λ to obtain an overall service performance index. In the evaluation procedure, DMs give the value α, based on the concept of an α-cut, with respect to the criteria's weights and alternative performance ratings.

This work proposes to use confidence level in a new perspective, which is confidence level, symbolized as μ, will be integrated into measuring areas spreading based on the RMMS model to assist the ranking FNs procedure, as shown in Figure 1. First, h experts in the group of DMs, $D = \{D_1, \ldots, D_e, \ldots, D_h\}$ are asked to specify their confidence μ_{D_e}, representing their confidence for alternatives to obtain $\mu = \frac{\sum_{e}^{h} \mu_{D_e}}{h}$, $\mu_{D_e} \in [0,\ 1]$. The greater the μ, the more assured is the decision-maker on the alternative.

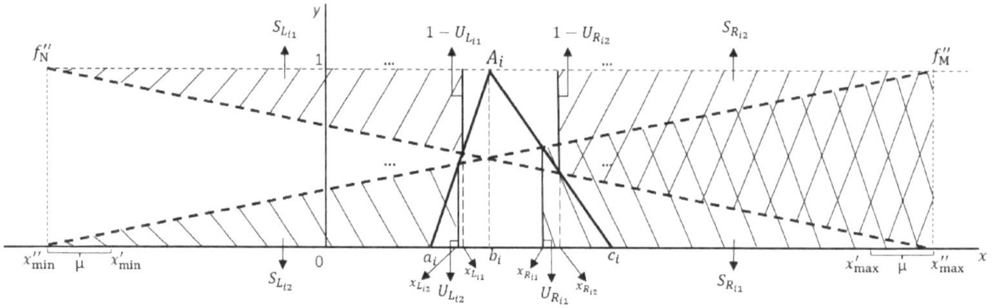

Figure 1. Spread area-based RMMS ranking method.

Since DMs' confidence in an alternative will influence their confidence level in other alternatives, the confidence level μ, calculated by the average of all DMs' evaluation, should be engaged simultaneously with the immensity of the RMMS concept. Accordingly, value μ is integrated by shifting the RMMS's infimum element to the left, provided that the new infimum element is obtained as $x''_{\min} = x'_{\min} - \mu$. Similarly, the average value of μ will be integrated by shifting the RMMS's infimum element to the right, provided that the new supremum element is obtained as $x''_{\max} = x'_{\max} + \mu$.

The coordinates of the intersection of the A_i with the relative maximizing set M'' and the relative minimizing set N'' can be seen in Figure 1 and are determined as the following equations.

$$x_{L_{i1}} = \frac{bx''_{\max} - ax''_{\min}}{b - a - x''_{\min} + x''_{\max}} \tag{9}$$

$$x_{L_{i2}} = \frac{bx''_{\min} - ax''_{\max}}{x''_{\min} - x''_{\max} + b - a} \tag{10}$$

$$x_{R_{i1}} = \frac{bx''_{\min} - cx''_{\max}}{b - c + x''_{\min} - x''_{\max}} \tag{11}$$

$$x_{R_{i2}} = \frac{cx''_{\min} - bx''_{\max}}{c - b + x''_{\min} - x''_{\max}} \tag{12}$$

The first left spread area $S_{L_{i1}}$ is defined as follows.

$$\begin{aligned}
&S_{L_{i1}}(A_i) = \int_{x''_{\min}}^{x_{L_{i1}}} 1dx - \int_{x''_{\min}}^{x_{L_{i1}}} f''_N(x)dx \\
&S_{L_{i1}}(A_i) = x_{L_{i1}} - x''_{\min} - \int_{x''_{\min}}^{x_{L_{i1}}} \left(\frac{x - x''_{\max}}{x''_{\min} - x''_{\max}} \right) dx \\
&= x_{L_{i1}} - x''_{\min} - \left(\frac{x^2}{2(x''_{\min} - x''_{\max})} - \frac{xx''_{\max}}{x''_{\min} - x''_{\max}} \right) \Bigg|_{x'''_{\min}}^{x_{L_{i1}}} \\
&= x_{L_{i1}} - x''_{\min} - \left(\frac{x^2_{L_{i1}} - 2x_{L_{i1}} x''_{\max}}{2(x''_{\min} - x''_{\max})} - \frac{x''^{\,2}_{\min} - 2x''_{\min} x''_{\max}}{2(x''_{\min} - x''_{\max})} \right) \\
&= \frac{(x_{L_{i1}} - x''_{\min})(2x''_{\min} - 2x''_{\max} - x_{L_{i1}} - x''_{\min} - 2x''_{\max})}{2(x''_{\min} - x''_{\max})} = \frac{(x_{L_{i1}} - x''_{\min})(x''_{\min} - 4x''_{\max} - x_{L_{i1}})}{2(x''_{\min} - x''_{\max})}
\end{aligned} \tag{13}$$

If the first left spread area $S_{L_{i1}}$ is larger, the fuzzy number A_i is larger. The second left spread area $S_{L_{i2}}$ is defined as Equation (14); and if $S_{L_{i2}}$ is larger, the fuzzy number

A_i is also larger. The first right spread area $S_{R_{i1}}$ is defined as Equation (15); but if $S_{R_{i1}}$ is larger, the fuzzy number A_i is smaller. Finally, the second right spread area $S_{R_{i2}}$ is defined as Equation (16); and if $S_{R_{i2}}$ is larger, the fuzzy number A_i is also smaller. Therefore, the above four areas must be considered when ranking FNs. The detailed derivation for Equations (14)–(16) is placed in Appendix A.

$$S_{L_{i2}}(A_i) = \int_{x''_{\min}}^{x_{L_{i2}}} f''_M(x)dx = \frac{\left(x_{L_{i2}} - x''_{\min}\right)^2}{2(x''_{\max} - x''_{\min})} \tag{14}$$

$$S_{R_{i1}}(A_i) = \int_{x_{R_{i1}}}^{x''_{\max}} f''_M(x)dx = \frac{(x''_{\max} + x_{R_{i1}})(x''_{\max} - x_{R_{i1}} - 2x''_{\min})}{2(x''_{\max} - x''_{\min})} \tag{15}$$

$$S_{R_{i2}}(A_i) = \int_{x''_{R_{i2}}}^{x''_{\max}} 1dx - \int_{x''_{R_{i2}}}^{x''_{\max}} f''_N(x)dx = \frac{(x''_{\max} - x_{R_{i2}})(2x''_{\min} - x''_{\max} + x_{R_{i2}})}{2(x''_{\min} - x''_{\max})} \tag{16}$$

Finally, the ranking value of each A_i is determined as Equation (17) to classify FNs. An FN is more prominent if its value is larger.

$$V(A_i) = \frac{1}{4}\left(S_{L_1}(A_i) - S_{R_1}(A_i) + S_{L_2}(A_i) - S_{R_2}(A_i)\right) \tag{17}$$

3.4. *The Hybrid DEMATEL-ANP Based Fuzzy PROMETHEE II Model*

3.4.1. DEMATEL

The DEMATEL method is first used to demonstrate the interrelationships between criteria and produce the influential network relationship map. The constructing equations of the classical DEMATEL model can be summarized as follows [67].

Assume that h experts in a decision group $D = \{D_1, D_2, \ldots, D_h\}$ are asked to indicate the direct effect of factor (criterion) C_i has on factor (criterion) C_j in a system with m factors (criteria) $C = \{C_1, C_2, \ldots, C_m\}$ using an integer scale of No Effect (0), Low Effect (1), Medium Low Effect (2), Medium Effect (3), Medium High Effect (4), High Effect (5) and Extremely Strong Effect (6). Next, the individual direct-influence matrix $Z_e = \left[z^e_{ij}\right]_{m\times m}$ provided by the eth expert can be constructed, where all main diagonal components are equal to zero and z^e_{ij} represent the respondent's evaluation of DM on the degree to which criterion C_i affects C_j.

Step 1. Generating the group direct-influence matrix. By aggregating h DMs' judgments, the group direct-influence matrix $Z = \left[z_{ij}\right]_{m\times m}$ can be constructed by

$$Z = \frac{1}{h}\sum_{e=1}^{h} z_{ij}, i,j = 1,2,\ldots,m. \tag{18}$$

Step 2. Acquiring the normalized direct-influence matrix. At this step, the normalized direct-influence matrix by the eth expert is $X_e = \left[x^e_{ij}\right]_{m\times m}, e = 1,2,\ldots h$.

The following equations calculate the average matrix X

$$X = \frac{(x^1 \oplus x^2 \oplus \ldots \oplus x^h)}{h} \tag{19}$$

$$x_{ij} = \frac{\sum_{e=1}^{h} x^e_{ij}}{h} \tag{20}$$

Step 3. Computing the total-influence matrix T. The total-influence matrix $T = \left[t_{ij}\right]_{m\times m}$ is computed as the summation of the direct impacts and all the indirect impacts by Equation (21)

$$T = X + X^2 + X^3 + \ldots + X^h = X(I - X)^{-1}, \tag{21}$$

when $h \to \infty$ in the identity matrix, known as I.

Step 4. Setting up a threshold value and producing the causal diagram.

The sum of columns and the sum of rows are symbolized as R and D, respectively, within the total-relation matrix $T = [t_{ij}], \{i, j \in 1, 2, \ldots, m\}$ by the following formulas:

$$D = [d_i]_{m\times 1} = \left[\sum_{j=1}^{m} t_{ij}\right]_{m\times 1} \tag{22}$$

$$R = [r_j]_{1\times m} = \left[\sum_{i=1}^{m} t_{ij}\right]_{1\times m} \tag{23}$$

The horizontal axis vector (D + R) called "Prominence" demonstrates the power of influence degree that is given and received by the criteria. The vertical axis vector ($D - R$) named "Relation" shows the system's criteria effect. If ($D - R$) is positive, the criterion C_j influences other criteria and can be grouped into a causal group; if ($D + R$) is negative, the criterion C_j is being influenced by the other criteria and can be grouped into an effect group. A causal diagram can be produced by mapping the (D + R, D − R) dataset, yielding valuable assessment perception. A threshold value can be defined to screen out the negligible factors [68,69]. In this work, factors that have a value higher than the average value of the "Prominence" (D + R) and/or ($D - R$) is positive are selected to use in the next step.

3.4.2. ANP

Next, the present work applied the ANP method to produce the weights of the criteria. The generalized ANP process from previous studies is summarized as follows [30,70,71]. In this work, a set of importance scales [28] is adopted to weight each criterion using linguistic values, including 1—*Identically Important (II)*, 3—*Moderately Important (MI)*, 5—*Highly Important (HI)*, 7—*Very Highly Importance (VHI)*, 9—*Extremely Important (EI)*, and 2, 4, 6, 8 are the median values. Reciprocal values are used for inverse comparison.

Step 1. Obtaining Pairwise Comparison Matrix (PCM). Assume that h experts in a decision group $D = \{D_1, D_2, \ldots, D_h\}$ are responsible for evaluating criteria $C = \{C_1, C_2, \ldots, C_m\}$ that are screened through the previous step. The PCM is generated by comparing the ith row with the jth column. The weights of components are formed as shown in matrix A. The diagonal components with identical importance are illustrated by 1.

$$A = [a_{ij}]_{m\times m}\begin{bmatrix} 1 & a_{12} & \ldots & a_{1m} \\ a_{21} & 1 & & \\ \vdots & & 1 & \\ a_{m1} & & & 1 \end{bmatrix}$$

As there are several DMs, the pairwise comparison values from different DMs may vary. Experts can decide together, or each assessment can be integrated into a PCM by the geometric mean GM as in Equation (24).

$$GM = \sqrt[j]{i_1 i_2 i_3 \ldots i_j} \tag{24}$$

Step 2. Computing eigenvectors and the unweighted supermatrix. In this step, eigenvector E_i is obtained through Equation (25), which is computed by each row's average.

$$E_i = \frac{1}{m}\sum_{j=1}^{m} a_{ij} \tag{25}$$

Then, the eigenvectors of each matrix are consolidated to form the unweighted matrix.

Step 3. Examining the consistency. To guarantee consistency among the judgments of the DMs, it is necessary to test the consistency by three metrics, including Consistency Measure (*CM*), Consistency Index (*CI*), and Consistency Ratio (*CR*).

The general form for *CM* values is obtained through Equation (26).

$$CM_j = \frac{a_j \times E}{E_j}, \tag{26}$$

where $j = 1,2,3,\ldots, m$, a_j is the corresponding row of the comparison matrix, E is Eigenvector and E_j represents the corresponding component in E.

Then, $\lambda_{\max}$ is obtained by the average of the *CM* vector. The *CI* is calculated as shown in Equation (27).

$$CI = \frac{\lambda_{\max} - m}{m - 1} \tag{27}$$

Next, a random index, as listed in Table 1 [28], is computed following the order of the PCM. Consequently, the Consistency Ratio *CR* is obtained by Equation (28).

$$CR = \frac{CI}{RI} \tag{28}$$

Table 1. Random Index.

Order	**1**	**2**	**3**	**4**	**5**	**6**	**7**	**8**	**9**	**10**
R.I	0	0	0.52	0.89	1.11	1.25	1.35	1.40	1.45	1.49

The value of $CR \leq 0.1$ is in the satisfactory range; otherwise, the pairwise comparison is required to be revised.

Step 4. Obtain the weighted supermatrix. A weighted supermatrix is obtained to evaluate the relation between criteria. Then, the unweighted matrix is converted into a weighted supermatrix to make the sum of each column 1, called column stochastic.

Step 5. Determining stable weights by obtaining limit supermatrix. The values produced from the previous step are elevated to the power of $2k$ until the values are firmly established, where k is an arbitrarily large number. The final priorities can be determined using the normalization function on each block of the limit matrix. The most significant value represents the most critical criterion among other criteria. The stable weights w constructed from this step are utilized in the following steps.

3.4.3. Fuzzy PROMETHEE-Based Ranking Method

The same group of h experts $D = \{D_1, D_2, \ldots, D_h\}$ will assess n alternatives $A = \{A_1, A_2, \ldots, A_n\}$ under m criteria $C = \{C_1, C_2, \ldots, C_m\}$ that are screened through the previous steps. Let $f_{ij}^e = (a_{ij}^e, b_{ij}^e, c_{ij}^e)$, $i = 1,2\ldots,n$, $j = 1,2\ldots,m$, $e = 1,2\ldots,h$, be the rating assigned to an alternative A_i under the criterion C_j by a decision-maker D_e. Criteria chosen from the earlier steps are first categorized into the cost-benefit framework as qualitative benefit criteria, C_j, $j = 1,2\ldots,k$, quantitative benefit criteria, C_j, $j = k+1,\ldots,k'$, cost qualitative criteria, C_j, $j = k'+1,\ldots,k''$, and cost quantitative criteria C_j, $j = k''+1,\ldots,m$. The fuzzy PROMETHEE II process is summarized as follows [72,73].

Step 1. Constructing the fuzzy decision matrix. Aggregated rating $f_{ij} = (a_{ij}, b_{ij}, c_{ij})$ is:

$$f_{ij} = \left(\frac{1}{h}\right) \otimes \left(f_{ij1} \oplus \ldots \oplus f_{ije} \oplus \ldots \oplus f_{ijh}\right) \tag{29}$$

where $a_{ij} = \sum\limits_{e=1}^{h} \frac{a_{ije}}{h}$, $b_{ij} = \sum\limits_{e=1}^{h} \frac{b_{ije}}{h}$, $c_{ij} = \sum\limits_{e=1}^{h} \frac{c_{ije}}{h}$.

Step 2. Computing the normalized matrix. The normalization is completed using the Chu and Nguyen [63] approach. The ranges of normalized TFNs belong to [0, 1]. Suppose $l_{ij} = (al_{ij}, bl_{ij}, cl_{ij})$ is the value of an alternative A_i, $i = 1, 2, \ldots, n$, versus a benefit (B) criterion or a cost (C) criterion. The normalized value l_{ij} can be as

$$l_{ij} = \left(\frac{al_{ij} - al_j^*}{y_j^*}, \frac{bl_{ij} - al_j^*}{y_j^*}, \frac{cl_{ij} - al_j^*}{y_j^*} \right), j \in \mathrm{B}, \tag{30}$$

$$l_{ij} = \left(\frac{cl_j^* - cl_{ij}}{y_j^*}, \frac{cl_j^* - bl_{ij}}{y_j^*}, \frac{cl_j^* - al_{ij}}{y_j^*} \right), j \in \mathrm{C}, \tag{31}$$

where $al_j^* = \min_i al_{ij}$, $cl_j^* = \max_i cl_{ij}$, $y_j^* = cl_j^* - al_j^*$, $i = 1, 2, \ldots, n$, $j = k' + 1, \ldots, k''$ and $j = k'' + 1, \ldots, m$, $l_{ij} = (al_{ij}, bl_{ij}, cl_{ij})$.

Step 3. Calculating the evaluative differences. Pairwise comparison is made by calculating the evaluative differences of ith alternative with respect to other alternatives. The intensity of the fuzzy preference $P_j(A_i, A_{i'})$ of an alternative A_i over $A_{i'}$ is obtained by Equations (32) and (33), based on Equation (3)

$$P'_j(C_j(A_i) - C_j(A_{i'})) = P'_j(A_i, A_{i'}) \tag{32}$$

$$= l_{ij} - l_{ij'} = (al_{ij}, bl_{ij}, cl_{ij}) - (al'_{ij}, bl'_{ij}, cl'_{ij}) = (al_{ij} - cl'_{ij}, bl_{ij} - bl'_{ij}, cl_{ij} - al'_{ij}) \tag{33}$$

where P_j is the fuzzy preference function for the jth criterion and $C_j(A_i)$ is the evaluation of alternative A_i corresponding to criterion C_j.

Step 4. Determining the preference function. To avoid the complexity and be in a more practicable form, the simplified fuzzy preference function is applied in this study as in Equations (34) and (35).

$$P'_j(A_i, A_{i'}) = 0 \text{ if } C_j(A_i) \leq C_j(A_{i'}) \tag{34}$$

$$P'_j(A_i, A_{i'}) = (C_j(A_i) - C_j(A_{i'})) \text{ if } C_j(A_i) > C_j(A_{i'}) \tag{35}$$

Step 5. Reckoning the aggregated fuzzy preference function. Calculate the aggregated fuzzy preference function considering the criteria weights computed from the ANP method.

$$\pi'(A_i, A_{i'}) = \sum_{j=1}^{m} w_j P'_j(A_i, A_{i'}) / \sum_{j=1}^{m} w_j \tag{36}$$

The higher $\pi'(A_i, A_{i'})$ is, the stronger preference for the ith alternative will be.

Step 6. Determining the fuzzy leaving flow $\varphi'^{+}(A_i)$ and the fuzzy entering flow $\varphi'^{-}(A_i)$

The fuzzy leaving flow of A_i is determined as

$$\varphi'^{+}(A_i) = \frac{1}{n-1} \sum_{\substack{i' = 1 \\ i' \neq 1}}^{n} \pi'(A_i, A_{i'}) \tag{37}$$

The fuzzy entering flow of A_i is determined as

$$\varphi'^{-}(A_i) = \frac{1}{n-1} \sum_{\substack{i' = 1 \\ i' \neq 1}}^{n} \pi'(A_{i'}, A_i) \tag{38}$$

Step 7. Calculating the fuzzy net outranking flow for each alternative

$$\varphi'(A_i) = \varphi'^{+}(A_i) - \varphi'^{-}(A_i) \quad (39)$$

Step 8. Defuzzifying the fuzzy net outranking flow value and obtaining the ranking of alternatives. In this step, the spread area-based RMMS model is proposed to apply to assist defuzzification and obtain the final ranking using Equations (12)–(20). An FN is more prominent if its value $V(A_i)$ is more significant.

4. Numerical Comparison and Consistency Test

In this section, various examples of comparisons are established to investigate the effectiveness of the proposed method. The first example illustrates the ranking orders of the method compared with the methods of Wang et al. [54] and Nejad and Mashinchi [59]. We used FNs in Examples 2–4 from Nejad and Mashinchi [59], and then different situations were generated through the addition of new FNs for testing the consistency of the ranking results, as shown in Table 2. In Situation (1), methods from both Nejad and Mashinchi and Wang et al. produce $A_1 = A_2 \prec A_3$, but the proposed method can discriminate between three FNs with the order $A_1 \prec A_2 \prec A_3$. Furthermore, the ranking order is $A_1 \prec A_2 \prec A_3$, and either $A_4 = (-3, -2, -1)$ is added (see Situation (1.1)) or $A_4 = (8.75, 9.5, 11)$ is added (see Situation (1.2)). In Situation (2), the proposed method yields the same ranking, $A_1 \prec A_2$, as that of the method of Nejad and Mashinchi when either $A_4 = (-1.5, -0.8, -0.6)$ or $A_4 = (1.15, 2.5, 3.15)$ is added. However, the method of Wang et al. highlights the inconsistency and produces $A_1 = A_2$ in Situation (2.2). In Situation (3), the proposed method yields the same ranking $A_1 \succ A_2$ as that of Nejad and Mashinchi when $A_4 = (-5, -4, -3, -1)$ or $A_4 = (6, 6, 7, 8)$ is added, but the method of Wang et al. compensates for the inconsistency and produces $A_1 = A_2$ in Situation (3.2). The first comparison demonstrates the usefulness of the proposed method in discriminating FNs.

Table 2. Modified comparison based on Examples 2, 3, and 4 from Nejad and Mashinchi [59].

Situations	Methods	Results	Results after Adding New FNs	
(1)			(1.1) $A_4 = (-3, -2, -1)$	(1.2) $A_4 = (8.75, 9.5, 11)$
$A_1 = (2, 3, 5, 6)$ $A_2 = (1, 4, 7)$ $A_3 = (4, 5, 7)$	[54]	$A_1 = A_2 \prec A_3$	$A_2 \prec A_1 \prec A_3$	$A_1 = A_2 \prec A_3$
	[59]	$A_1 = A_2 \prec A_3$	$A_1 = A_2 \prec A_3$	$A_1 = A_2 \prec A_3$
	Proposed method	$A_1 \prec A_2 \prec A_3$	$A_1 \prec A_2 \prec A_3$	$A_1 \prec A_2 \prec A_3$
(2)			(2.1) $A_4 = (-1.5, -0.8, -0.6)$	(2.2) $A_4 = (1.15, 2.5, 3.15)$
$A_1 = (0.2, 0.5, 0.8)$ $A_2 = (0.4, 0.5, 0.6)$	[54]	$A_1 \prec A_2$	$A_1 \prec A_2$	$A_1 = A_2$
	[59]	$A_1 \prec A_2$	$A_1 \prec A_2$	$A_1 \prec A_2$
	Proposed method	$A_1 \prec A_2$	$A_1 \prec A_2$	$A_1 \prec A_2$
(3)		$A_1 \succ A_2$	(3.1) $A_4 = (-5, -4, -3, -1)$	(3.2) $A_4 = (6, 6, 7, 8)$
$A_1 = (1, 2, 5)$ $A_2 = (1, 2, 2, 4)$	[54]	$A_1 \succ A_2$	$A_1 \succ A_2$	$A_1 = A_2$
	[59]		$A_1 \succ A_2$	$A_1 \succ A_2$
	Proposed method	$A_1 \succ A_2$	$A_1 \succ A_2$	$A_1 \succ A_2$

Second, three sets of FNs are created to further examine the proposed method's stability and credibility, as shown in Table 3. In all previous situations, the method of Wang et al. is ineffective in distinguishing FNs. For example, in Situation (1.1), the method of Nejad and Mashinchi yields an FN ranking, $A_1 \prec A_2 \prec A_3 \prec A_4$, but yields $A_1 = A_2 = A_3 = A_4$ in cases (1) and (1.2), indicating inconsistency, but the proposed method yields $A_1 \prec A_2 \prec A_3 \prec A_4$ in all Situations (1), (1.1), and (1.2). Similarly, in Situations (2) and (2.2), the ranking order obtained using the method of Nejad and Mashinchi is $A_1 = A_2 = A_3$; however, when $A_4 = (-7, -5, -3, -2)$ is added, the order changes to $A_1 \prec A_2 \prec A_3$, as in Situation (2.1); whereas the suggested method persistently ranks in the following order: $A_1 \prec A_2 \prec A_3$. In Situations (3) and (3.2), both the proposed method and the method of Nejad and Mashinchi yield a ranking order of $A_1 \prec A_2$; however, in (3.1), when $A_3 = (-4, -2.5, -1.5)$ is added, the method of Nejad and Mashinchi yields $A_1 \succ A_2$; however, the proposed method yields a persistent rank order of $A_1 \prec A_2$. Hence, the second comparison has demonstrated the effectiveness of the proposed method in discriminating FNs compared to Wang et al.'s technique and the consistency compared with the method of Nejad and Mashinchi.

Table 3. Comparison with Wang et al.'s [54] and Nejad and Mashinchi [59].

Situations	Methods	Results	Results after Adding New FNs	
(1)			(1.1) $A_5 = (-5, -4, -3)$	(1.2) $A_5 = (8, 9, 10)$
$A_1 = (3,3,3)$ $A_2 = (3,3,6)$ $A_3 = (3,3,8)$ $A_4 = (3,3,6,8)$	[54]	$A_1 = A_2 = A_3 = A_4$	$A_1 = A_2 = A_3 = A_4$	$A_1 = A_2 = A_3 = A_4$
	[59]	$A_1 = A_2 = A_3 = A_4$	$A_1 \prec A_2 \prec A_3 \prec A_4$	$A_1 = A_2 = A_3 = A_4$
	Proposed method	$A_1 \prec A_2 \prec A_3 \prec A_4$	$A_1 \prec A_2 \prec A_3 \prec A_4$	$A_1 \prec A_2 \prec A_3 \prec A_4$
(2)			(2.1) $A_4 = (-7, -5, -3, -2)$	(2.2) $A_4 = (7, 9, 11, 12)$
$A_1 = (3,3,3)$ $A_2 = (3,3,6)$ $A_3 = (3,3,5,6)$	[54]	$A_1 = A_2 = A_3$	$A_1 = A_2 = A_3$	$A_1 = A_2 = A_3$
	[59]	$A_1 = A_2 = A_3$	$A_1 \prec A_2 \prec A_3$	$A_1 = A_2 = A_3$
	Proposed method	$A_1 \prec A_2 \prec A_3$	$A_1 \prec A_2 \prec A_3$	$A_1 \prec A_2 \prec A_3$
(3)			(3.1) $A_3 = (-4, -2.5, -1.5)$	(3.2) $A_3 = (6, 7.8, 8.5)$
$A_1 = (2,2,7)$ $A_2 = (2,4,4)$	[54]	$A_1 = A_2$	$A_1 = A_2$	$A_1 = A_2$
	[59]	$A_1 \prec A_2$	$A_1 \succ A_2$	$A_1 \prec A_2$
	Proposed method	$A_1 \prec A_2$	$A_1 \prec A_2$	$A_1 \prec A_2$

Additionally, a consistency test is designed to examine the reliability of the proposed method, as shown in Tables 4 and 5. In Example 1, the result is $A_1 \prec A_2 \prec A_3$, $A_1 \prec A_2 \prec A_3$ for all assumed various μ values. In Example 2, when $A_4 = (8, 9, 10)$ is added, the classifying order remains the same as $A_1 \prec A_2 \prec A_3$ for all $0.1 \prec \mu \prec 1$. Finally, in Example 3, when $A_4 = (-3, -2, -1)$ is added, the proposed method consistently yields an order of $A_1 \prec A_2 \prec A_3$ for all tested values of μ. The results of the numerical comparison demonstrate the credibility and effectiveness of the suggested ranking method based on spread area-based RMMS.

Table 4. Numerical comparison with Chu and Nguyen [63].

Situations	Methods	Results	Results after Adding New FNs	
(1)			(1.1) $A_3 = (1,4,5)$	(1.2) $A_3 = (-3,-2,-1)$
$A_1 = (1,3,5)$ $A_2 = (2,3,4)$	[63]	$A_1 = A_2$	$A_1 = A_2$	$A_1 = A_2$
	Proposed method	$A_1 \prec A_2$	$A_1 \prec A_2$	$A_1 \prec A_2$
(2)			(2.1) $A_3 = (2,3,7)$	(2.2) $A_3 = (-4,-2,-2)$
$A_1 = (2,2,4)$ $A_2 = (2,2,6)$	[63]	$A_1 = A_2$	$A_1 = A_2$	$A_1 = A_2$
	Proposed method	$A_1 \prec A_2$	$A_1 \prec A_2$	$A_1 \prec A_2$

Table 5. A consistency test with various values of μ in different examples.

μ	Examples		
	(1) Three FNs $A_1 = (2,3,5,6)$, $A_2 = (1,4,7)$ $A_3 = (4,5,7)$	**(2) Add an FN to the Right Side** $A_1 = (2,3,5,6)$, $A_2 = (1,4,7)$ $A_3 = (4,5,7)$, $A_4 = (8,9,10)$	**(3) Add an FN to the Left Side** $A_1 = (2,3,5,6)$, $A_2 = (1,4,7)$ $A_3 = (4,5,7)$, $A_4 = (-3,-2,-1)$
0.1	$A_1 \prec A_2 \prec A_3$	$A_1 \prec A_2 \prec A_3$	$A_1 \prec A_2 \prec A_3$
0.2	$A_1 \prec A_2 \prec A_3$	$A_1 \prec A_2 \prec A_3$	$A_1 \prec A_2 \prec A_3$
0.3	$A_1 \prec A_2 \prec A_3$	$A_1 \prec A_2 \prec A_3$	$A_1 \prec A_2 \prec A_3$
0.4	$A_1 \prec A_2 \prec A_3$	$A_1 \prec A_2 \prec A_3$	$A_1 \prec A_2 \prec A_3$
0.5	$A_1 \prec A_2 \prec A_3$	$A_1 \prec A_2 \prec A_3$	$A_1 \prec A_2 \prec A_3$
0.6	$A_1 \prec A_2 \prec A_3$	$A_1 \prec A_2 \prec A_3$	$A_1 \prec A_2 \prec A_3$
0.7	$A_1 \prec A_2 \prec A_3$	$A_1 \prec A_2 \prec A_3$	$A_1 \prec A_2 \prec A_3$
0.8	$A_1 \prec A_2 \prec A_3$	$A_1 \prec A_2 \prec A_3$	$A_1 \prec A_2 \prec A_3$
0.9	$A_1 \prec A_2 \prec A_3$	$A_1 \prec A_2 \prec A_3$	$A_1 \prec A_2 \prec A_3$
1.0	$A_1 \prec A_2 \prec A_3$	$A_1 \prec A_2 \prec A_3$	$A_1 \prec A_2 \prec A_3$

5. Numerical Example

Suppose 4 DMs $(D_h, h = 1,2,3,4)$ of an accelerator must establish criteria and analyze the criteria's effect on a technology-based acceleration program. To achieve this goal, the methods DEMATEL and ANP are performed. Assume $(C_m, m = 1,2,\ldots,19)$ are the qualitative criteria and quantitative criteria under consideration, as shown in Figure A1 (see Appendix B for details). Assuming that DMs have reached a consensus, the effects of criteria on each other are indicated using a scale of *No Effect (1)*, *Low Effect (2)*, *Medium Low Effect (3)*, *Medium Effect (4)*, *Medium High Effect (5)*, *High Effect (6)*, and *Extremely Strong Effect (7)*. After each DM rates the alternatives, the aggregating direct-relation matrix is determined using Equation (18) and is shown in Table A1 (see Appendix C for details).

Subsequently, values of the normalized direct-relation matrix are obtained using Equations (19) and (20) and are shown in Table A2 (see Appendix D for details). Finally, the total-relation matrix is attained using Equation (21), as shown in Table A3 (see Appendix E for details). Next, the prominence $(D + R)$ and relation $(D - R)$ values are calculated using Equations (22) and (23). Thereafter, the threshold value is set, which determines the filtered factors. The causal relationship and notable factors are displayed in Table 6 and Figure 2. According to Table 6, "(C_6) demand validation" has the greatest $(D + R)$ value and is the most critical factor, followed by "(C_7) customer affordability" and "(C_8) market demographic". All these factors need to be evaluated in the initial steps when building a product or service. Additionally, the $(D - R)$ values of "(C_3) prior startup experience", "(C_1) sales", and "(C_2) product development cost" demonstrate that these criteria have net influences on other factors. Other medium value factors that are selected when proceeding to the next steps are "(C_9) concept maturity", "(C_{10}) product maturity", "(C_{11}) value proposition", "(C_{13}) technology experience", "(C_{15}) growth strategy", "(C_{18}) creativity", and "(C_{19}) negotiation".

Table 6. Prominence and Relation value of criteria.

	D	R	$D+R$	$D-R$
C_1	1.4660	1.0098	2.476	0.4562
C_2	1.3166	0.9326	2.249	0.3840
C_3	2.8245	2.0101	4.835	0.8144
C_4	1.3291	1.9605	3.290	−0.6314
C_5	1.5740	2.1332	3.707	−0.5593
C_6	3.0201	3.1850	6.205	−0.1649
C_7	2.9359	3.1138	6.050	−0.1778
C_8	2.9104	3.1088	6.019	−0.1985
C_9	2.5804	2.8768	5.457	−0.2964
C_{10}	2.3358	2.7069	5.043	−0.3711
C_{11}	2.2284	2.6253	4.854	−0.3969
C_{12}	1.3718	1.9929	3.365	−0.6211
C_{13}	2.6701	2.9201	5.590	−0.2500
C_{14}	1.3602	2.0277	3.388	−0.6675
C_{15}	2.1349	2.5318	4.667	−0.3969
C_{16}	1.6294	2.1784	3.808	−0.5490
C_{17}	1.8855	2.3726	4.258	−0.4871
C_{18}	2.4492	2.7714	5.221	−0.3222
C_{19}	2.5088	2.8181	5.327	−0.3093
		Average	4.516	

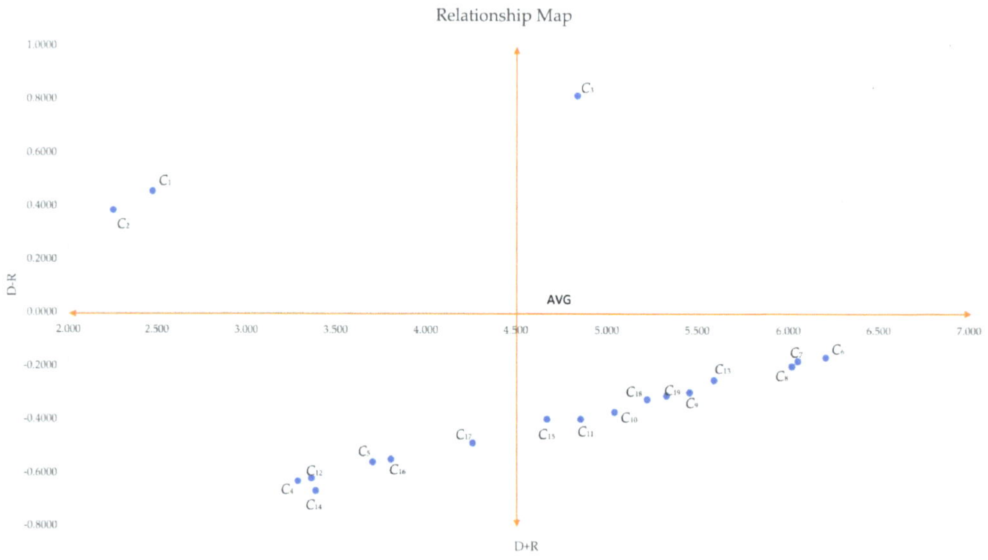

Figure 2. Causal Diagram.

Next, the pairwise comparison must be carefully evaluated by DMs according to the criteria. In this study, the statistical software Super Decisions was used for the analysis. Super Decisions is a decision-support program that implements AHP and ANP to calculate the weights of the dimensions and tests the expert's competency. After obtaining the integrated PCM, the values are entered into the software to compute CR values. First, the integrated matrix is computed with respect to each criterion, including the Consistency Ratio $CR \leq 0.1$, as shown in Equations (24)–(28) (see Tables A4–A16 in Appendix F for details). Then, the unweighted supermatrix and weighted matrix are created, as shown in Tables 7 and 8. Finally, the limited matrix with the stable weights and the final weight order can be determined, as shown in Tables 9 and 10. According to Table 10, "(C_8) market demographics" has the highest value with 0.1253, followed by "(C_6) demand validation" with 0.1196 and "(C_3) prior startup experience" with 0.0940. The lowest weight value is "(C_{11}) value proposition" with 0.0215.

Table 7. The unweighted supermatrix.

	C1S	C2 PDC	C3 PSE	C6 DV	C7 CA	C8 MD	C9 CM	C10 PM	C11 VP	C13 TE	C15 GS	C18 Cre	C19 Neg
C1S	0.03614	0.02318	0.17330	0.03728	0.03882	0.03972	0.14571	0.15206	0.03883	0.17549	0.03695	0.04388	0.18769
C2 PDC	0.01653	0.04581	0.10342	0.04958	0.04628	0.04749	0.10732	0.10631	0.02190	0.10314	0.01747	0.01689	0.10480
C3 PSE	0.03338	0.15625	0.04742	0.17850	0.17685	0.17666	0.03819	0.04569	0.03989	0.04565	0.03391	0.03338	0.04480
C6 DV	0.01423	0.18950	0.13997	0.15548	0.14938	0.13554	0.14084	0.13822	0.01601	0.15156	0.01444	0.01552	0.13387
C7 CA	0.16286	0.07300	0.06280	0.08022	0.08345	0.08509	0.06857	0.06773	0.14256	0.04789	0.16540	0.16458	0.04802
C8 MD	0.12483	0.16978	0.06602	0.18612	0.18717	0.18934	0.06726	0.06685	0.10806	0.06635	0.12572	0.12298	0.06639
C9 CM	0.03999	0.06738	0.02107	0.10666	0.10801	0.11045	0.02145	0.02110	0.03401	0.02129	0.03741	0.03359	0.02160
C10 PM	0.07094	0.07526	0.08248	0.08041	0.08405	0.08596	0.08810	0.08796	0.06341	0.08393	0.06744	0.06902	0.08430
C11 VP	0.02344	0.01298	0.02558	0.01390	0.01398	0.01415	0.02695	0.02594	0.06117	0.02575	0.02421	0.02427	0.02577
C13 TE	0.22517	0.05499	0.06987	0.03349	0.03341	0.03477	0.08177	0.07892	0.21790	0.07524	0.22318	0.22273	0.07842
C15 GS	0.06504	0.01957	0.05668	0.01696	0.01707	0.01769	0.05688	0.05184	0.06853	0.05801	0.06724	0.06790	0.05799
C18 Cre	0.11064	0.09415	0.02974	0.03832	0.03826	0.03718	0.03954	0.03792	0.10767	0.03890	0.10966	0.10762	0.03941
C19 Neg	0.07680	0.01817	0.12165	0.02307	0.02327	0.02597	0.11742	0.11946	0.08005	0.10681	0.07698	0.07764	0.10694

Table 8. The weighted supermatrix.

	C1 S	C2 PDC	C3 PSE	C6 DV	C7CA	C8 MD	C9 CM	C10 PM	C11 VP	C13 TE	C15 GS	C18 Cre	C19 Neg
C1 S	0.03614	0.02318	0.17330	0.03728	0.03882	0.03972	0.14571	0.15206	0.03883	0.17549	0.03695	0.04388	0.18769
C2 PDC	0.01653	0.04581	0.10342	0.04958	0.04628	0.04749	0.10732	0.10631	0.02190	0.10314	0.01747	0.01689	0.10480
C3 PSE	0.03338	0.15625	0.04742	0.17850	0.17685	0.17666	0.03819	0.04569	0.03989	0.04565	0.03391	0.03338	0.04480
C6 DV	0.01423	0.18950	0.13997	0.15548	0.14938	0.13554	0.14084	0.13822	0.01601	0.15156	0.01444	0.01552	0.13387
C7CA	0.16286	0.07300	0.06280	0.08022	0.08345	0.08509	0.06857	0.06773	0.14256	0.04789	0.16540	0.16458	0.04802
C8 MD	0.12483	0.16978	0.06602	0.18612	0.18717	0.18934	0.06726	0.06685	0.10806	0.06635	0.12572	0.12298	0.06639
C9 CM	0.03999	0.06738	0.02107	0.10666	0.10801	0.11045	0.02145	0.02110	0.03401	0.02129	0.03741	0.03359	0.02160
C10 PM	0.07094	0.07526	0.08248	0.08041	0.08405	0.08596	0.08810	0.08796	0.06341	0.08393	0.06744	0.06902	0.08430
C11 VP	0.02344	0.01298	0.02558	0.01390	0.01398	0.01415	0.02695	0.02594	0.06117	0.02575	0.02421	0.02427	0.02577
C13 TE	0.22517	0.05499	0.06987	0.03349	0.03341	0.03477	0.08177	0.07892	0.21790	0.07524	0.22318	0.22273	0.07842
C15 GS	0.06504	0.01957	0.05668	0.01696	0.01707	0.01769	0.05688	0.05184	0.06853	0.05801	0.06724	0.06790	0.05799
C18 Cre	0.11064	0.09415	0.02974	0.03832	0.03826	0.03718	0.03954	0.03792	0.10767	0.03890	0.10966	0.10762	0.03941
C19 Neg	0.07680	0.01817	0.12165	0.02307	0.02327	0.02597	0.11742	0.11946	0.08005	0.10681	0.07698	0.07764	0.10694

Table 9. The limited supermatrix.

	C1 S	C2 PDC	C3 PSE	C6 DV	C7 CA	C8 MD	C9 CM	C10 PM	C11 VP	C13 TE	C15 GS	C18 Cre	C19 Neg
C1 S	0.08852	0.08852	0.08852	0.08852	0.08852	0.08852	0.08852	0.08852	0.08852	0.08852	0.08852	0.08852	0.08852
C2 PDC	0.06374	0.06374	0.06374	0.06374	0.06374	0.06374	0.06374	0.06374	0.06374	0.06374	0.06374	0.06374	0.06374
C3 PSE	0.09400	0.09400	0.09400	0.09400	0.09400	0.09400	0.09400	0.09400	0.09400	0.09400	0.09400	0.09400	0.09400
C6 DV	0.11965	0.11965	0.11965	0.11965	0.11965	0.11965	0.11965	0.11965	0.11965	0.11965	0.11965	0.11965	0.11965
C7 CA	0.08917	0.08917	0.08917	0.08917	0.08917	0.08917	0.08917	0.08917	0.08917	0.08917	0.08917	0.08917	0.08917
C8 MD	0.12532	0.12532	0.12532	0.12532	0.12532	0.12532	0.12532	0.12532	0.12532	0.12532	0.12532	0.12532	0.12532
C9 CM	0.05665	0.05665	0.05665	0.05665	0.05665	0.05665	0.05665	0.05665	0.05665	0.05665	0.05665	0.05665	0.05665
C10 PM	0.08054	0.08054	0.08054	0.08054	0.08054	0.08054	0.08054	0.08054	0.08054	0.08054	0.08054	0.08054	0.08054
C11 VP	0.02149	0.02149	0.02149	0.02149	0.02149	0.02149	0.02149	0.02149	0.02149	0.02149	0.02149	0.02149	0.02149
C13 TE	0.09150	0.09150	0.09150	0.09150	0.09150	0.09150	0.09150	0.09150	0.09150	0.09150	0.09150	0.09150	0.09150
C15 GS	0.04306	0.04306	0.04306	0.04306	0.04306	0.04306	0.04306	0.04306	0.04306	0.04306	0.04306	0.04306	0.04306
C18 Cre	0.05593	0.05593	0.05593	0.05593	0.05593	0.05593	0.05593	0.05593	0.05593	0.05593	0.05593	0.05593	0.05593
C19 Neg	0.07044	0.07044	0.07044	0.07044	0.07044	0.07044	0.07044	0.07044	0.07044	0.07044	0.07044	0.07044	0.07044

Table 10. Final weight order.

Criteria	Symbol	Values	Ranking
(C_8) Market Demographic	*C8 MD*	0.1253	1
(C_6) Demand Validation	*C6 DV*	0.1196	2
(C_3) Prior Startup Experience	*C3 PSE*	0.0940	3
(C_{13}) Technology Experience	*C13 TE*	0.0915	4
(C_7) Customer affordability	*C7 CA*	0.0892	5
(C_1) Sales	*C1 S*	0.0885	6
(C_{10}) Product Maturity	*C10 PM*	0.0805	7
(C_{19}) Negotiation	*C19 Neg*	0.0704	8
(C_2) Product Development Cost	*C2 PDC*	0.0637	9
(C_9) Concept Maturity	*C9 CM*	0.0567	10
(C_{18}) Creativity	*C18 Cre*	0.0559	11
(C_{15}) Growth Strategy	*C15 GS*	0.0431	12
(C_{11}) Value Proposition	*C11 VP*	0.0215	13

Finally, the fuzzy PROMETHEE II-based spread area ranking method is applied. Suppose the same DM group assesses four technology-based startup projects $(A_n, n = 1,2,3,4)$ under 13 criteria that are screened during the previous steps. The ratings of the alternatives over qualitative criteria and quantitative criteria are shown in Tables A17 and A18 (see Sections G and H, respectively, for details). Subsequently, the mean ratings are calculated using Equation (29), as shown in Table 11, and the alternatives' normalized gradings versus quantitative criteria are produced using Equations (30) and (31), as shown in Table 12. The confidence level ratings on alternatives are also collected to produce μ value, as shown in Table 13.

The aggregated fuzzy preference is attained using Equations (32)–(36), as shown in Table 14. Subsequently, the fuzzy leaving flow $\varphi'^{+}(A_i)$, the fuzzy entering flow $\varphi'^{-}(A_i)$, and the fuzzy net outranking flow for each alternative are computed using Equations (37)–(39), as presented in Table 15. Using the proposed spread area-based RMMS model, the fuzzy net outranking flow of each alternative is defuzzified using Equations (9)–(17) and yields values of A_1 (−0.0519), A_2 (0.0905), A_3 (0.0594) and A_4 (−0.0980) as presented in Table 16. The final ranking of four startup projects $A_4 < A_1 < A_3 < A_2$ indicates that startup project A_2 has the highest comprehensive potential, followed by startup project A_3.

Table 11. The average ratings of the alternatives over qualitative criteria.

C_n	Average Rating											
	A_1			A_2			A_3			A_4		
	(aj_1, bj_1, cj_1)			(aj_2, bj_2, cj_2)			(aj_3, bj_3, cj_3)			(aj_4, bj_4, cj_4)		
C_6	0.500	0.650	0.750	0.600	0.750	0.850	0.500	0.650	0.763	0.388	0.538	0.675
C_7	0.750	0.900	1.000	0.538	0.688	0.788	0.538	0.688	0.788	0.388	0.538	0.675
C_8	0.425	0.575	0.700	0.425	0.575	0.700	0.538	0.688	0.788	0.350	0.500	0.650
C_9	0.425	0.575	0.700	0.613	0.763	0.863	0.650	0.800	0.900	0.325	0.463	0.613
C_{10}	0.500	0.650	0.763	0.563	0.713	0.813	0.500	0.650	0.750	0.388	0.538	0.675
C_{11}	0.463	0.613	0.725	0.438	0.575	0.688	0.375	0.500	0.638	0.425	0.575	0.700
C_{13}	0.213	0.313	0.463	0.650	0.800	0.900	0.650	0.800	0.900	0.350	0.500	0.650
C_{15}	0.213	0.313	0.463	0.650	0.800	0.900	0.650	0.800	0.900	0.350	0.500	0.650
C_{18}	0.388	0.538	0.675	0.500	0.650	0.775	0.613	0.763	0.863	0.188	0.288	0.438
C_{19}	0.500	0.650	0.750	0.388	0.538	0.675	0.425	0.575	0.700	0.350	0.500	0.650

Table 12. The average ratings of the alternatives over quantitative criteria.

C_n	Average Rating											
	A_1			A_2			A_3			A_4		
	(al_1, bl_1, cl_1)			(al_2, bl_2, cl_2)			(al_3, bl_3, cl_3)			(al_4, bl_4, cl_4)		
C_1	0.250	0.375	0.500	0.750	0.875	1.000	0.500	0.625	0.750	0.000	0.125	0.250
C_2	0.752	0.877	1.000	0.000	0.125	0.248	0.501	0.627	0.749	0.750	0.875	1.000
C_3	0.000	0.125	0.250	0.750	0.875	1.000	0.375	0.500	0.625	0.375	0.500	0.625

Table 13. Confidence level μ from DMs.

	A_1	A_2	A_3	A_4		A_1	A_2	A_3	A_4	μ
D_1	0.6	0.8	0.7	0.6	D_3	0.7	0.7	0.8	0.5	0.6625
D_2	0.6	0.8	0.7	0.5	D_4	0.5	0.8	0.7	0.6	

Table 14. The aggregated fuzzy TNs preference.

	A_1			A_2			A_3			A_4		
A_1	-	-	-	0.0321	0.0479	0.0637	0.0002	0.0160	0.0318	0.0164	0.0771	0.1301
A_2	0.0863	0.1594	0.1998	-	-	-	0.0118	0.0574	0.1030	0.0628	0.1825	0.2857
A_3	0.0289	0.1020	0.1659	0.0161	0.0320	0.0478	-	-	-	0.0373	0.1336	0.2118
A_4	0.0117	0.0352	0.0587	0.0321	0.0479	0.0637	0.0002	0.0160	0.0318	-	-	-

Table 15. The fuzzy TNs net outranking flow for each alternative.

	$\phi+$			$\phi-$			ϕ		
A_1	0.0162	0.0470	0.0752	0.0423	0.0989	0.1415	−0.1253	−0.0519	0.0329
A_2	0.0536	0.1331	0.1962	0.0268	0.0426	0.0584	−0.0048	0.0905	0.1694
A_3	0.0275	0.0892	0.1418	0.0040	0.0298	0.0555	−0.0281	0.0594	0.1378
A_4	0.0147	0.0330	0.0514	0.0388	0.1310	0.2092	−0.1945	−0.0980	0.0126

Table 16. Defuzzification and ranking of the alternatives.

		Φ		S_{L1}	S_{L2}	S_{R1}	S_{R2}	$V(A_i)$	Ranking
A_1	−0.1253	−0.0519	0.0329	0.9364	0.1733	0.6221	0.6364	−0.0519	3
A_2	−0.0048	0.0905	0.1694	1.1217	0.2415	0.6852	0.4956	0.0905	1
A_3	−0.0281	0.0594	0.1378	1.0830	0.2263	0.6719	0.5262	0.0594	2
A_4	−0.1945	−0.0980	0.0126	0.8599	0.1462	0.6058	0.6724	−0.0980	4

The utilization of the DEMATEL-ANP-based fuzzy PROMETHEE II provides a comprehensive procedure for ranking alternatives. The DEMATEL investigated the cause–effect relationships between criteria and filtered out the nonsignificant criteria. Subsequently, ANP helped to determine the criteria weights because it permits criterion dependency. Finally, the final ranking was generated by the fuzzy-based PROMETHEE II method, which includes a proposed ranking model to enhance consistency and discrimination ability. The numerical results demonstrated the feasibility of the hybrid model for various decision-making management applications.

6. Conclusions

Language has naturally evolved to reflect human judgment and fuzzy ranking is required to turn assessments into decision-making. An extension on ranking FNs using spread area-based RMMS was proposed to improve the applicability and differentiation of the methods of Wang et al. [54], Nejad and Mashinchi [59], and Chu and Nguyen [63]. The algorithm and equations were derived by implementing a ranking method. Comparative examples demonstrated the strengths of the proposed method in discriminating fuzzy numbers and consistency ranking. Finally, the suggested ranking method was integrated into a hybrid DEMATEL-ANP-based fuzzy PROMETHEE II model to inspect the interrelationships among factors, obtain critical criteria weights, and organize startups for a comprehensive decision-making procedure. The numerical example has illustrated the feasibility of the hybrid fuzzy MCDM method.

In future studies, the proposed fuzzy ranking method can be amalgamated into different MCDM methods to further investigate its validity and apply the method to various practices in entrepreneurial problems, such as project selections, business investment evaluation, accelerator evaluation, risk management, performance evaluation, and other areas where decision-making involves subjective judgment and uncertainty. Hybrid fuzzy ranking methods enable comprehensive evaluation and prioritization of project proposals or initiatives by considering multiple criteria and incorporating fuzzy logic, aiding decision-makers in selecting projects aligned with their strategic objectives. In addition, fuzzy ranking methods can aid in evaluating and comparing different accelerators based on their offerings, mentorship quality, network strength, success rate, and other relevant criteria. This helps entrepreneurs make informed decisions about which accelerator program would best suit their needs and increase their chances of success. The fuzzy ranking approach adds a layer of flexibility to handle uncertain or imprecise data in investment decision-making.

Author Contributions: Conceptualization, H.T.N. and T.-C.C.; methodology, H.T.N. and T.-C.C.; validation, H.T.N. and T.-C.C.; formal analysis, H.T.N.; investigation, H.T.N. and T.-C.C.; resources, T.-C.C.; data curation, H.T.N.; writing—original draft preparation, H.T.N.; writing—review and editing, H.T.N. and T.-C.C.; visualization, H.T.N.; supervision, T.-C.C.; project administration, T.-C.C. All authors have read and agreed to the published version of the manuscript.

Funding: This work was supported in part by the National Science and Technology Council, Taiwan, under Grant MOST 111-2410-H-218-004.

Institutional Review Board Statement: Not applicable.

Informed Consent Statement: Not applicable.

Data Availability Statement: Not applicable.

Acknowledgments: The authors would like to thank the anonymous reviewers for their constructive comments which improved the presentation of this work. This work was supported in part by the National Science and Technology Council, Taiwan, under Grant MOST 111-2410-H-218-004.

Conflicts of Interest: The authors declare no conflict of interest.

Appendix A

The derivation of Equation (A1) for the second left spread area $S_{L_{i2}}$ is presented as follows.

$$\begin{aligned}
S_{L_{i2}}(A_i) &= \int_{x''_{\min}}^{x_{L_{i2}}} f''_M(x)dx \\
&= \int_{x''_{\min}}^{x_{L_{i2}}} \left(\frac{x-x''_{\min}}{x''_{\max}-x''_{\min}}\right)dx \\
&= \left(\frac{x^2}{2(x''_{\max}-x''_{\min})} - \frac{2xx''_{\min}}{2(x''_{\max}-x''_{\min})}\right)\Bigg|_{x''_{\min}}^{x_{L_{i2}}} = \frac{x^2_{L_{i2}}-2x_{L_{i2}}x''_{\min}}{2(x''_{\max}-x''_{\min})} - \frac{(-x''^{\,2}_{\min})}{2(x''_{\max}-x''_{\min})} \\
&= \frac{x^2_{L_{i2}}-2x_{L_{i2}}x''_{\min}+x''^{\,2}_{\min}}{2(x''_{\max}-x''_{\min})} = \frac{(x_{L_{i2}}-x''_{\min})^2}{2(x''_{\max}-x''_{\min})}
\end{aligned} \tag{A1}$$

The derivation of Equation (A2) for the first right spread area $S_{R_{i1}}$ is presented as follows.

$$\begin{aligned}
S_{R_{i1}}(A_i) &= \int_{x_{R_{i1}}}^{x''_{\max}} f''_M(x)dx \\
&= \int_{x_{R_{i1}}}^{x''_{\max}} \left(\frac{x-x''_{\min}}{x''_{\max}-x''_{\min}}\right)dx \\
&= \left(\frac{x^2}{2(x''_{\max}-x''_{\min})} - \frac{2xx''_{\min}}{2(x''_{\max}-x''_{\min})}\right)\Bigg|_{x_{R_{i1}}}^{x''_{\max}} = \frac{(x''^{\,2}_{\max}-2x''_{\max}x''_{\min})}{2(x''_{\max}-x''_{\min})} - \frac{x^2_{R_{i1}}-2x_{R_{i1}}x''_{\min}}{2(x''_{\max}-x''_{\min})} \\
&= \frac{(x''_{\max}-x_{R_{i1}})(x''_{\max}+x_{R_{i1}})-2x''_{\min}(x''_{\max}+x_{R_{i1}})}{2(x''_{\max}-x''_{\min})} = \frac{(x''_{\max}+x_{R_{i1}})(x''_{\max}-x_{R_{i1}}-2x''_{\min})}{2(x''_{\max}-x''_{\min})}
\end{aligned} \tag{A2}$$

The derivation of Equation (A3) for the second right spread area $S_{R_{i2}}$ is presented as follows.

$$\begin{aligned}
S_{R_{i2}}(A_i) &= \int_{x''_{R_{i2}}}^{x''_{\max}} 1dx - \int_{x''_{R_{i2}}}^{x''_{\max}} f''_N(x)dx \\
&= x''_{\max} - x_{R_{i2}} - \int_{x_{R_{i2}}}^{x''_{\max}} \left(\frac{x-x''_{\max}}{x''_{\min}-x''_{\max}}\right)dx \\
&= x''_{\max} - x_{R_{i2}} - \left(\frac{x^2}{2(x''_{\min}-x''_{\max})} - \frac{xx''_{\max}}{x''_{\min}-x''_{\max}}\right)\Bigg|_{x_{R_{i2}}}^{x''_{\max}} \\
&= x''_{\max} - x_{R_{i2}} - \left(\frac{x''^{\,2}_{\max}-2x''^{\,2}_{\max}}{2(x''_{\min}-x''_{\max})} - \frac{x^2_{R_{i2}}-2x_{R_{i2}}x''_{\max}}{2(x''_{\min}-x''_{\max})}\right) \\
&= \frac{\left(x''_{\max}-x_{R_{i2}}\right)\left(2\left(x''_{\min}-x''_{\max}\right)+\left(x''_{\max}+x_{R_{i2}}\right)\right)}{2\left(x''_{\min}-x''_{\max}\right)} = \frac{\left(x''_{\max}-x_{R_{i2}}\right)\left(2x''_{\min}-x''_{\max}+x_{R_{i2}}\right)}{2\left(x''_{\min}-x''_{\max}\right)}
\end{aligned} \tag{A3}$$

Appendix B

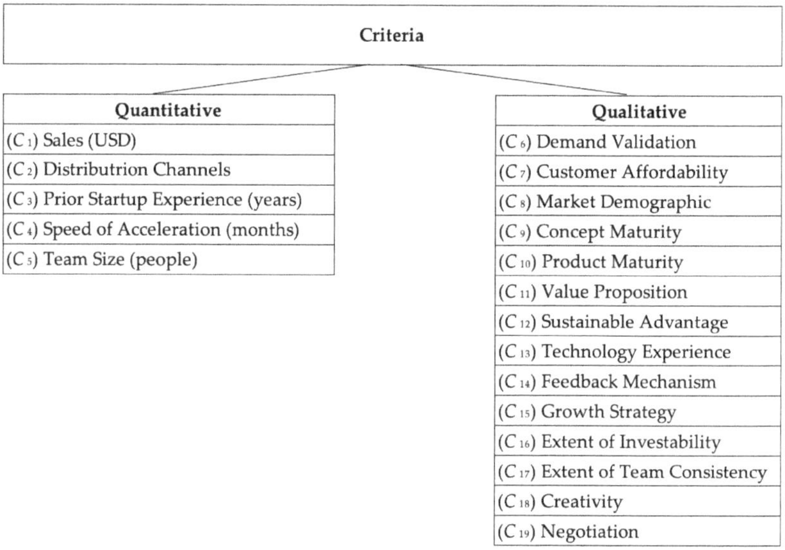

Figure A1. Structure of criteria (Yin and Luo, 2018 [5]; Mariño-Garrido et al., 2020 [14]).

Appendix C

Table A1. The aggregating direct-relation matrix of decision-makers.

	C_1	C_2	C_3	C_4	C_5	C_6	C_7	C_8	C_9	C_{10}	C_{11}	C_{12}	C_{13}	C_{14}	C_{15}	C_{16}	C_{17}	C_{18}	C_{19}
C_1	0	2	1.5	4	5	1	1	1	1.5	1	2.75	4	4	3	2	5	3	3	3
C_2	5.5	0	1.25	5	4	1.25	1	1.75	1.25	2	1	4	2	3	1	4	2	2	2
C_3	6	6	0	6	6	4	5	4	3	3	4	6	5	6	6	5	6	5	5
C_4	4	3	1	0	4	1	1.75	1	2	1	2	3	1	5	2	3	3	4	2
C_5	3	4	1	4	0	1.75	2	2	1	2	1	4	4	6	1	2	3	5	4
C_6	6	6	4	5.75	6	0	4.25	4.5	6	6	6	6	4	5.75	6	6	5.75	5.25	3.75
C_7	6	6	3	6	6	3.75	0	3.25	6	5.75	5.75	6	6	6	5.25	5.5	6	4.75	3.75
C_8	5.75	6	4	5.75	5.25	3.5	4.75	0	5.5	5.25	6	5.5	4	6	5.75	6	5.75	5	3.75
C_9	6	6	5	6	6	1.5	1	2.5	0	6	5	6	3.25	6	6	6	5	4	4
C_{10}	6	6	5	6	5.75	1	1	2.25	2	0	5	6	3	6	5	5	6	3	4
C_{11}	5.25	6	4	6	6	2	2	1.75	2.75	3	0	6	5	5	3	6	4	3	3
C_{12}	4	4	1	4.75	4	1	1	1	1	2	2	0	2	3	2	4	4	2	3
C_{13}	4	6	3	6	4	4	2	4	4.75	5	3	6	0	6	6	5	5	6	6
C_{14}	5	5	2	3	2	2.25	2	1.5	1.75	1.75	3	5	1	0	1	2	2	1	3
C_{15}	6	6	2	6	6	1	2.75	2	1.75	3	5	6	2	6	0	6	6	2	3
C_{16}	3	4	3	5	6	2	1.75	1	2	3	2	4	3	6	1	0	3	2	2
C_{17}	5	6	2	5	5	1.75	2	2	3	2	4	4	3	6	2	5	0	3	2
C_{18}	4.75	5.75	3	4	3	2.75	3	3	4	5	5	6	2	6	6	6	5	0	5
C_{19}	5	6	3	6	4	4	4	4	4	4	5	5	2	5	5	6	6	3	0

Appendix D

Table A2. The normalized direct-relation matrix.

	C_1	C_2	C_3	C_4	C_5	C_6	C_7	C_8	C_9	C_{10}	C_{11}	C_{12}	C_{13}	C_{14}	C_{15}	C_{16}	C_{17}	C_{18}	C_{19}
C_1	0	0.0206	0.0155	0.0412	0.0515	0.0103	0.0103	0.0103	0.0155	0.0103	0.0284	0.0412	0.0412	0.0309	0.0206	0.0515	0.0309	0.0309	0.0309
C_2	0.0567	0	0.0129	0.0515	0.0412	0.0129	0.0103	0.0180	0.0129	0.0206	0.0103	0.0412	0.0206	0.0309	0.0103	0.0412	0.0206	0.0206	0.0206
C_3	0.0619	0.0619	0	0.0619	0.0619	0.0412	0.0515	0.0412	0.0309	0.0309	0.0412	0.0619	0.0515	0.0619	0.0619	0.0515	0.0619	0.0515	0.0515
C_4	0.0412	0.0309	0.0103	0	0.0412	0.0103	0.0180	0.0103	0.0206	0.0103	0.0206	0.0309	0.0103	0.0515	0.0206	0.0309	0.0309	0.0412	0.0206
C_5	0.0309	0.0412	0.0103	0.0412	0	0.0180	0.0206	0.0206	0.0103	0.0206	0.0103	0.0412	0.0412	0.0619	0.0103	0.0206	0.0309	0.0515	0.0412
C_6	0.0619	0.0619	0.0412	0.0593	0.0619	0	0.0438	0.0464	0.0619	0.0619	0.0619	0.0619	0.0412	0.0593	0.0619	0.0619	0.0593	0.0541	0.0387
C_7	0.0619	0.0619	0.0309	0.0619	0.0619	0.0387	0	0.0335	0.0619	0.0593	0.0593	0.0619	0.0619	0.0619	0.0541	0.0567	0.0619	0.0490	0.0387
C_8	0.0593	0.0619	0.0412	0.0593	0.0541	0.0361	0.0490	0	0.0567	0.0541	0.0619	0.0567	0.0412	0.0619	0.0593	0.0619	0.0593	0.0515	0.0387
C_9	0.0619	0.0619	0.0515	0.0619	0.0619	0.0155	0.0103	0.0258	0	0.0619	0.0515	0.0619	0.0335	0.0619	0.0619	0.0619	0.0515	0.0412	0.0412
C_{10}	0.0619	0.0619	0.0515	0.0619	0.0593	0.0103	0.0103	0.0232	0.0206	0	0.0515	0.0619	0.0309	0.0619	0.0515	0.0515	0.0619	0.0309	0.0412
C_{11}	0.0541	0.0619	0.0412	0.0619	0.0619	0.0206	0.0206	0.0180	0.0284	0.0309	0	0.0619	0.0515	0.0515	0.0309	0.0619	0.0412	0.0309	0.0309
C_{12}	0.0412	0.0412	0.0103	0.0490	0.0412	0.0103	0.0103	0.0103	0.0103	0.0206	0.0206	0	0.0206	0.0309	0.0206	0.0412	0.0412	0.0206	0.0309
C_{13}	0.0412	0.0619	0.0309	0.0619	0.0412	0.0412	0.0206	0.0412	0.0490	0.0515	0.0309	0.0619	0	0.0619	0.0619	0.0515	0.0515	0.0619	0.0619
C_{14}	0.0515	0.0515	0.0206	0.0309	0.0206	0.0232	0.0206	0.0155	0.0180	0.0180	0.0309	0.0515	0.0103	0	0.0103	0.0206	0.0206	0.0103	0.0309
C_{15}	0.0619	0.0619	0.0206	0.0619	0.0619	0.0103	0.0284	0.0206	0.0180	0.0309	0.0515	0.0619	0.0206	0.0619	0	0.0619	0.0619	0.0206	0.0309
C_{16}	0.0309	0.0412	0.0309	0.0515	0.0619	0.0206	0.0180	0.0103	0.0206	0.0309	0.0206	0.0412	0.0309	0.0619	0.0103	0	0.0309	0.0206	0.0206
C_{17}	0.0515	0.0619	0.0206	0.0515	0.0515	0.0180	0.0206	0.0206	0.0309	0.0206	0.0412	0.0412	0.0309	0.0619	0.0206	0.0515	0	0.0309	0.0206
C_{18}	0.0490	0.0593	0.0309	0.0412	0.0309	0.0284	0.0309	0.0309	0.0412	0.0515	0.0515	0.0619	0.0206	0.0619	0.0619	0.0619	0.0515	0	0.0515
C_{19}	0.0515	0.0619	0.0309	0.0619	0.0412	0.0412	0.0412	0.0412	0.0412	0.0412	0.0515	0.0515	0.0206	0.0515	0.0515	0.0619	0.0619	0.0309	0

Appendix E

Table A3. The total-relation matrix.

	C_1	C_2	C_3	C_4	C_5	C_6	C_7	C_8	C_9	C_{10}	C_{11}	C_{12}	C_{13}	C_{14}	C_{15}	C_{16}	C_{17}	C_{18}	C_{19}
C_1	0.0681	0.0911	0.0507	0.1117	0.1165	0.0406	0.0418	0.0417	0.0532	0.0539	0.0754	0.1097	0.0820	0.1040	0.0655	0.1147	0.0891	0.0780	0.0781
C_2	0.1152	0.0620	0.0442	0.1134	0.1000	0.0391	0.0381	0.0450	0.0462	0.0580	0.0532	0.1018	0.0581	0.0954	0.0504	0.0979	0.0727	0.0631	0.0629
C_3	0.1930	0.1963	0.0692	0.1986	0.1895	0.0979	0.1110	0.1010	0.1050	0.1151	0.1353	0.1951	0.1320	0.2009	0.1486	0.1774	0.1747	0.1419	0.1419
C_4	0.1022	0.0939	0.0424	0.0646	0.1001	0.0372	0.0459	0.0383	0.0542	0.0495	0.0642	0.0936	0.0488	0.1156	0.0607	0.0893	0.0830	0.0821	0.0635
C_5	0.1045	0.1161	0.0487	0.1166	0.0708	0.0507	0.0543	0.0545	0.0524	0.0675	0.0635	0.1154	0.0841	0.1374	0.0606	0.0910	0.0938	0.1004	0.0917
C_6	0.2028	0.2062	0.1154	0.2064	0.1997	0.0614	0.1072	0.1095	0.1383	0.1502	0.1615	0.2051	0.1286	0.2088	0.1554	0.1962	0.1806	0.1503	0.1363
C_7	0.1983	0.2019	0.1034	0.2044	0.1952	0.0969	0.0626	0.0957	0.1360	0.1451	0.1556	0.2009	0.1449	0.2067	0.1453	0.1873	0.1790	0.1430	0.1338
C_8	0.1952	0.2009	0.1124	0.2009	0.1873	0.0941	0.1096	0.0627	0.1307	0.1396	0.1577	0.1950	0.1254	0.2056	0.1493	0.1912	0.1759	0.1442	0.1326
C_9	0.1810	0.1836	0.1125	0.1862	0.1783	0.0671	0.0657	0.0797	0.0652	0.1341	0.1349	0.1828	0.1071	0.1881	0.1389	0.1748	0.1541	0.1226	0.1237
C_{10}	0.1690	0.1713	0.1054	0.1737	0.1641	0.0575	0.0606	0.0718	0.0791	0.0670	0.1258	0.1703	0.0975	0.1753	0.1205	0.1537	0.1528	0.1050	0.1152
C_{11}	0.1562	0.1659	0.0934	0.1685	0.1614	0.0655	0.0677	0.0652	0.0846	0.0953	0.0727	0.1653	0.1140	0.1607	0.0988	0.1582	0.1293	0.1028	0.1029
C_{12}	0.1037	0.1051	0.0431	0.1140	0.1026	0.0378	0.0393	0.0391	0.0452	0.0596	0.0648	0.0649	0.0595	0.0985	0.0614	0.1006	0.0942	0.0645	0.0739
C_{13}	0.1672	0.1894	0.0965	0.1911	0.1631	0.0941	0.0785	0.0975	0.1172	0.1301	0.1213	0.1880	0.0765	0.1933	0.1446	0.1709	0.1593	0.1453	0.1463
C_{14}	0.1139	0.1143	0.0532	0.0973	0.0834	0.0500	0.0491	0.0441	0.0528	0.0576	0.0749	0.1143	0.0505	0.0667	0.0526	0.0819	0.0752	0.0545	0.0738
C_{15}	0.1589	0.1607	0.0711	0.1633	0.1571	0.0529	0.0721	0.0642	0.0714	0.0909	0.1184	0.1600	0.0822	0.1648	0.0632	0.1535	0.1435	0.0885	0.0981
C_{16}	0.1067	0.1181	0.0692	0.1289	0.1331	0.0533	0.0525	0.0452	0.0619	0.0774	0.0734	0.1175	0.0770	0.1404	0.0606	0.0716	0.0952	0.0737	0.0738
C_{17}	0.1379	0.1492	0.0659	0.1415	0.1357	0.0558	0.0599	0.0598	0.0781	0.0755	0.1011	0.1297	0.0846	0.1527	0.0777	0.1331	0.0749	0.0909	0.0815
C_{18}	0.1638	0.1759	0.0910	0.1614	0.1443	0.0770	0.0827	0.0822	0.1032	0.1224	0.1324	0.1772	0.0915	0.1816	0.1359	0.1703	0.1496	0.0782	0.1285
C_{19}	0.1689	0.1808	0.0923	0.1833	0.1570	0.0904	0.0939	0.0932	0.1056	0.1148	0.1343	0.1701	0.0941	0.1755	0.1281	0.1727	0.1613	0.1115	0.0810

Appendix F

Table A4. Comparison Matrix of 13 criteria with respect to criterion 1.

	C_1	C_2	C_3	C_6	C_7	C_8	C_9	C_{10}	C_{11}	C_{13}	C_{15}	C_{18}	C_{19}
C_1	1	2	3	3	1/4	1/5	1/4	1/5	3	1/7	1/4	1/3	1/2
C_2	1/2	1	1/3	3	1/5	1/8	1/4	1/5	1/4	1/8	1/3	1/8	1/5
C_3	1/3	3	1	3	1/5	1/4	2	1/2	3	1/6	1/3	1/3	1/6
C_6	1/3	1/3	1/3	1	1/5	1/8	1/2	1/5	1/3	1/8	1/6	1/8	1/3
C_7	4	5	5	5	1	3	6	3	6	1/2	4	2	3
C_8	5	8	4	8	1/3	1	5	2	5	1/3	2	2	3
C_9	4	4	1/2	2	1/6	1/5	1	1/2	2	1/8	1/2	1/7	1/5
C_{10}	5	5	2	5	1/3	1/2	2	1	3	1/6	1/2	1/2	2
C_{11}	1/3	4	1/3	3	1/6	1/5	1/2	1/3	1	1/8	1/2	1/5	1/4
C_{13}	7	8	6	8	2	3	8	6	8	1	4	2	4
C_{15}	4	3	3	6	1/4	1/2	2	2	2	1/4	1	1/2	1/2
C_{18}	3	8	3	8	1/2	1/2	7	5	5	1/2	2	1	2
C_{19}	2	5	6	3	1/3	1/3	5	4	4	1/4	2	1/2	1

Inconsistency: 0.08328

Table A5. Comparison Matrix of 13 criteria with respect to criterion 2.

	C_1	C_2	C_3	C_6	C_7	C_8	C_9	C_{10}	C_{11}	C_{13}	C_{15}	C_{18}	C_{19}
C_1	1	1/6	1/5	1/7	1/8	1/8	1/6	1/8	3	1/2	3	1/4	3
C_2	6	1	1/3	1/5	1/2	1/4	1/3	1/2	3	1/2	4	1/2	4
C_3	5	3	1	3	3	1/2	2	4	7	2	5	2	6
C_6	7	5	1/3	1	5	3	2	3	9	5	7	3	7
C_7	8	2	1/3	1/5	1	1/6	3	1/2	5	2	4	1/2	4
C_8	8	4	2	1/3	6	1	3	2	8	6	4	2	6
C_9	6	3	1/2	1/2	1/3	1/3	1	1/2	4	2	4	1/3	5
C_{10}	8	2	1/4	1/3	2	1/2	2	1	6	1/3	4	1/2	4
C_{11}	1/3	1/3	1/7	1/9	1/5	1/8	1/4	1/6	1	1/3	1/2	1/7	1/3
C_{13}	2	2	1/2	1/5	1/2	1/6	1/2	3	3	1	3	1/2	2
C_{15}	1/3	1/4	1/5	1/7	1/4	1/4	1/4	1/4	2	1/3	1	1/6	2
C_{18}	4	2	1/2	1/3	2	1/2	3	2	7	2	6	1	4
C_{19}	1/3	1/4	1/6	1/7	1/4	1/6	1/5	1/4	3	1/2	1/2	1/4	1

Inconsistency: 0.09659

Table A6. Comparison Matrix of 13 criteria with respect to criterion 3.

	C_1	C_2	C_3	C_6	C_7	C_8	C_9	C_{10}	C_{11}	C_{13}	C_{15}	C_{18}	C_{19}
C_1	1	3	3	1/2	3	2	8	5	6	3	2	4	3
C_2	1/3	1	3	1/2	2	2	4	3	3	2	4	2	1/2
C_3	1/3	1/3	1	1/3	3	1/2	2	1/3	2	1/3	1/2	3	1/4
C_6	2	2	3	1	1/2	2	4	2	4	3	2	5	2
C_7	1/3	1/2	1/3	2	1	1/2	2	1/3	2	1/2	3	2	1/3
C_8	1/2	1/2	2	1/2	2	1	3	1/2	3	1/2	2	3	1/2
C_9	1/8	1/4	1/2	1/4	1/2	1/3	1	1/4	1/2	1/3	1/2	1/2	1/4
C_{10}	1/5	1/3	3	1/2	3	2	4	1	2	3	1/2	4	1/3
C_{11}	1/6	1/3	1/2	1/4	1/2	1/3	2	1/2	1	1/2	1/3	1/2	1/5
C_{13}	1/3	1/2	3	1/3	2	2	3	1/3	2	1	3	1/2	2
C_{15}	1/2	1/4	2	1/2	1/3	1/2	2	2	3	1/3	1	3	1/3
C_{18}	1/4	1/2	1/3	1/5	1/2	1/3	2	1/4	2	2	1/3	1	1/2
C_{19}	1/3	2	4	2	3	2	4	3	5	1/2	3	2	1
							Inconsistency: 0.09391						

Table A7. Comparison Matrix of 13 criteria with respect to criterion 6.

	C_1	C_2	C_3	C_6	C_7	C_8	C_9	C_{10}	C_{11}	C_{13}	C_{15}	C_{18}	C_{19}
C_1	1	1/6	1/5	1/7	1/6	1/7	1/4	1/5	3	2	3	3	4
C_2	6	1	1/3	1/4	1/2	1/4	1/3	1/3	3	1/2	4	1/2	3
C_3	5	3	1	4	3	1/2	2	4	6	3	4	6	7
C_6	7	4	1/4	1	4	1/2	2	2	7	8	6	6	7
C_7	6	2	1/3	1/4	1	1/2	1/3	2	5	3	4	3	2
C_8	7	4	2	2	2	1	3	2	8	5	8	6	7
C_9	4	3	1/2	1/2	3	1/3	1	2	7	3	7	3	5
C_{10}	5	3	1/4	1/2	1/2	1/2	1/2	1	5	3	4	3	5
C_{11}	1/3	1/3	1/6	1/7	1/5	1/8	1/7	1/5	1	1/3	1/2	1/4	1/2
C_{13}	1/2	2	1/3	1/8	1/3	1/5	1/3	3	3	1	4	1/2	1/2
C_{15}	1/3	1/4	1/4	1/6	1/4	1/8	1/7	1/4	2	1/4	1	1/3	1/2
C_{18}	1/3	2	1/6	1/6	1/3	1/6	1/3	1/3	4	2	3	1	3
C_{19}	1/4	1/3	1/7	1/7	1/2	1/7	1/5	1/5	2	2	2	1/3	1
							Inconsistency: 0.09784						

Table A8. Comparison Matrix of 13 criteria with respect to criterion 7.

	C_1	C_2	C_3	C_6	C_7	C_8	C_9	C_{10}	C_{11}	C_{13}	C_{15}	C_{18}	C_{19}
C_1	1	1/5	1/5	1/4	1/6	1/7	1/4	1/5	3	2	3	3	4
C_2	5	1	1/4	1/4	1/2	1/4	1/3	1/4	3	1/2	4	1/2	3
C_3	5	4	1	4	2	1/2	2	4	6	3	4	6	7
C_6	4	4	1/4	1	4	1/2	2	2	7	8	6	6	7
C_7	6	2	1/2	1/4	1	1/2	1/3	2	5	3	4	3	2
C_8	7	4	2	2	2	1	3	2	8	5	8	6	7
C_9	4	3	1/2	1/2	3	1/3	1	2	7	3	7	3	5
C_{10}	5	4	1/4	1/2	1/2	1/2	1/2	1	5	3	4	3	5
C_{11}	1/3	1/3	1/6	1/7	1/5	1/8	1/7	1/5	1	1/3	1/2	1/4	1/2
C_{13}	1/2	2	1/3	1/8	1/3	1/5	1/3	3	3	1	4	1/2	1/2
C_{15}	1/3	1/4	1/4	1/6	1/4	1/8	1/7	1/4	2	1/4	1	1/3	1/2
C_{18}	1/3	2	1/6	1/6	1/3	1/6	1/3	1/3	4	2	3	1	3
C_{19}	1/4	1/3	1/7	1/7	1/2	1/7	1/5	1/5	2	2	2	1/3	1
Inconsistency: 0.09426													

Table A9. Comparison Matrix of 13 criteria with respect to criterion 8.

	C_1	C_2	C_3	C_6	C_7	C_8	C_9	C_{10}	C_{11}	C_{13}	C_{15}	C_{18}	C_{19}
C_1	1	1/5	1/5	1/4	1/6	1/7	1/4	1/5	3	2	3	3	4
C_2	5	1	1/4	1/4	1/2	1/4	1/3	1/4	3	1/2	4	1/2	3
C_3	5	4	1	4	2	1/2	2	4	6	3	4	6	7
C_6	4	4	1/4	1	4	1/2	2	2	7	5	5	4	5
C_7	6	2	1/2	1/4	1	1/2	1/3	2	5	3	4	3	2
C_8	7	4	2	2	2	1	3	2	8	5	8	6	7
C_9	4	3	1/2	1/2	3	1/3	1	2	7	3	7	3	5
C_{10}	5	4	1/4	1/2	1/2	1/2	1/2	1	5	3	4	3	5
C_{11}	1/3	1/3	1/6	1/7	1/5	1/8	1/7	1/5	1	1/3	1/2	1/4	1/2
C_{13}	1/2	2	1/3	1/5	1/3	1/5	1/3	3	3	1	4	1/2	1/2
C_{15}	1/3	1/4	1/4	1/5	1/4	1/8	1/7	1/4	2	1/4	1	1/2	1/3
C_{18}	1/3	2	1/6	1/4	1/3	1/6	1/3	1/3	4	2	2	1	2
C_{19}	1/4	1/3	1/7	1/5	1/2	1/7	1/5	1/5	2	2	3	1/2	1
Inconsistency: 0.09230													

Table A10. Comparison Matrix of 13 criteria with respect to criterion 9.

	C_1	C_2	C_3	C_6	C_7	C_8	C_9	C_{10}	C_{11}	C_{13}	C_{15}	C_{18}	C_{19}
C_1	1	3	3	1/2	3	2	6	3	4	2	2	4	2
C_2	1/3	1	3	1/2	2	2	5	3	3	2	4	2	1/2
C_3	1/3	1/3	1	1/3	1/2	1/2	2	1/3	2	1/3	1/2	3	1/4
C_6	2	2	3	1	1/2	2	4	2	4	3	2	5	2
C_7	1/3	1/2	2	2	1	1/2	2	1/3	2	1/2	3	2	1/3
C_8	1/2	1/2	2	1/2	2	1	3	1/2	3	1/2	2	3	1/2
C_9	1/6	1/5	1/2	1/4	1/2	1/3	1	1/4	1/2	1/3	1/2	1/2	1/4
C_{10}	1/3	1/3	3	1/2	3	2	4	1	2	3	1/2	4	1/3
C_{11}	1/4	1/3	1/2	1/4	1/2	1/3	2	1/2	1	1/2	1/3	1/2	1/5
C_{13}	1/2	1/2	3	1/3	2	2	3	1/3	2	1	3	1/2	2
C_{15}	1/2	1/4	2	1/2	1/3	1/2	2	2	3	1/3	1	3	1/3
C_{18}	1/4	1/2	1/3	1/5	1/2	1/3	2	1/4	2	2	1/3	1	1/2
C_{19}	1/2	2	4	2	3	2	4	3	5	1/2	3	2	1

Inconsistency: 0.09890

Table A11. Comparison Matrix of 13 criteria with respect to criterion 10.

	C_1	C_2	C_3	C_6	C_7	C_8	C_9	C_{10}	C_{11}	C_{13}	C_{15}	C_{18}	C_{19}
C_1	1	3	3	1/2	3	2	7	3	4	3	2	4	2
C_2	1/3	1	3	1/2	2	2	5	3	3	2	4	2	1/2
C_3	1/3	1/3	1	1/3	1/2	1/2	2	1/3	2	1/3	1/2	3	1/4
C_6	2	2	3	1	1/2	2	4	2	4	3	2	4	2
C_7	1/3	1/2	2	2	1	1/2	2	1/3	2	1/2	3	2	1/3
C_8	1/2	1/2	2	1/2	2	1	3	1/2	3	1/2	2	3	1/2
C_9	1/7	1/5	1/2	1/4	1/2	1/3	1	1/4	1/2	1/3	1/2	1/2	1/4
C_{10}	1/3	1/3	3	1/2	3	2	4	1	2	3	1/2	4	1/3
C_{11}	1/4	1/3	1/2	1/4	1/2	1/3	2	1/2	1	1/3	1/3	1/2	1/5
C_{13}	1/3	1/2	3	1/3	2	2	3	1/3	3	1	3	1/2	2
C_{15}	1/2	1/4	2	1/2	1/3	1/2	2	2	3	1/3	1	3	1/3
C_{18}	1/4	1/2	1/3	1/4	1/2	1/3	2	1/4	2	2	1/3	1	1/4
C_{19}	1/2	2	4	2	3	2	4	3	5	1/2	3	4	1

Inconsistency: 0.09964

Table A12. Comparison Matrix of 13 criteria with respect to criterion 11.

	C_1	C_2	C_3	C_6	C_7	C_8	C_9	C_{10}	C_{11}	C_{13}	C_{15}	C_{18}	C_{19}
C_1	1	2	3	3	1/4	1/2	1/2	1/4	1/3	1/3	1/2	1/3	1/2
C_2	1/2	1	1/2	2	1/4	1/3	1/2	1/4	1/2	1/7	1/4	1/5	1/3
C_3	1/3	2	1	3	1/5	1/4	2	1/2	3	1/6	1/3	1/3	1/6
C_6	1/3	1/2	1/3	1	1/5	1/8	1/2	1/5	1/3	1/8	1/6	1/4	1/3
C_7	4	4	5	5	1	2	5	3	3	1/2	3	2	2
C_8	2	3	4	8	1/2	1	5	2	1/2	1/3	2	2	3
C_9	2	2	1/2	2	1/5	1/5	1	1/2	2	1/8	1/2	1/7	1/5
C_{10}	4	4	2	5	1/3	1/2	2	1	1/3	1/6	1/2	1/2	2
C_{11}	3	2	1/3	3	1/3	2	1/2	3	1	1/5	1/2	1/3	1/2
C_{13}	3	7	6	8	2	3	8	6	5	1	4	2	4
C_{15}	2	4	3	6	1/3	1/2	2	2	2	1/4	1	1/2	1/2
C_{18}	3	5	3	4	1/2	1/2	7	5	3	1/2	2	1	2
C_{19}	2	3	6	3	1/2	1/3	5	4	2	1/4	2	1/2	1
Inconsistency: 0.09801													

Table A13. Comparison Matrix of 13 criteria with respect to criterion 13.

	C_1	C_2	C_3	C_6	C_7	C_8	C_9	C_{10}	C_{11}	C_{13}	C_{15}	C_{18}	C_{19}
C_1	1	4	4	1/2	3	2	7	4	6	3	2	4	3
C_2	1/4	1	3	1/2	2	2	4	3	3	2	4	2	1/2
C_3	1/4	1/3	1	1/3	3	1/2	2	1/3	2	1/2	1/2	2	1/2
C_6	2	2	3	1	3	2	4	2	4	3	2	5	2
C_7	1/3	1/2	1/3	1/3	1	1/2	2	1/3	2	1/2	3	2	1/3
C_8	1/2	1/2	2	1/2	2	1	3	1/2	3	1/2	2	3	1/2
C_9	1/7	1/4	1/2	1/4	1/2	1/3	1	1/4	1/2	1/3	1/2	1/2	1/4
C_{10}	1/4	1/3	3	1/2	3	2	4	1	2	3	1/2	4	1/3
C_{11}	1/6	1/3	1/2	1/4	1/2	1/3	2	1/2	1	1/2	1/3	1/2	1/5
C_{13}	1/3	1/2	2	1/3	2	2	3	1/3	2	1	3	1/2	2
C_{15}	1/2	1/4	2	1/2	1/3	1/2	2	2	3	1/3	1	3	1/3
C_{18}	1/4	1/2	1/2	1/5	1/2	1/3	2	1/4	2	2	1/3	1	1/2
C_{19}	1/3	2	2	2	3	2	4	3	5	1/2	3	2	1
Inconsistency: 0.09084													

Table A14. Comparison Matrix of 13 criteria with respect to criterion 15.

	C_1	C_2	C_3	C_6	C_7	C_8	C_9	C_{10}	C_{11}	C_{13}	C_{15}	C_{18}	C_{19}
C_1	1	5	4	3	1/4	1/4	1/3	1/4	4	1/5	1/4	1/3	1/2
C_2	1/5	1	1/3	3	1/5	1/7	1/3	1/5	1/4	1/8	1/4	1/6	1/4
C_3	1/4	3	1	3	1/5	1/4	2	1/2	3	1/6	1/3	1/3	1/6
C_6	1/3	1/3	1/3	1	1/5	1/8	1/2	1/5	1/3	1/8	1/6	1/8	1/3
C_7	4	5	5	5	1	3	6	3	6	1/2	4	2	3
C_8	4	7	4	8	1/3	1	5	2	5	1/3	2	2	3
C_9	3	3	1/2	2	1/6	1/5	1	1/2	2	1/8	1/2	1/7	1/5
C_{10}	4	5	2	5	1/3	1/2	2	1	3	1/6	1/2	1/2	2
C_{11}	1/4	4	1/3	3	1/6	1/5	1/2	1/3	1	1/8	1/2	1/5	1/4
C_{13}	5	8	6	8	2	3	8	6	8	1	4	2	4
C_{15}	4	4	3	6	1/4	1/2	2	2	2	1/4	1	1/2	1/2
C_{18}	3	6	3	8	1/2	1/2	7	5	5	1/2	2	1	2
C_{19}	2	4	6	3	1/3	1/3	5	4	4	1/4	2	1/2	1
Inconsistency: 0.08784													

Table A15. Comparison Matrix of 13 criteria with respect to criterion 18.

	C_1	C_2	C_3	C_6	C_7	C_8	C_9	C_{10}	C_{11}	C_{13}	C_{15}	C_{18}	C_{19}
C_1	1	2	3	3	1/4	1/5	1/5	1/4	4	1/5	1/4	1/3	1/2
C_2	1/2	1	1/3	3	1/5	1/7	1/3	1/5	1/4	1/8	1/4	1/6	1/4
C_3	1/3	3	1	3	1/5	1/4	2	1/2	3	1/6	1/3	1/3	1/6
C_6	1/3	1/3	1/3	1	1/4	1/7	1/2	1/5	1/3	1/7	1/5	1/6	1/4
C_7	4	5	5	4	1	3	6	3	6	1/2	4	2	3
C_8	5	7	4	7	1/3	1	5	2	5	1/3	2	2	3
C_9	5	3	1/2	2	1/6	1/5	1	1/2	2	1/8	1/2	1/7	1/5
C_{10}	5	5	2	5	1/3	1/2	2	1	3	1/6	1/2	1/2	2
C_{11}	1/4	4	1/3	3	1/6	1/5	1/2	1/3	1	1/8	1/2	1/5	1/4
C_{13}	5	8	6	7	2	3	8	6	8	1	4	2	4
C_{15}	4	4	3	5	1/4	1/2	2	2	2	1/4	1	1/2	1/2
C_{18}	3	6	3	6	1/2	1/2	7	5	5	1/2	2	1	2
C_{19}	2	4	6	4	1/3	1/3	5	4	4	1/4	2	1/2	1
Inconsistency: 0.08708													

Table A16. Comparison Matrix of 13 criteria with respect to criterion 19.

	C_1	C_2	C_3	C_6	C_7	C_8	C_9	C_{10}	C_{11}	C_{13}	C_{15}	C_{18}	C_{19}
C_1	1	4	4	2	3	2	5	4	6	3	2	4	3
C_2	1/4	1	3	1/2	2	2	5	3	3	2	4	2	1/2
C_3	1/4	1/3	1	1/3	3	1/2	2	1/3	2	1/3	1/2	2	1/2
C_6	1/2	2	3	1	3	2	4	2	4	3	2	5	2
C_7	1/3	1/2	1/3	1/3	1	1/2	2	1/3	2	1/2	3	2	1/3
C_8	1/2	1/2	2	1/2	2	1	3	1/2	3	1/2	2	3	1/2
C_9	1/5	1/5	1/2	1/4	1/2	1/3	1	1/4	1/2	1/3	1/2	1/2	1/4
C_{10}	1/4	1/3	3	1/2	3	2	4	1	2	3	1/2	4	1/3
C_{11}	1/6	1/3	1/2	1/4	1/2	1/3	2	1/2	1	1/2	1/3	1/2	1/5
C_{13}	1/3	1/2	3	1/3	2	2	3	1/3	2	1	3	1/2	2
C_{15}	1/2	1/4	2	1/2	1/3	1/2	2	2	3	1/3	1	3	1/3
C_{18}	1/4	1/2	1/2	1/5	1/2	1/3	2	1/4	2	2	1/3	1	1/2
C_{19}	1/3	2	2	2	3	2	4	3	5	1/2	3	2	1
Inconsistency: 0.09114													

Appendix G

Table A17. Rating of Alternative Qualitative Criteria—Linguistic Values.

DMs	Alternatives	Qualitative Criteria									
		C_6	C_7	C_8	C_9	C_{10}	C_{11}	C_{12}	C_{13}	C_{14}	C_{15}
D_1	A_1	H	EH	H	M	H	H	VP	VH	H	H
	A_2	VH	H	H	VH	H	H	VH	H	VH	M
	A_3	H	H	VH	VH	H	VH	VH	H	VH	H
	A_4	M	H	M	M	H	M	M	VP	EP	M
D_2	A_1	H	EH	M	M	H	H	P	VH	M	H
	A_2	H	VH	M	VH	H	H	VH	H	VH	H
	A_3	M	H	H	VH	H	M	VH	H	VH	M
	A_4	H	M	M	M	M	M	M	VP	P	M
D_3	A_1	H	EH	M	H	VH	H	P	VH	M	H
	A_2	EH	H	H	VH	H	H	VH	H	M	M
	A_3	VH	H	H	VH	H	P	VH	H	H	H
	A_4	M	M	M	M	M	H	M	P	P	M
D_4	A_1	H	EH	H	H	M	M	P	VH	M	H
	A_2	H	H	M	H	EH	P	VH	H	M	M
	A_3	H	VH	H	VH	H	P	VH	H	VH	M
	A_4	M	M	M	P	M	H	M	P	P	M

Appendix H

Table A18. Rating of Alternative versus Quantitative Criteria.

Alternatives	Quantitative Criteria								
	C_1			C_2			C_3		
A_1	2001	2500	3000	101	150	200	3	4	5
A_2	4001	4500	5000	401	450	500	9	10	11
A_3	3001	3500	4000	201	250	300	6	7	8
A_4	1001	1500	2000	101	150	200	6	7	8

References

1. Peterson, R.; Valliere, D. Entrepreneurship and National Economic Growth: The European Entrepreneurial Deficit. *Eur. J. Int. Manag.* **2008**, *2*, 471. [CrossRef]
2. Vandenberg, P.; Hampel-Milagrosa, A.; Helble, M. Financing of Tech Startups in Selected Asian Countries. Tokyo *ADBI* 2020. Available online: https://www.adb.org/publications/financing-tech-startups-selected-asian-countries (accessed on 24 May 2020).
3. Stam, E. Entrepreneurial Ecosystems and Regional Policy: A Sympathetic Critique. *Eur. Plan. Stud.* **2015**, *23*, 1759–1769. [CrossRef]
4. Chang, C. Portfolio Company Selection Criteria: Accelerators vs. Venture Capitalists. CMC Senior Theses. 2013. Available online: http://scholarship.claremont.edu/cmc_theses/566 (accessed on 6 May 2013).
5. Yin, B.; Luo, J. How Do Accelerators Select Startups? Shifting Decision Criteria Across Stages. *IEEE Trans. Eng. Manag.* **2018**, *65*, 574–589. [CrossRef]
6. Amezcua, A.S.; Grimes, M.G.; Bradley, S.W.; Wiklund, J. Organizational Sponsorship and Founding Environments: A Contingency View on the Survival of Business-Incubated Firms, 1994–2007. *Acad. Manag. J.* **2013**, *56*, 1628–1654. [CrossRef]
7. Lin, M.; Chen, Z.; Chen, R.; Fujita, H. Evaluation of Startup Companies Using Multicriteria Decision Making Based on Hesitant Fuzzy Linguistic Information Envelopment Analysis Models. *Int. J. Intell. Syst.* **2021**, *36*, 2292–2322. [CrossRef]
8. Kahraman, C.; Onar, S.C.; Oztaysi, B. Fuzzy Multicriteria Decision-Making: A Literature Review. *Int. J. Comput. Intell. Syst.* **2015**, *8*, 637. [CrossRef]
9. Liu, J.; Yin, Y. An Integrated Method for Sustainable Energy Storing Node Optimization Selection in China. *Energy Convers. Manag.* **2019**, *199*, 112049. [CrossRef]
10. Drori, I.; Wright, M. *Accelerators: Characteristics, Trends and the New Entrepreneurial Ecosystem*; Edward Elgar Publishing: Cheltenham, UK, 2018. [CrossRef]
11. Seed-DB. Seed-DB Charts and Tables. Available online: https://www.seed-db.com/accelerators (accessed on 19 April 2023).
12. 500startups. 2021. Available online: https://500.co/accelerators/500-global-flagship-accelerator-program (accessed on 22 December 2021).
13. Butz, H.; Mrożewski, M.J. The Selection Process and Criteria of Impact Accelerators. An Exploratory Study. *Sustainability* **2021**, *13*, 6617. [CrossRef]
14. Garrido, T.M.; de Lema, D.G.P.; Duréndez, A. Assessment Criteria for Seed Accelerators in Entrepreneurial Project Selections. *Int. J. Entrep. Innov.* **2020**, *24*, 53. [CrossRef]
15. Majumdar, R.; Mittal, A. Startup Financing: Some Evidence from the Indian Venture Capital Industry. *FIIB Bus. Rev.* **2023**, 23197145221142109. [CrossRef]
16. Sreenivasan, A.; Suresh, M. Agility Adaptability and Alignment in Start-Ups. *J. Sci. Technol. Policy Manag.* 2023. [CrossRef]
17. Honoré, F.; Ganco, M. Entrepreneurial Teams' Acquisition of Talent: Evidence from Technology Manufacturing Industries Using a Two-sided Approach. *Strateg. Manag. J.* **2020**, *44*, 141–170. [CrossRef]
18. Zavadskas, E.K.; Turskis, Z.; Kildienė, S. State of art surveys of overviews on MCDM/MADM methods. *Technol. Econ. Dev. Econ.* **2014**, *20*, 165–179. [CrossRef]
19. Kumar, A.; Sah, B.; Singh, A.R.; Deng, Y.; He, X.; Kumar, P.; Bansal, R.C. A Review of Multi Criteria Decision Making (MCDM) towards Sustainable Renewable Energy Development. *Renew. Sust. Energ. Rev.* **2017**, *69*, 596–609. [CrossRef]
20. Stojčić, M.; Zavadskas, E.; Pamučar, D.; Stević, Ž.; Mardani, A. Application of MCDM Methods in Sustainability Engineering: A Literature Review 2008–2018. *Symmetry* **2019**, *11*, 350. [CrossRef]
21. Jamwal, A.; Agrawal, R.; Sharma, M.; Kumar, V. Review on Multi-Criteria Decision Analysis in Sustainable Manufacturing Decision Making. *Int. J. Sustain. Eng.* **2020**, *14*, 202–225. [CrossRef]
22. Gabus, A.; Fontela, E. *World Problems, an Invitation to Further Thought within the Framework of DEMATEL*; Battelle Geneva Research Center: Geneva, Switzerland, 1972; pp. 1–8.
23. Fontela, E.; Gabus, A. *The DEMATEL Observer*; DEMATEL 1976 report; Battelle Geneva Research Center: Geneva, Switzerland, 1976.

24. Moraga, J.A.; Quezada, L.E.; Palominos, P.I.; Oddershede, A.M.; Silva, H.A. A Quantitative Methodology to Enhance a Strategy Map. *Int. J. Prod. Econ.* **2020**, *219*, 43–53. [CrossRef]
25. Altuntas, F.; Gok, M.S. The Effect of COVID-19 Pandemic on Domestic Tourism: A DEMATEL Method Analysis on Quarantine Decisions. *Int. J. Hosp. Manag.* **2021**, *92*, 102719. [CrossRef] [PubMed]
26. Wang, Y.; Qi, L.; Dou, R.; Shen, S.; Hou, L.; Liu, Y.; Yang, Y.; Kong, L. An Accuracy-Enhanced Group Recommendation Approach Based on DEMATEL. *Pattern Recognit. Lett.* **2023**, *167*, 171–180. [CrossRef]
27. Si, S.-L.; You, X.-Y.; Liu, H.-C.; Zhang, P. DEMATEL Technique: A Systematic Review of the State-of-the-Art Literature on Methodologies and Applications. *Math. Probl. Eng.* **2018**, *2018*, 1–33. [CrossRef]
28. Saaty, T.L.; Vargas, L.G. *The Analytic Network Process, Decision Making with the Analytic Network Process*; Springer: Berlin, Germany, 2013; pp. 1–40.
29. Saaty, T.L. *The Analytic Hierarchy Process: Planning, Priority Setting, Resource Allocation*; McGraw-Hill International Book Company: New York, NY, USA, 1980.
30. Saaty, T.L. *The Analytic Network Process: Decision Making with Dependence and Feedback*, 2nd ed.; RWS Publications: Pittsburgh, PA, USA, 2001.
31. Saaty, T.L. *Theory and Applications of the Analytic Network Process: Decision Making with Benefits, Opportunities, Costs, and Risks*; RWS Publications: Pittsburgh, PA, USA, 2005.
32. Yu, D.; Kou, G.; Xu, Z.; Shi, S. Analysis of Collaboration Evolution in AHP Research: 1982–2018. *Int. J. Inf. Technol. Decis. Mak.* **2021**, *20*, 7–36. [CrossRef]
33. Rahiminezhad Galankashi, M.; Mokhatab Rafiei, F.; Ghezelbash, M. Portfolio Selection: A Fuzzy-ANP Approach. *Financ. Innov* **2020**, *6*, 17. [CrossRef]
34. Saputro, K.E.A.; Hasim; Karlinasari, L.; Beik, I.S. Evaluation of Sustainable Rural Tourism Development with an Integrated Approach Using MDS and ANP Methods: Case Study in Ciamis, West Java, Indonesia. *Sustainability* **2023**, *15*, 1835. [CrossRef]
35. Kadoić, N. Characteristics of the Analytic Network Process, a Multi-Criteria Decision-Making Method. *Croat. Oper. Res. Rev.* **2018**, *9*, 235–244. [CrossRef]
36. Zheng, Y.; He, Y.; Xu, Z.; Pedrycz, W. Assessment for Hierarchical Medical Policy Proposals Using Hesitant Fuzzy Linguistic Analytic Network Process. *Knowl. Based Syst.* **2018**, *161*, 254–267. [CrossRef]
37. Brans, J.P. Lingenierie de La Decision. Elaboration Dinstruments Daide a La Decision. Methode PROMETHEE. In *Laide a la Decision: Nature, Instruments et Perspectives Davenir*; Nadeau, R., Landry, M., Eds.; Presses de Universite Laval: Québec, QC, Canada, 1982; pp. 183–214.
38. Brans, J.P.; Vincke, P. Note—A Preference Ranking Organisation Method. *Manage Sci.* **1985**, *31*, 647–656. [CrossRef]
39. Brans, J.-P. The Space of Freedom of the Decision Maker Modelling the Human Brain. *Eur. J. Oper. Res.* **1996**, *92*, 593–602. [CrossRef]
40. Seikh, M.R.; Mandal, U. Interval-Valued Fermatean Fuzzy Dombi Aggregation Operators and SWARA Based PROMETHEE II Method to Bio-Medical Waste Management. *Expert Syst. Appl.* **2023**, *226*, 120082. [CrossRef]
41. Khorasaninejad, E.; Fetanat, A.; Hajabdollahi, H. Prime Mover Selection in Thermal Power Plant Integrated with Organic Rankine Cycle for Waste Heat Recovery Using a Novel Multi Criteria Decision Making Approach. *Appl. Therm. Eng.* **2016**, *102*, 1262–1279. [CrossRef]
42. Govindan, K.; Kannan, D.; Shankar, M. Evaluation of Green Manufacturing Practices Using a Hybrid MCDM Model Combining DANP with PROMETHEE. *Int. J. Prod. Res.* **2014**, *53*, 6344–6371. [CrossRef]
43. Hua, Z.; Jing, X. A Generalized Shapley Index-Based Interval-Valued Pythagorean Fuzzy PROMETHEE Method for Group Decision-Making. *Soft Comput.* **2023**, *27*, 6629–6652. [CrossRef]
44. Zadeh, L.A. Fuzzy Sets. *Inf. Control.* **1965**, *8*, 338–353. [CrossRef]
45. Al-Tahan, M.; Hoskova-Mayerova, S.; Al-Kaseasbeh, S.; Tahhan, S.A. Linear Diophantine Fuzzy Subspaces of a Vector Space. *Mathematics* **2023**, *11*, 503. [CrossRef]
46. Ardil, C. Aircraft Supplier Selection Using Multiple Criteria Group Decision Making Process with Proximity Measure Method for Determinate Fuzzy Set Ranking Analysis. *Int. J. Ind. Syst. Eng.* **2023**, *17*, 127–135.
47. Karmaker, C.L.; Aziz, R.A.; Palit, T.; Bari, A.B.M.M. Analyzing Supply Chain Risk Factors in the Small and Medium Enterprises under Fuzzy Environment: Implications towards Sustainability for Emerging Economies. *Sustain. Technol. Entrep.* **2023**, *2*, 100032. [CrossRef]
48. Kahraman, C.; Öztaysi, B.; Onar, S.C. A Comprehensive Literature Review of 50 Years of Fuzzy Set Theory. *Int. J. Comput. Intell. Syst.* **2016**, *9* (Suppl. S1), 3. [CrossRef]
49. Jain, R. Decision-making in the Presence of Fuzzy Variables. *IEEE Trans. Syst. Man Cybern. Syst.* **1976**, *SMC-6*, 698–703. [CrossRef]
50. Dubois, D.; Prade, H. Ranking Fuzzy Numbers in the Setting of Possibility Theory. *Inf. Sci.* **1983**, *30*, 183–224. [CrossRef]
51. Chen, S.-H. Ranking Fuzzy Numbers with Maximizing Set and Minimizing Set. *Fuzzy Sets Syst.* **1985**, *17*, 113–129. [CrossRef]
52. Fortemps, P.; Roubens, M. Ranking and Defuzzification Methods Based on Area Compensation. *Fuzzy Sets Syst.* **1996**, *82*, 319–330. [CrossRef]
53. Deng, Y.; Zhenfu, Z.; Qi, L. Ranking Fuzzy Numbers with an Area Method Using Radius of Gyration. *Comput. Math. Appl.* **2006**, *51*, 1127–1136. [CrossRef]

54. Wang, Z.-X.; Liu, Y.-J.; Fan, Z.-P.; Feng, B. Ranking L–R Fuzzy Number Based on Deviation Degree. *Inf. Sci.* **2009**, *179*, 2070–2077. [CrossRef]
55. Chen, S.-M.; Chen, J.-H. Fuzzy Risk Analysis Based on Ranking Generalized Fuzzy Numbers with Different Heights and Different Spreads. *Expert Syst. Appl.* **2009**, *36*, 6833–6842. [CrossRef]
56. Nguyen, H.T.; Chu, T.-C. Using a Fuzzy Multiple Criteria Decision-Making Method to Evaluate Personal Diversity Perception to Work in a Diverse Workgroup. *J. Intell. Fuzzy Syst.* **2021**, *41*, 1407–1428. [CrossRef]
57. Wang, Y.-M.; Luo, Y. Area Ranking of Fuzzy Numbers Based on Positive and Negative Ideal Points. *Comput. Math. Appl.* **2009**, *58*, 1769–1779. [CrossRef]
58. Asady, B. The Revised Method of Ranking LR Fuzzy Number Based on Deviation Degree. *Expert Syst. Appl.* **2010**, *37*, 5056–5060. [CrossRef]
59. Nejad, A.M.; Mashinchi, M. Ranking Fuzzy Numbers Based on the Areas on the Left and the Right Sides of Fuzzy Number. *Comput. Math. Appl.* **2011**, *61*, 431–442. [CrossRef]
60. Yu, V.F.; Chi, H.T.X.; Shen, C. Ranking Fuzzy Numbers Based on Epsilon-Deviation Degree. *Appl. Soft Comput.* **2013**, *13*, 3621–3627. [CrossRef]
61. Chutia, R. Ranking of Fuzzy Numbers by Using Value and Angle in the Epsilon-Deviation Degree Method. *Appl. Soft Comput.* **2017**, *60*, 706–721. [CrossRef]
62. Ghasemi, R.; Nikfar, M.; Roghanian, E. A Revision on Area Ranking and Deviation Degree Methods of Ranking Fuzzy Numbers. *Sci. Iran.* **2015**, *22*, 1142–1154.
63. Chu, T.-C.; Nguyen, H.T. Ranking Alternatives with Relative Maximizing and Minimizing Sets in a Fuzzy MCDM Model. *Int. J. Fuzzy Syst.* **2019**, *21*, 1170–1186. [CrossRef]
64. Kaufman, A.; Gupta, M.M. *Introduction to Fuzzy Arithmetic*; Van Nostrand Reinhold Company: New York, NY, USA, 1991.
65. Zadeh, L.A. Outline of a New Approach to the Analysis of Complex Systems and Decision Processes. *IEEE Trans. Syst. Man Cybern. Syst.* **1973**, *SMC-3*, 28–44. [CrossRef]
66. Yeh, C.-H.; Kuo, Y.-L. Evaluating Passenger Services of Asia-Pacific International Airports. *Transp. Res. Part E Logist. Transp. Rev.* **2003**, *39*, 35–48. [CrossRef]
67. Huang, C.-Y.; Shyu, J.Z.; Tzeng, G.-H. Reconfiguring the Innovation Policy Portfolios for Taiwan's SIP Mall Industry. *Technovation* **2007**, *27*, 744–765. [CrossRef]
68. Tzeng, G.; Chiang, C.; Li, C. Evaluating Intertwined Effects in E-Learning Programs: A Novel Hybrid MCDM Model Based on Factor Analysis and DEMATEL. *Expert Syst. Appl.* **2007**, *32*, 1028–1044. [CrossRef]
69. Chien, K.-F.; Wu, Z.-H.; Huang, S.-C. Identifying and Assessing Critical Risk Factors for BIM Projects: Empirical Study. *Autom. Constr.* **2014**, *45*, 1–15. [CrossRef]
70. Lin, W.-R.; Wang, Y.-H.; Hung, Y.-M. Analyzing the Factors Influencing Adoption Intention of Internet Banking: Applying DEMATEL-ANP-SEM Approach. *PLoS ONE* **2020**, *15*, e0227852. [CrossRef]
71. Farman, H.; Javed, H.; Jan, B.; Ahmad, J.; Ali, S.; Khalil, F.N.; Khan, M. Analytical Network Process Based Optimum Cluster Head Selection in Wireless Sensor Network. *PLoS ONE* **2017**, *12*, e0180848. [CrossRef]
72. Geldermann, J.; Spengler, T.; Rentz, O. Fuzzy Outranking for Environmental Assessment. Case Study: Iron and Steel Making Industry. *Fuzzy Sets Syst.* **2000**, *115*, 45–65. [CrossRef]
73. Maity, S.R.; Chakraborty, S. Tool Steel Material Selection Using PROMETHEE II Method. *Int. J. Adv. Manuf. Technol.* **2015**, *78*, 1537–1547. [CrossRef]

axioms

Article

A New Perspective for Multivariate Time Series Decision Making through a Nested Computational Approach Using Type-2 Fuzzy Integration

Martha Ramirez and Patricia Melin *

Division of Graduate Studies and Research, Tijuana Institute of Technology, Tijuana 22414, Mexico; martharmz@gmail.com
* Correspondence: pmelin@tectijuana.mx

Abstract: The integration of key indicators from the results of the analysis of time series represents a constant challenge within organizations; this could be mainly due to the need to establish the belonging of each indicator within a process, geographic region or category. This paper thus illustrates how both primary and secondary indicators are relevant for decision making, and why they need to be integrated by making new final fuzzy indicators. Thus, our proposal consists of a type-2 fuzzy integration of multivariate time series, such as OECD country risk classification, inflation, population and gross national income (GNI) by using multiple type-1 fuzzy inference systems to perform time series classification tasks. Our contribution consists of the proposal to integrate multiple nested type-1 fuzzy inference systems using a type-2 fuzzy integration. Simulation results show the advantages of using the proposed method for the fuzzy classification of multiple time series. This is done in order so the user can have tools that allow them to understand the environment and generate comparative analyses of multiple information sources, and finally use it during the process prior to decision making considering the main advantage of modeling the inherent uncertainty.

Keywords: type-2 fuzzy system; time series; decision making

MSC: 03B52; 03E72; 62P30

Citation: Ramirez, M.; Melin, P. A New Perspective for Multivariate Time Series Decision Making through a Nested Computational Approach Using Type-2 Fuzzy Integration. *Axioms* **2023**, *12*, 385. https://doi.org/10.3390/axioms12040385

Academic Editor: Ta-Chung Chu

Received: 17 March 2023
Revised: 8 April 2023
Accepted: 10 April 2023
Published: 17 April 2023

1. Introduction

Analytical purposes about demographics, financial, industry and labor market statistics indicators, among others, are commonly accumulated over time and represent a significant part of the decision-making process. Furthermore, the analysis of historical information [1–3] makes it possible to use the collected information to issue early warnings on the current and future measurement of indicators (variables).

For governments, a key aspect that must end poverty is sustainable development, through which better health conditions are created and prosperity is fostered, in addition to considering improvements in education and social conditions. Sustainable development is found in the main goals and policies of societies all over the world [4]. On the other hand, many organizations have been recording the fulfillment of their goals, objectives and performance indicators for decades.

A common aspect of the historical analysis of variables (indicators) is that it is necessary to classify these indicators into categories, geographical regions or topics in order to achieve an optimal composition of each group. Because of the individual characteristics and based on the membership of the determined group, it could place each indicator in more than one category, which means that there is uncertainty in class membership.

Due to a lack of comprehensive measures in some time series, making comparisons in terms of multiple variables (indicators) is a complex task, as is the case with most country statistics. This, in some cases, causes summaries of the information to be made and composite measures to be constructed, for which reason a frequent problem for the analyst is the selection of an adequate weighting for each of the considered indicators [5,6]. The ideal objective is for individuals to assign weights according to their preferences and based on experience, though in most cases it is not possible because there is not enough information available [7].

Thus, our motivation comes from the need for a computational intelligence model for handling uncertainty in decision making, through comparisons in terms of nested fuzzy classification of multiple variables (indicators) instead a typical model of Multi-Criteria Decision Making (MCDM), with the common understanding that it evaluates the criteria using an aggregation method function which returns a binary output; no preference is represented by a 0 and the strongest preference is represented by a 1. So, we can rely our proposal on the theory of fuzzy reasoning that is an inference procedure that derives conclusions from a set of fuzzy if-then rules and known facts [8,9] by modeling vagueness and unreliability of information, where an interval also represents the degree of membership (consists of two limits between 0 and 1) of the function.

Therefore, the main contribution of this paper consists of the combination of multiple fuzzy systems [10,11] to perform integration of time series' analysis results, which slightly simulate the cognitive functioning of the human brain when the person makes a decision and is focusing on achieving a management of uncertainty in this type of decisive process. It consists of a type-2 fuzzy integration of multivariate time series such as OECD country risk classification, inflation, population and gross national income (GNI) by using time series classification tasks multiple type-1 fuzzy systems.

This approach differs from most existing methods and computational models in the literature by combining multiple nested type-1 fuzzy systems using a type-2 fuzzy integration for comparisons in terms of multiple variables (indicators), which represents a great advantage of our method when managing uncertainty in decision making using linguistic variables.

This paper consists of the following sections. In Sections 2 and 3, we show the literature review and theoretical aspects, respectively. In Section 4, the problem is described. The methodology used is clarified in Section 5. The experimental and discussion of results are presented in Sections 6 and 7, respectively. Ultimately, in Section 8, the final conclusions are outlined.

2. Literature Review

In recent decades, attention has been paid to the design of decision-making systems, mainly those that consider multiple criteria weighted by a group of experts; that is, they establish the importance or select certain criteria based on their knowledge, experience or intuition. There is a challenge in establishing an appropriate hierarchy among the multiple criteria [12,13].

There are numerous mathematical techniques for Multi-Criteria Decision Making (MCDM) [14]: Simple Additive Weightage (SAW), Technique for Order Preference by Similarity to Ideal Solution (TOPSIS), Weighted Aggregated Sum Product Assessment (WASPAS), Preference Ranking Organization Method for Enrichment Evaluation (PROMETHEE) Elimination Et Choice Translating Reality (ELECTRE), Linear programming, Goal programming, LINMAP, Measurement of alternative and ranking according to the compromise solution (MARCOS) and Lexicographic, among others. Some of these models have been contemplated in different areas such as the humanities, administration, politics or engineering.

In the last decade, researchers have proposed hybrid models that combine mathematical models such as those mentioned above with general aspects of the fuzzy set theory proposed by Zadeh in the 1960s. In [15,16], the authors investigate an outranking approach with ELECTRE II and MARCOS methods, respectively, for group decision making in 2-tuple linguistic fuzzy context. Furthermore, other authors applied TOPSIS approach to modeling problems based on interval-valued probabilistic linguistic q-rung orthopair fuzzy sets in [17]. The Intuitionistic fuzzy set theory is used by the authors in [18] to choose the most appropriate energy alternative among a set of renewable energy alternatives; to map (fuzzy ranking) the linguistic judgements of the MCDM problems [19,20]; and to introduce a new aggregation and ranking method based on the WASPAS and TOPSIS methods [21]. In [22], authors proposed a new interval type-2 fuzzy (IT2F) MCDM method based on the analytic hierarchy process (AHP) and TOPSIS to the selection of a maintenance strategy for an industrial asset. In [23,24], the authors investigate the multiple attribute group decision-making problems in which the attribute values and the weights take the form of trapezoidal interval type-2 fuzzy sets.

Limitation of the current models validates the results, since they use different methodologies to establish the importance or hierarchy of the criteria or indicators [25], as they are dynamic or uncertain since they depend on the global environment. As far as intelligent techniques are concerned [26–28], the review of the literature indicates that there is no MCDM model that contemplates multiple type-1 and type-2 fuzzy systems to carry out the evaluation of the impact of using or ignoring criteria through a fuzzy integration that can model the results obtained.

3. Basic Concepts

In this section, we show a summary about the theory considered during the development of our model, covering mainly the use of bio-inspired methods, as is the case with fuzzy systems.

3.1. Type-2 Fuzzy Systems

When we refer to fuzzy logic we can take into account two relevant aspects; if it is type 1 we seek to model the vagueness in linguistic concepts, while in type 2 we intend to model uncertainty, mainly that which affects the decision-making process [29] and is inherent to the information attribute, appearing in several different ways.

We can start from the composition of a fuzzy inference system: a fuzzy rule base; a database containing the type and parameters of the considered the membership functions, which will be very useful when generating fuzzy rules; and finally a reasoning mechanism, that is a procedure to infer which rules apply to the given values to reach a result, which are mostly fuzzy sets. Therefore, it is necessary to use a defuzzification method to extract a crisp value that represents a fuzzy set [30].

There are no changes in the basic concepts of the different fuzzy sets; both type-1 and type-2 fuzzy use the fuzzy if-then rules in the antecedent or consequent. The difference is that the uncertainty in type-2 membership functions is modeled, since they contain type-2 fuzzy sets (which contain type 1 fuzzy sets). In other words, it consists of a representation of the uncertainty by means of a crisp output due to the perturbation, once the reduced type set is deblurred to produce a crisp type 2 output, by finding the centroid of the type-reduced set. This means the equivalent of finding the weighted average of the outputs of all the type-1 fuzzy logic systems that are embedded in the type-2 fuzzy logic system, where the weights correspond to the memberships in the type-reduced set.

So, the amount of uncertainty in a system can be reduced by using type-2 fuzzy logic as it offers better capabilities to handle linguistic uncertainties by modeling vagueness and unreliability of information, and also an interval represents the degree of membership (consists of two limits between 0 and 1) of the function [31].

It is possible to mathematically express an Interval Type-2 Fuzzy Set as Equation (1)

$$J_x = \left\{((x,u)) \mid u \in \left[\underline{\mu}_A(x),\ \overline{\mu}_A(x)\right]\right\} \tag{1}$$

where $\underline{\mu}_A(x)$ and $\overline{\mu}_A(x)$ correspond to the limits of the fuzzy set, frequently known as lower and upper membership functions, correspondingly.

The mathematical expression of the Footprint of Uncertainty (FOU) is presented as Equation (2)

$$\text{FOU} \in \left[\underline{\mu}_A(x),\ \overline{\mu}_A(x)\right] \tag{2}$$

where the $\underline{\mu}_A(x)$ and $\overline{\mu}_A(x)$ are the lower and upper membership functions, respectively [32].

We can highlight that a similar process is carried out in an Interval Type-2 Mamdani FIS as in a Type-1, with the difference that lies in the activation forces of upper and lower rules. We rely on the fuzzy logic version of modus ponens to compute the inference calculation, as can be seen in the equation in Equation (3)

$$R^l \; : IF\ x_1\ \ is\ \widetilde{F}_1^l\ and \ldots and\ x_p\ is\ \widetilde{F}_1^l\ ,\ THEN\ y\ is\ \widetilde{G}^l \tag{3}$$

where $l = 1, \ldots, M$.

3.2. Multi-Criteria Decision Making

Multiple-criteria decision making or multiple-criteria decisions belong to the area of operations research (OR), where its purpose is to allow a quantitative analysis to be carried out first for the solution of complex problems in a public, private or social organization.

Generally, regardless of the person or work environment, it is considered that for decision making many criteria should be evaluated systematically and formally, through an analysis that includes multiple criteria, such as cost, price or measurement of the quality. A simple example is portfolio management, where obtaining high returns is a priority, but reducing risks is also required. In the service industry, customer satisfaction and the cost of providing the service are two of the significant criteria that must be weighed [33].

In artificial intelligence, the generalization of logical connectives is mainly used when a system must decide. It is possible that the system has a multiple-criteria decision problem, which means that the system has numerous criteria. To simulate the environment in an information system, it is required to have a general understanding of the environment and that the sources of information are reliable. Unfortunately, when the information is provided by a single source (by a sensor or an expert) it is often not reliable enough. Thus, when the information is provided by several sensors (or experts), it must be combined to improve the reliability and precision of the data [34–36].

An indicator (criterion) is presumed to be useful if its predictions result in a smaller loss when compared to a prediction where that indicator was ignored. In the case of vulnerability indicators, these are used as useful early warning indicators for policymakers when faced with severe recessions. Hence the importance of considering international events when assessing a country's vulnerabilities. In a global economy, the vulnerabilities of countries accumulate, and are potentially transmitted between them [37].

There are multiple factors for the development of a country. By carrying out a ranking process, it is possible to compare the strengths and weaknesses of each nation. Therefore, it is necessary to identify a correct classification mechanism by which it is possible to perform a comparative analysis [38].

4. Problem Description

The Organization for Economic Co-operation and Development (OECD) is an international organization that works to build better policies for better lives. Together with governments, policy makers and citizens, OECD works on establishing evidence-based international standards and finding solutions to a range of social, economic and environmental challenges.

As part of the civil and governmental actions, global comparable data are available to uncover the strengths of the OECD and other leading economies, which makes it possible to analyze multiple historical trends such as inflation, population and gross national income (GNI), among others, though this is an arduous task that will probably take a long time, so in the meantime it is necessary to have tools that allow these variables to be associated, provide new insights into areas of policy interest and inform the global panorama to decision-makers. Furthermore, indicators are pointers; they do not address causal relationships. Moreover, the validity of a set of indicators depends on its use [39].

For this case, four datasets were selected for each of the 38 OECD member countries, and no data preprocessing was performed (Table 1):

Table 1. OECD member countries.

Country Name	Country Code	Country Name	Country Code
Australia	AUS	Japan	JPN
Austria	AUT	Republic of Korea, Rep.	KOR
Belgium	BEL	Latvia	LVA
Canada	CAN	Lithuania	LTU
Chile	CHL	Luxembourg	LUX
Colombia	COL	Mexico	MEX
Costa Rica	CRI	Netherlands	NLD
Czech Republic	CZE	New Zealand	NZL
Denmark	DNK	Norway	NOR
Estonia	EST	Poland	POL
Finland	FIN	Portugal	PRT
France	FRA	Slovak Republic	SVK
Germany	DEU	Slovenia	SVN
Greece	GRC	Spain	ESP
Hungary	HUN	Sweden	SWE
Iceland	ISL	Switzerland	CHE
Ireland	IRL	Turkey	TUR
Israel	ISR	United Kingdom	GBR
Italy	ITA	United States	USA

The first dataset consists of six attributes (Table 2) for 61 instances corresponding to the total annual population, from 1960 to 2020 (Figure 1) [40].

Table 2. Annual total population: attributes of the time series.

Code Attribute	Attribute Name
att1	Country
att2	IDCountry
att3	Criterion
att4	IDCriterion
att5	IDYear
att6	ValueCriterion

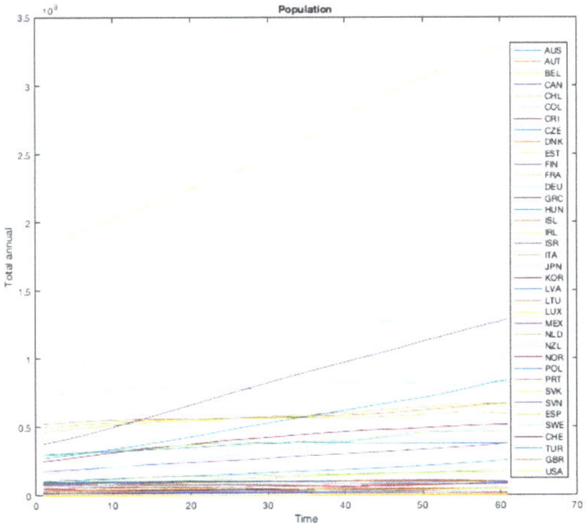

Figure 1. OECD member countries: total annual population.

In the second dataset, it consists of six attributes (Table 3) for 15 instances corresponding to GNI, from 2006 to 2020 (Figure 2) [41].

Table 3. GNI: attributes of the time series.

Code Attribute	Attribute Name
att1	Country
att2	IDCountry
att3	Criterion
att4	IDCriterion
att5	IDYear
att6	ValueCriterion

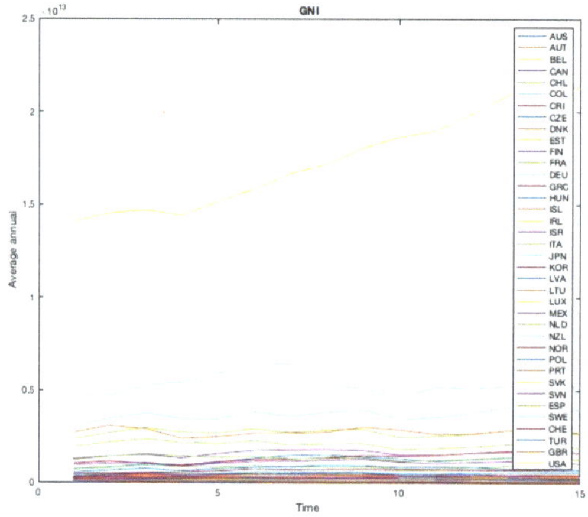

Figure 2. OECD members: average annual GNI.

The third dataset consists of six attributes (Table 4) for 28 instances corresponding to the inflation, from 1993 to 2020 (Figure 3) [42].

Table 4. Inflation: attributes of the time series.

Code Attribute	Attribute Name
att1	Country
att2	IDCountry
att3	Criterion
att4	IDCriterion
att5	IDYear
att6	ValueCriterion

The fourth dataset consists of four attributes (Table 5) for 34 instances corresponding to OECD country risk, from 1987 to 2020 [43].

Table 5. OECD country risk: attributes of the time series.

Code Attribute	Attribute Name
att1	Country
att2	IDCountry
att3	IDYear
att4	ValueCriterion

High-income OECD member countries have been not classified; these belong to category 0.

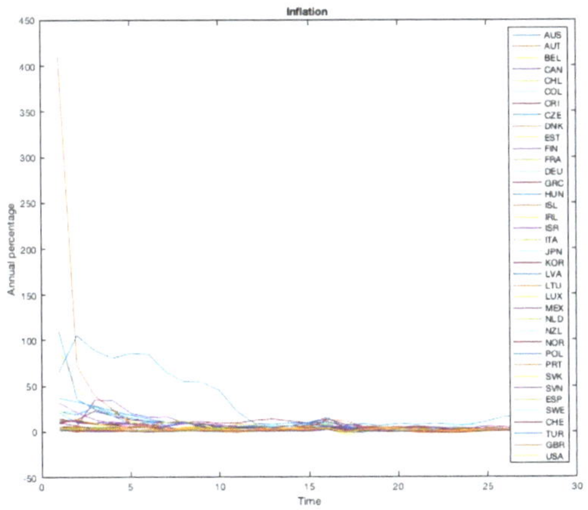

Figure 3. OECD members: inflation annual percentage.

5. Proposed Method

We presented a computational model that comprises three levels. In the first, a time series dataset is selected; and for the second, type-1 fuzzy inference systems are used to classify a set of countries by weighting: population, GNI, inflation and OECD country risk time series values (based on the time series values a class is assigned to each country). Finally, in the third, the results obtained in the previous levels are used as inputs of a type-2 fuzzy inference system to integrate the results and obtain an indicator or global result (Figure 4).

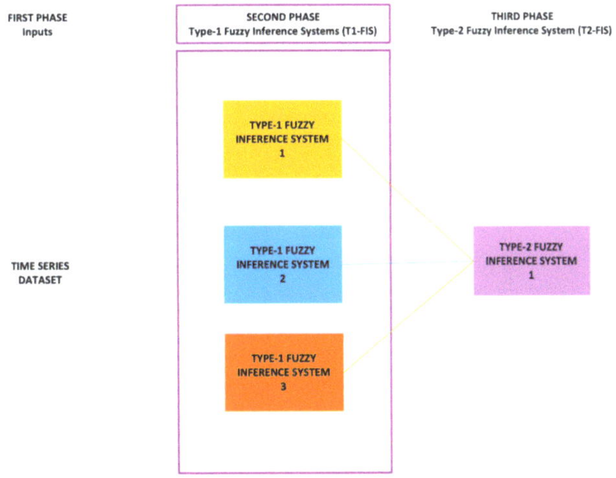

Figure 4. Illustration of the proposed method.

Type-1 fuzzy systems used to integrate the population and GNI time series are the Mamdani type, consisting of two inputs and one output (Table 6), triangular membership functions type: Low (LW), Medium (MM), High (HH), 9 rules and the centroid defuzzification method. To select these values, the parameters of the membership function and fuzzy rules were thoroughly tested (Figure 5).

Table 6. First Type-1 fuzzy system population and GNI variables.

Variables Type	Variables Name	Membership Functions
Input	Population	LW, MM, HH
	GNI	LW, MM, HH
Output	Criterion 1	LW, MM, HH

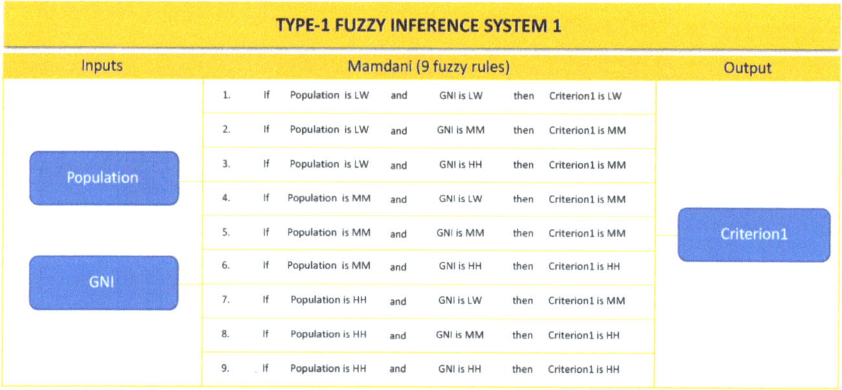

Figure 5. First Type-1 fuzzy system population and GNI variables.

The second type-1 fuzzy system is the Mamdani type, and is used to integrate GNI time series values; it consists of three inputs and one output (Table 7), triangular membership functions type (Lower income (L), Lower Middle income (LM), Upper Middle income (UM), High income (H), 10 rules and the centroid defuzzification method. To select these values, the parameters of the membership function and fuzzy rules were thoroughly tested (Figure 6).

Table 7. Second Type-1 fuzzy system GNI variables.

Variables Type	Variables Name	Membership Functions
Inputs	GNI1	L, LM, UM, H
	GNI2	L, LM, UM, H
	GNI3	L, LM, UM, H
	GNI4	L, LM, UM, H
Output	Criterion 2	L, LM, UM, H

Type-1 fuzzy systems used to integrate the inflation and OECD country risk time series consist of two inputs and one output, triangular membership functions type: Low (LW), Medium (MM) and High (HH) (Table 8). It is Mamdani type, with 9 rules and the centroid defuzzification method. To select these values, the parameters of the membership function and fuzzy rules were tested thoroughly (Figure 7).

Table 8. Third Type-1 fuzzy system inflation and OECD country risk variables.

Variables Type	Variables Name	Membership Functions
Input	Inflation	LW, MM, HH
	OECD cr	LW, MM, HH
Output	Criterion 3	LW, MM, HH

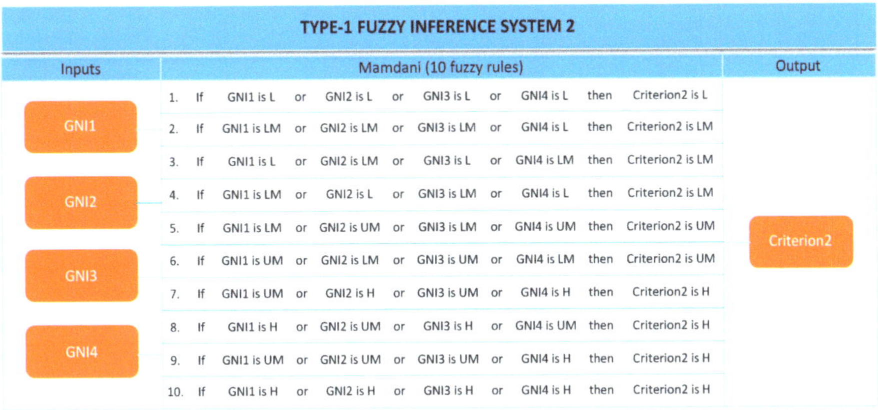
TYPE-1 FUZZY INFERENCE SYSTEM 2
Inputs
Mamdani (10 fuzzy rules)
Output
GNI1
GNI2
GNI3
GNI4
1. If GNI1 is L or GNI2 is L or GNI3 is L or GNI4 is L then Criterion2 is L
2. If GNI1 is LM or GNI2 is LM or GNI3 is LM or GNI4 is L then Criterion2 is LM
3. If GNI1 is L or GNI2 is LM or GNI3 is L or GNI4 is LM then Criterion2 is LM
4. If GNI1 is LM or GNI2 is L or GNI3 is LM or GNI4 is L then Criterion2 is LM
5. If GNI1 is LM or GNI2 is UM or GNI3 is LM or GNI4 is UM then Criterion2 is UM
6. If GNI1 is UM or GNI2 is LM or GNI3 is UM or GNI4 is LM then Criterion2 is UM
7. If GNI1 is UM or GNI2 is H or GNI3 is UM or GNI4 is H then Criterion2 is H
8. If GNI1 is H or GNI2 is UM or GNI3 is H or GNI4 is UM then Criterion2 is H
9. If GNI1 is UM or GNI2 is UM or GNI3 is UM or GNI4 is H then Criterion2 is H
10. If GNI1 is H or GNI2 is H or GNI3 is H or GNI4 is H then Criterion2 is H
Criterion2

Figure 6. Second Type-1 fuzzy system GNI variables.

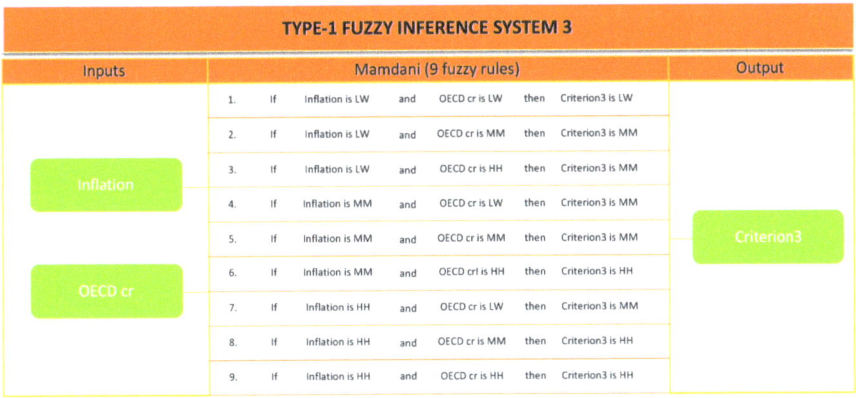
TYPE-1 FUZZY INFERENCE SYSTEM 3
Inputs
Mamdani (9 fuzzy rules)
Output
Inflation
OECD cr
1. If Inflation is LW and OECD cr is LW then Criterion3 is LW
2. If Inflation is LW and OECD cr is MM then Criterion3 is MM
3. If Inflation is LW and OECD cr is HH then Criterion3 is MM
4. If Inflation is MM and OECD cr is LW then Criterion3 is MM
5. If Inflation is MM and OECD cr is MM then Criterion3 is MM
6. If Inflation is MM and OECD crl is HH then Criterion3 is HH
7. If Inflation is HH and OECD cr is LW then Criterion3 is MM
8. If Inflation is HH and OECD cr is MM then Criterion3 is HH
9. If Inflation is HH and OECD cr is HH then Criterion3 is HH
Criterion3

Figure 7. Third Type-1 fuzzy system inflation and OECD country risk variables.

Type-2 fuzzy system used to integrate the results of the type-1 fuzzy systems consists of three inputs and one output, triangular membership functions type Low (LW), Medium (MM) and High (HH) (Table 9). It is the Mamdani type, with 27 rules and the centroid defuzzification method. To select these values, the parameters of the membership function and fuzzy rules were tested thoroughly (Figure 8).

Table 9. Type-2 fuzzy indicators.

Variables Type	Variables	Membership Functions
Input	Criterion 1	LW, MM, HH
	Criterion 2	LW, MM, HH
	Criterion 3	LW, MM, HH
Output	Criteria	LW, MM, HH

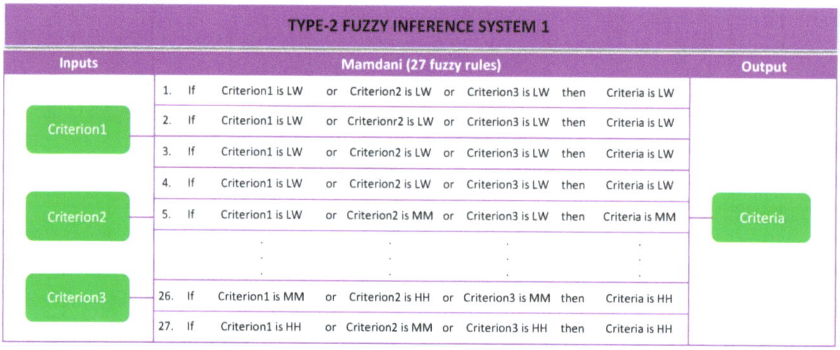

Figure 8. Type-2 fuzzy criteria integration.

It is necessary to establish that after the type reduction, to calculate the final output of the type-2 fuzzy system, the weights of the outputs of the fuzzy rules are averaged, where the membership functions of the rules are the weights as Equation (4)

$$y(t) = \frac{\mu_1 y_1(t) + \mu_2 y_2(t) + \cdots + \mu_{27} y_{27}(t)}{\mu_1 + \mu_2 + \cdots + \mu_{27}} \tag{4}$$

where $y_i(t)$ are the outputs of the rules, i = 1, ... 27, μ_i are the membership function values at the outputs of the rules, i = 1, ... 27 and y(t) is the total output.

6. Experimental Results

For the case of calculating the level of increase for each variable (time series), we first subtract the immediate previous value of the time series from the initial value. After obtaining the result, it is divided by the initial value. Subsequently, with the result of calculating the level of increase in the population and the GNI for each of the calendar years, we classify the variables of the level of increase in the population and the GNI by using a Mamdani type-1 fuzzy inference system with two inputs and one output, triangular membership functions: Low (LW), Medium (MM) and High (HH) (Tables 10 and 11).

Table 10. First Type-1 FIS: Input-Output Parameters.

Variables	Membership Functions	Parameter a	Parameter b	Parameter c
Input 1 Population	LW	0.000	0.100	0.300
	MM	0.200	0.600	1.200
	HH	0.900	2.500	3.000
Input 2 GNI	LW	0.000	0.100	0.300
	MM	0.200	0.600	1.200
	HH	0.900	2.500	3.000
Output 1 Criterion 1	LW	0.000	0.600	1.200
	MM	0.900	1.600	2.200
	HH	2.000	2.500	3.000

By using a Mamdani fuzzy type-1 Inference System with four inputs and one output, as well as triangular membership functions based on the GNI variable, the indicator level was reached for each of the countries according to the corresponding classification for every calendar year, Lower income (L), Lower Middle income (LM), Upper Middle income (UM) and High income (H), seeking to integrate the results with respect to the class assigned to each country, into a new classification label (Tables 12 and 13).

Table 11. First Type-1 FIS: Output increase level.

Country Code	Criterion 1	Country Code	Criterion 1
AUS	MM	JPN	MM
AUT	LW	KOR	LW
BEL	LW	LVA	LW
CAN	LW	LTU	LW
CHL	MM	LUX	MM
COL	MM	MEX	MM
CRI	MM	NLD	MM
CZE	MM	NZL	MM
DNK	LW	NOR	LW
EST	MM	POL	MM
FIN	LW	PRT	LW
FRA	LW	SVK	LW
DEU	MM	SVN	MM
GRC	MM	ESP	MM
HUN	MM	SWE	MM
ISL	LW	CHE	LW
IRL	MM	TUR	MM
ISR	MM	GBR	MM
ITA	LW	USA	LW

Table 12. Type-1 FIS: Input-Output membership function.

Variables	Membership Functions	Parameter a	Parameter b	Parameter c
Input1 GNI1	L	0	0.002	0.010
	LM	0.008	0.144	0.244
	UM	0.234	0.274	0.915
	HM	0.910	1.2	1.5
Input 2 GNI2	L	0	0.002	0.011
	LM	0.009	0.144	0.245
	UM	0.235	0.274	0.916
	HM	0.911	1.2	1.5
Input 3 GNI3	L	0	0.002	0.012
	LM	0.010	0.144	0.246
	UM	0.236	0.274	0.917
	HM	0.912	1.2	1.5
Input 4 GNI4	L	0	0.002	0.013
	LM	0.011	0.144	0.247
	UM	0.237	0.274	0.918
	HM	0.914	1.2	1.5
Output 1 Criterion 2	L	0	0.002	0.010
	LM	0.008	0.144	0.250
	UM	0.230	0.280	0.890
	HM	0.790	1.2	1.5

Then, a third Mamdani type-1 fuzzy inference system was used to classify the level of increase of inflation and OECD country risk. It consists of two inputs and one output, as well as triangular membership functions: Low (LW), Medium (MM) and High (HG) (Tables 14 and 15).

Table 13. Output first Type-1 FIS Criterion 2.

Country Code	Criterion 2	Country Code	Criterion 2
AUS	H	JPN	H
AUT	H	KOR	H
BEL	H	LVA	H
CAN	H	LTU	UM
CHL	UM	LUX	H
COL	UM	MEX	UM
CRI	UM	NLD	H
CZE	H	NZL	H
DNK	H	NOR	H
EST	H	POL	UM
FIN	H	PRT	H
FRA	H	SVK	UM
DEU	H	SVN	H
GRC	UM	ESP	H
HUN	UM	SWE	H
ISL	H	CHE	H
IRL	H	TUR	UM
ISR	H	GBR	H
ITA	H	USA	H

Table 14. Third Type-1 FIS: Input-Output parameters.

Variables	Membership Functions	Parameter a	Parameter b	Parameter c
	LW	−3.000	0.100	0.300
Input 1 Inflation	MM	0.200	0.600	1.200
	HH	0.900	2.500	3.000
	LW	−3.000	0.100	0.300
Input 2 OECD cr	MM	0.200	0.600	1.200
	HH	0.900	2.500	3.000
	LW	−3.000	0.600	1.200
Output 1 Criterion 3	MM	0.900	1.600	2.200
	HH	2.000	2.500	3.000

Table 15. Third Type-1 FIS Criterion 3.

Country Code	Criterion 3	Country Code	Criterion 3
AUS	LW	JPN	LW
AUT	LW	KOR	LW
BEL	LW	LVA	LW
CAN	LW	LTU	LW
CHL	LW	LUX	LW
COL	LW	MEX	LW
CRI	LW	NLD	LW
CZE	LW	NZL	LW
DNK	LW	NOR	LW
EST	LW	POL	LW
FIN	LW	PRT	LW
FRA	LW	SVK	LW
DEU	LW	SVN	LW
GRC	LW	ESP	LW
HUN	LW	SWE	LW
ISL	LW	CHE	LW
IRL	LW	TUR	LW
ISR	LW	GBR	LW
ITA	LW	USA	LW

Finally, once the classification results of the level of increase of the variables population, GNI, inflation and OECD country risk from one period to another are obtained by using three type-1 fuzzy systems, these results are integrated by using a type-2 fuzzy system with triangular membership functions, where the three inputs correspond to the classification of each of the type-1 fuzzy systems, for the purpose of obtaining an output that represents the final classification of the level of increase of the variables Low (LW), Medium (MM) or High (HH), as appropriate (Tables 16 and 17).

Table 16. Fist Type-2 FIS: Input-Output parameters.

Variables	Membership Function	a	b	c	Lower Scale	Lower Lag	
	LW	0.1223	0.6223	1.2980	1.000	0.2000	0.2000
Input 1 Criterion 1	MM	0.894	1.544	2.166	1.000	0.2000	0.2000
	HH	2.000	2.500	3.000	1.000	0.2000	0.2000
	LW	0.1223	0.6223	1.2980	1.000	0.2000	0.2000
Input 2 Criterion 2	MM	0.894	1.544	2.166	1.000	0.2000	0.2000
	HH	2.000	2.500	3.000	1.000	0.2000	0.2000
	LW	0.1223	0.6223	1.2980	1.000	0.2000	0.2000
Input 3 Criterion 3	MM	0.894	1.544	2.166	1.000	0.2000	0.2000
	HH	2.000	2.500	3.000	1.000	0.2000	0.2000
	LW	0.000	0.500	1.200	1.000	0.2000	0.2000
Output 1 Criteria	MM	1.000	1.500	2.200	1.000	0.2000	0.2000
	HH	2.000	2.500	3.000	1.000	0.2000	0.2000

Table 17. Output Type-2 FIS variables (Criteria).

Country Code	Criteria	Country Code	Criteria
AUS	MM	JPN	MM
AUT	MM	KOR	MM
BEL	MM	LVA	MM
CAN	MM	LTU	MM
CHL	MM	LUX	MM
COL	MM	MEX	LW
CRI	MM	NLD	MM
CZE	MM	NZL	MM
DNK	MM	NOR	MM
EST	MM	POL	MM
FIN	MM	PRT	MM
FRA	MM	SVK	MM
DEU	MM	SVN	MM
GRC	MM	ESP	MM
HUN	MM	SWE	MM
ISL	MM	CHE	MM
IRL	MM	TUR	LW
ISR	MM	GBR	MM
ITA	MM	USA	MM

Finally, we are presenting a comparison of the classification results obtained for each country using type-1 fuzzy systems and the final classification obtained using the type-2 fuzzy system as an integrator of all the criteria evaluated (Table 18).

Table 18. Comparison of Type-1 (T1) and Type-2 (T2) fuzzy systems results (FIS).

Country Code	T1 FIS1 Criterion 1	T1 FIS2 Criterion 2	T1 FIS3 Criterion 3	T2 FIS1 Criteria
AUS	MM	H	LW	MM
AUT	LW	H	LW	MM
BEL	LW	H	LW	MM
CAN	LW	H	LW	MM
CHL	MM	UM	LW	MM
COL	MM	UM	LW	MM
CRI	MM	UM	LW	MM
CZE	MM	H	LW	MM
DNK	LW	H	LW	MM
EST	MM	H	LW	MM
FIN	LW	H	LW	MM
FRA	LW	H	LW	MM
DEU	MM	H	LW	MM
GRC	MM	UM	LW	MM
HUN	MM	UM	LW	MM
ISL	LW	H	LW	MM
IRL	MM	H	LW	MM
ISR	MM	H	LW	MM
ITA	LW	H	LW	MM
JPN	MM	H	LW	MM
KOR	LW	H	LW	MM
LVA	LW	H	LW	MM
LTU	LW	UM	LW	MM
LUX	MM	H	LW	MM
MEX	MM	UM	LW	LW
NLD	MM	H	LW	MM
NZL	MM	H	LW	MM
NOR	LW	H	LW	MM
POL	MM	UM	LW	MM
PRT	LW	H	LW	MM
SVK	LW	UM	LW	MM
SVN	MM	H	LW	MM
ESP	MM	H	LW	MM
SWE	MM	H	LW	MM
CHE	LW	H	LW	MM
TUR	MM	UM	LW	LW
GBR	MM	H	LW	MM
USA	LW	H	LW	MM

7. Discussion of Results

The main goal of this work lies in achieving the separation of the results obtained using each type-1 fuzzy system, with the idea of making decisions based on the integrated results through the type-2 fuzzy system. In this understanding, these results show that it is possible to integrate utilizing type-2 fuzzy systems with the outputs of type-1 fuzzy systems, with which it is possible to identify countries with similar trends and provide an overview of the performance of multiple variables in different countries concurrently.

Since most member countries of the OECD have been classified by international organizations, in the case of the type-1 fuzzy system classification by GNI and classification of the OECD risk variables, the vast majority of the countries obtained similar a classification.

Similarly, the final classification obtained using type-2 fuzzy concentrates most of the results in the middle range, with 36 countries classified as MM, with the exception of two countries that obtained low (LW) classification. This is because both criterion 1 and criterion 2, which represent the increase in population and gross national income, respectively, obtained different weights compared to the rest of the countries with similar classifications in the results of type-1 fuzzy systems. This means that by integrating the results using the type-2 fuzzy system, it is possible to separate into new classes elements that belong to the same group, since in this case the difference is mainly marked in criterion 2, which models the increase in gross national income of the last four years for each country. Consequently, the results of type-2 fuzzy are slightly similar.

8. Conclusions

We have presented in this work a model for the classification of time series of population, GNI, inflation and OECD country risk using multiple type-1 fuzzy systems and a type-2 fuzzy system as integrator.

The simulation results have shown the countries with similar indicators. The results have shown that it is possible to use type-2 fuzzy systems to find the final country key indicator (criteria) based on the trend and similarity of their primary indicators. Therefore, the combination of nested fuzzy models to perform integration of time series' analysis results slightly simulate the cognitive functioning of the human brain when the person makes a decision, and focuses on achieving a management of uncertainty in this type of decisive process, with this representing the main contribution of this work.

By carrying out the experiments, we identified some of the advantages of using type-2 fuzzy integration for classification problems, particularly applied as a decision-support tool, as it is possible to achieve results for a specific place or area by having the data grouped based on their similarity or groups of elements. Furthermore, by incorporating the type-2 fuzzy system, it was possible to observe the improvement in the integration of the outputs of type-1 fuzzy systems, since by offering threshold values between 0 and 1, elements that apparently belong to the same group (in other words, that obtained the same classification) can be separated into a new cluster due to differences that only an expert in the function could detect through exhaustive analysis. This will depend on the problem to be solved to decide whether to use three phases of the proposed method simultaneously, or to work with each one of them separately.

As future work, we could design a model consisting of multiple type-2 fuzzy inference systems, with the aim of performing tests with other types of membership functions, seeking to extend the membership threshold of each function. On the other hand, we also intend to combine our proposal with the use of supervised neural networks to perform multi-variable prediction tasks, seeking to reach a greater number of global indicators. In addition, we are evaluating working with new datasets, with the idea of considering the relevant attributes within the time series by using several types of demographics and financial, industry and labor market statistics indicators, among others. On the other hand, it is also intended to combine our proposal with the use of supervised neural networks to perform variable prediction tasks, seeking to reach a greater number of global indicators.

Author Contributions: Conceptualization, M.R. and P.M.; methodology, P.M.; software, M.R.; validation, M.R. and P.M.; formal analysis, P.M.; investigation, M.R.; data curation, M.R.; writing—original draft preparation, M.R.; and writing—review and editing, P.M. All authors have read and agreed to the published version of the manuscript.

Funding: This research received no external funding.

Data Availability Statement: Not applicable.

Conflicts of Interest: The authors declare no conflict of interest.

References

1. Chacón, H.; Kesici, E.; Najafirad, P. Improving Financial Time Series Prediction Accuracy Using Ensemble Empirical Mode Decomposition and Recurrent Neural Networks. *IEEE Access* **2020**, *8*, 117133–117145. [CrossRef]
2. Hu, Y.; Sun, X.; Nie, X.; Li, Y.; Liu, L. An Enhanced LSTM for Trend Following of Time Series. *IEEE Access* **2019**, *7*, 34020–34030. [CrossRef]
3. Moghar, A.; Hamiche, M. Stock Market Prediction Using LSTM Recurrent Neural Network. *Procedia Comput. Sci.* **2020**, *170*, 1168–1173. [CrossRef]
4. Megyesiova, S.; Lieskovska, V. Analysis of the Sustainable Development Indicators in the OECD Countries. *Sustainability* **2018**, *10*, 4554. [CrossRef]

5. Zhao, S.; Dong, Y.; Martíne, L.; Pedrycz, W. Analysis of Ranking Consistency in Linguistic Multiple Attribute Decision Making: The Roles of Granularity and Decision Rules. *IEEE Trans. Fuzzy Syst.* **2022**, *30*, 2266–2278. [CrossRef]
6. Pan, X.; Wang, Y.; He, S.; Chin, K. A Dynamic Programming Algorithm Based Clustering Model and Its Application to Interval Type-2 Fuzzy Large-Scale Group Decision-Making Problem. *IEEE Trans. Fuzzy Syst.* **2022**, *30*, 108–120. [CrossRef]
7. Peiró-Palomino, J.; Picazo-Tadeo, A.J. OECD: One or Many? Ranking Countries with a Composite Well-Being Indicator. *Soc. Indic. Res.* **2018**, *139*, 847–869. [CrossRef]
8. Baskov, O.V.; Noghin, V.D. Type-2 Fuzzy Sets and Their Application in Decision-Making: General Concepts. *Sci. Tech. Inf. Proc.* **2022**, *49*, 283–291. [CrossRef]
9. Morente-Molinera, J.; Wang, Y.-W.; Gong, Z.; Morfeq, A.; Al-Hmouz, R.; Herrera-Viedma, E. Reducing Criteria in Multicriteria Group Decision-Making Methods Using Hierarchical Clustering Methods and Fuzzy Ontologies. *IEEE Trans. Fuzzy Syst.* **2022**, *30*, 1585–1598. [CrossRef]
10. Melin, P.; Monica, J.C.; Sanchez, D.; Castillo, O. Multiple Ensemble Neural Network Models with Fuzzy Response Aggregation for Predicting COVID-19 Time Series: The Case of Mexico. *Healthcare* **2020**, *8*, 181. [CrossRef]
11. Xu, T.T.; Qin, J.D. A New Representation Method for Type-2 Fuzzy Sets and Its Application to Multiple Criteria Decision Making. *Int. J. Fuzzy Syst.* **2023**, *25*, 1171–1190. [CrossRef]
12. Cheng, C.H.; Chen, M.Y.; Chang, J.R. Linguistic multi-criteria decision-making aggregation model based on situational ME-LOWA and ME-LOWGA operators. *Granul. Comput.* **2023**, *8*, 97–110. [CrossRef]
13. Putra, M.Y.; Arini; Fiade, A.; Malik, I.M. Fuzzy Multi-Criteria Decision Making for Optimization of Housing Construction Financing. In Proceedings of the 2021 Sixth International Conference on Informatics and Computing (ICIC), Jakarta, Indonesia, 3–4 November 2021; pp. 1–5. [CrossRef]
14. Thakkar, J.J. Introduction. In *Multi-Criteria Decision Making. Studies in Systems, Decision and Control*; Springer: Singapore, 2021; Volume 336, pp. 1–25. [CrossRef]
15. Akram, M.; Bibi, R.; Deveci, M. An outranking approach with 2-tuple linguistic Fermatean fuzzy sets for multi-attribute group decision-making. *Eng. Appl. Artif. Intell.* **2023**, *121*, 105992. [CrossRef]
16. Akram, M.; Khan, A.; Luqman, A.; Senapati, T.; Pamucar, D. An extended MARCOS method for MCGDM under 2-tuple linguistic q-rung picture fuzzy environment. *Eng. Appl. Artif. Intell.* **2023**, *120*, 105892. [CrossRef]
17. Xu, Y.; Liu, S.; Wang, S.; Shang, X. A novel two-stage TOPSIS approach based on interval-valued probabilistic linguistic q-rung orthopair fuzzy sets with its application to MAGDM problems. *Eng. Appl. Artif. Intell.* **2022**, *116*, 105413. [CrossRef]
18. Mishra, A.R.; Kumari, R.; Sharma, D.K. Intuitionistic fuzzy divergence measure-based multi-criteria decision-making method. *Neural Comput. Appl.* **2019**, *31*, 2279–2294. [CrossRef]
19. Chiao, K.P. Interval Type 2 Intuitionistic Fuzzy Sets Ranking Method Based on the General Graded Mean Integration Representation with Application to Multiple Criteria Decision Making. In Proceedings of the 2021 International Conference on Machine Learning and Cybernetics (ICMLC), Adelaide, Australia, 4–5 December 2021; pp. 1–6. [CrossRef]
20. Chiao, K.P. MCDM Prioritization Based on Interval Type 2 Intuitionistic Fuzzy Sets Ranking with Parametric General Graded Mean Integration Representation. In Proceedings of the 2022 International Conference on Fuzzy Theory and Its Applications (iFUZZY), Kaohsiung, Taiwan, 3–5 November 2022; pp. 1–6. [CrossRef]
21. Davoudabadi, R.; Mousavi, S.M.; Mohagheghi, V. A new last aggregation method of multi-attributes group decision making based on concepts of TODIM, WASPAS and TOPSIS under interval-valued intuitionistic fuzzy uncertainty. *Knowl. Inf. Syst.* **2020**, *62*, 1371–1391. [CrossRef]
22. Mathew, M.; Chakrabortty, R.K.; Ryan, M.J. Selection of an Optimal Maintenance Strategy Under Uncertain Conditions: An Interval Type-2 Fuzzy AHP-TOPSIS Method. *IEEE Trans. Eng. Manag.* **2022**, *69*, 1121–1134. [CrossRef]
23. Zhang, Z. Trapezoidal interval type-2 fuzzy aggregation operators and their application to multiple attribute group decision making. *Neural Comput. Appl.* **2018**, *29*, 1039–1054. [CrossRef]
24. Chen, Z.; Wan, S.; Dong, J. An efficiency-based interval type-2 fuzzy multi-criteria group decision making for makeshift hospital selection. *Appl. Soft Comput.* **2022**, *115*, 108243. [CrossRef]
25. Sharma, J.; Arora, M.; Sonia; Alsharef, A. An illustrative study on Multi Criteria Decision Making Approach: Analytical Hierarchy Process. In Proceedings of the 2022 2nd International Conference on Advance Computing and Innovative Technologies in Engineering (ICACITE), Greater Noida, India, 28–29 April 2022; pp. 2000–2005. [CrossRef]
26. Chimatapu, R.; Hagras, H.; Kern, M.; Owusu, G. Hybrid Deep Learning Type-2 Fuzzy Logic Systems for Explainable AI. In Proceedings of the 2020 IEEE International Conference on Fuzzy Systems (FUZZ-IEEE), Glasgow, UK, 19–24 July 2020; pp. 1–6. [CrossRef]
27. Kiani, M.; Andreu-Perez, J.; Hagras, H.; Filippetti, M.; Rigato, S. A Type-2 Fuzzy Logic Based Explainable Artificial Intelligence System for Developmental Neuroscience. In Proceedings of the 2020 IEEE International Conference on Fuzzy Systems (FUZZ-IEEE), Glasgow, UK, 19–24 July 2020; pp. 1–8. [CrossRef]
28. Rostam, M.; FazelZarandi, M.H. A new interval type-2 fuzzy reasoning method for classification systems based on normal forms of a possibility-based fuzzy measure. *Inf. Sci.* **2021**, *581*, 567–586. [CrossRef]
29. Melin, P.; Monica, J.C.; Sanchez, D.; Castillo, O. A new prediction approach of the COVID-19 virus pandemic behavior with a hybrid ensemble modular nonlinear autoregressive neural network. *Soft Comput.* **2023**, *27*, 2685–2694. [CrossRef] [PubMed]

30. Jang, J.-S.R. Fuzzy Inference Systems. In *Neuro-Fuzzy and Soft Computing: A Computational Approach to Learning and Machine Intelligence*; Prentice Hall: Upper Saddle River, NJ, USA, 1997; pp. 73–90. ISBN 0132610663.
31. Melin, P.; Ontiveros-Robles, E.; Castillo, O. Background and Theory. In *New Medical Diagnosis Models Based on Generalized Type-2 Fuzzy Logic*; SpringerBriefs in Applied Sciences and Technology; Springer: Cham, Switzerland, 2021; pp. 5–28. [CrossRef]
32. Ontiveros-Robles, E.; Castillo, O.; Melin, P. Towards asymmetric uncertainty modeling in designing General Type-2 Fuzzy classifiers for medical diagnosis. *Expert Syst. Appl.* **2021**, *183*, 115370. [CrossRef]
33. Azzabi, L.; Azzabi, D.; Kobi, A. The Multi-Criteria Approach Decision. In *The Multi-Criteria Approach for Decision Support*; International Series in Operations Research & Management Science; Springer: Cham, Switzerland, 2020; Volume 300, pp. 1–23. [CrossRef]
34. Melin, P.; Sánchez, D.; Castro, J.R.; Castillo, O. Design of Type-3 Fuzzy Systems and Ensemble Neural Networks for COVID-19 Time Series Prediction Using a Firefly Algorithm. *Axioms* **2022**, *11*, 410. [CrossRef]
35. Miramontes, I.; Melin, P. Interval Type-2 Fuzzy Approach for Dynamic Parameter Adaptation in the Bird Swarm Algorithm for the Optimization of Fuzzy Medical Classifier. *Axioms* **2022**, *11*, 485. [CrossRef]
36. Dombi, J.; Csiszár, O. Introduction—Aggregation and Intelligent Decision. In *Explainable Neural Networks Based on Fuzzy Logic and Multi-Criteria Decision Tools. Studies in Fuzziness and Soft Computing*; Springer: Cham, Switzerland, 2021; Volume 408, pp. ix–xi. [CrossRef]
37. Hermansen, M.; Röhn, O. Economic resilience: The usefulness of early warning indicators in OECD countries. *OECD J. Econ. Stud.* **2017**, *2016*, 9–35. [CrossRef]
38. Gupta, A.; Sharma, K. Ranking of Countries Using World Development Indicators: A Computational Approach. In Proceedings of the 2020 11th International Conference on Computing, Communication and Networking Technologies (ICCCNT), Kharagpur, India, 1–3 July 2020; pp. 1–4. [CrossRef]
39. OECD. *OECD Science, Technology and Industry Scoreboard 2017: The Digital Transformation*; OECD Publishing: Paris, France, 2017. [CrossRef]
40. The World Bank Data: Population, Total. Available online: https://data.worldbank.org/indicator/SP.POP.TOTL (accessed on 6 June 2022).
41. The World Bank Data: GNI per Capita. Available online: https://data.worldbank.org/indicator/NY.GNP.MKTP.CD (accessed on 6 June 2022).
42. The World Bank Data: Inflation. Available online: https://data.worldbank.org/indicator/FP.CPI.TOTL.ZG (accessed on 6 June 2022).
43. OECD: Country Risk Classifications of the Participants to the Arrangement on Officially Supported Export Credit. Available online: https://www.oecd.org/trade/topics/export-credits/documents/cre-crc-current-english.pdf (accessed on 1 February 2023).

MDPI AG
Grosspeteranlage 5
4052 Basel
Switzerland
Tel.: +41 61 683 77 34

Axioms Editorial Office
E-mail: axioms@mdpi.com
www.mdpi.com/journal/axioms

www.ingramcontent.com/pod-product-compliance
Lightning Source LLC
LaVergne TN
LVHW072343090526
838202LV00019B/2470
9783725834327